图解Spark
核心技术与案例实战

郭景瞻 编著

电子工业出版社
Publishing House of Electronics Industry
北京·BEIJING

内 容 简 介

本书以 Spark 2.0 版本为基础进行编写，全面介绍了 Spark 核心及其生态圈组件技术。主要内容包括 Spark 生态圈、实战环境搭建、编程模型和内部重要模块的分析，重点介绍了消息通信框架、作业调度、容错执行、监控管理、存储管理以及运行框架，同时还介绍了 Spark 生态圈相关组件，包括 Spark SQL 的即席查询、Spark Streaming 的实时流处理应用、MLbase/MLlib 的机器学习、GraphX 的图处理、SparkR 的数学计算和 Alluxio 的分布式内存文件系统等。

本书从 Spark 核心技术进行深入分析，重要章节会结合源代码解读其实现原理，围绕着技术原理介绍了相关典型实例，读者通过这些实例可以更加深入地理解 Spark 的运行机制。另外本书还应用了大量的图表进行说明，让读者能够更加直观地理解 Spark 相关原理。

本书不仅适合大数据、Spark 从业人员阅读，同时也适合大数据爱好者、架构师和软件开发人员阅读。通过本书，读者将能够很快地熟悉和掌握 Spark 大数据分析计算的利器，在生产中解决实际问题。

未经许可，不得以任何方式复制或抄袭本书之部分或全部内容。
版权所有，侵权必究。

图书在版编目（CIP）数据

图解 Spark：核心技术与案例实战 / 郭景瞻编著. —北京：电子工业出版社，2017.1
ISBN 978-7-121-30236-7

Ⅰ. ①图… Ⅱ. ①郭… Ⅲ. ①数据处理软件－图解 Ⅳ. ①TP274-64

中国版本图书馆 CIP 数据核字(2016)第 262015 号

责任编辑：安　娜
印　　刷：三河市良远印务有限公司
装　　订：三河市良远印务有限公司
出版发行：电子工业出版社
　　　　　北京市海淀区万寿路 173 信箱　邮编：100036
开　　本：787×980　1/16　印张：30.25　字数：570 千字
版　　次：2017 年 1 月第 1 版
印　　次：2018 年 4 月第 4 次印刷
定　　价：99.00 元

凡所购买电子工业出版社图书有缺损问题，请向购买书店调换。若书店售缺，请与本社发行部联系，联系及邮购电话：(010) 88254888，88258888。
质量投诉请发邮件至 zlts@phei.com.cn，盗版侵权举报请发邮件至 dbqq@phei.com.cn。
本书咨询联系方式：010-51260888-819，faq@phei.com.cn。

推荐序

移动互联网的兴起把我们带入了真正的大数据时代，各大互联网公司由于服务于海量用户，因此一般都储存了 EB 级的数据，而在一个典型的业务场景中，每次处理 TB 级的数据也是很常见的。同时由于竞争的加剧，互联网公司对业务的要求需要不断提高质量和降低响应速度，这就给大数据处理工具带来了非常大的挑战。目前的大数据生态圈以 Hadoop 为主，在数据存储、SQL 查询引擎、分布式计算、实时处理引擎和机器学习等方向先后诞生了一系列的开源项目，由于这些开源项目面向各自的领域，因而在设计的开发部署中，开发和运维的工程师就不得不同时面对不同的工具和环境，无形中大大增加了公司的成本和工程师的学习门槛。

Spark 就是在这样的背景下发展起来的，是目前为止唯一能够把交互式查询、实时处理、离线处理和机器学习无缝结合在一起的大数据产品。Spark 是基于内存的计算框架，其设计非常精巧，对于多次迭代的数据处理，例如机器学习和 SQL 查询，可以比 Hadoop 快很多倍，所以诞生不久即成为 Apache 的顶级项目，之后更是得到众多企业和开发者的拥护。一些大的互联网公司在过去几年虽然开发了自己的大数据处理框架，但最近几年也逐渐转向了 Spark。而对于那些对外提供云服务的厂商，Spark 更是成为一个标准配置环境，由此可见 Spark 的火热程度。

早期版本的 Spark 实现相当精简，然而随着版本的快速迭代和功能的不断增加，其实现已经变得相当复杂。由于分布式计算的复杂性，开发者和运维人员在实际使用过程中，经常会遇到集群甚至 Spark 本身相关的问题，在不了解工作原理的情况下很难快速定位和解决问题。本书作者在大数据领域深研究数年，自 Spark 诞生之日起就一直密切关注其发展，对其设计框架、运行机制和对外 API 都有较为深入的了解。作为京东大数据平台部门，我们需要解决业务部门在使用 Spark 过程中所遇到的各种复杂问题，在此过程中本书作者给予了我们莫大的帮助。

《图解 Spark：核心技术与案例实战》从生态系统讲起，先让读者对 Spark 生态圈有一个大概的了解。之后通过配置 Spark 开发环境，以及一个实际的例子告诉读者如何在 Spark 上快速开发。接下来作者详细介绍了 Spark 的编程模型和核心架构，这是本书的精华所在，也是真正了解 Spark 的必读内容，作者以平实的语言透彻地讲解了 RDD 含义、内部处理逻辑、任务执行的调度过程以及集群中 Driver 节点和 Worker 节点之间的交互等，至此读者可以清楚了解到 Spark

应用执行的完整过程。第 5 章详细介绍了存储原理及 Shuffle 过程，对于这些内容开发者往往不太在意，但对应用的整体性能有非常大的影响。第 6 章则以实际案例介绍了 Spark 的多种运行方式，读者从中可以了解到 Spark 是如何与自身资源管理框架、Yarn 集群或者 Mesos 集群进行交互的。

在对 Spark 应用有了整体的认识之后，作者又分别对 SQL 查询、流处理、机器学习、图计算和 SparkR 等核心子系统进行了深入解读，既介绍了各自的开发接口，又清楚地介绍了各个模块之间的关系。最后特别介绍了 Alluxio，其源于 Spark，可以为各种分布式系统提供抽象的文件存储服务。

大规模机器学习专家　京东大数据架构师　何云龙

前　言

为什么要写这本书

在过去的十几年里，随着计算机的普遍应用和互联网的普及，使得数据呈现爆发式增长，在这个背景下，Doug Cutting 在谷歌的两篇论文（GFS 和 MapReduce）的启发下开发了 Nutch 项目。2006 年 Hadoop 脱离了 Nutch，成为 Apache 的顶级项目，带动了大数据发展的新十年。在此期间，大数据开源产品如雨后春笋般层出不穷，特别是 2009 年由加州大学伯克利分校 AMP 实验室开发的 Spark，它以内存迭代计算的高效和各组件所形成一站式解决平台成为这些产品的翘楚。

Spark 在 2013 年 6 月成为 Apache 孵化项目，8 个月后成为其顶级项目，并于 2014 年 5 月发布了 1.0 版本，在 2016 年 7 月正式发布了 2.0 版本。在这个过程中，Spark 社区不断壮大，成为了最为活跃的大数据社区之一。作为大数据处理的"利器"，Spark 在发展过程中不断地演进，因此各个版本存在较大的差异。市面上关于 Spark 的书已经不少，但是这些书所基于的 Spark 版本稍显陈旧，另外在介绍 Spark 的时候，未能把原理、代码和实例相结合，于是便有了本书，本书能够在剖析 Spark 原理的同时结合实际案例，从而让读者能够更加深入理解和掌握 Spark。

在本书中，首先对 Spark 的生态圈进行了介绍，讲述了 Spark 的发展历程，同时也介绍 Spark 实战环境的搭建；接下来从 Spark 的编程模型、作业执行、存储原理和运行架构等方面讲解了 Spark 内部核心原理；最后对 Spark 的各组件进行详细介绍，这些组件包括 Spark SQL 的即席查询、Spark Streaming 的实时流处理应用、MLbase/MLlib 的机器学习、GraphX 的图处理、SparkR 的数学计算和 Alluxio 的分布式内存文件系统等。

读者对象

（1）大数据爱好者

随着大数据时代的来临，无论是传统行业、IT 行业还是互联网等行业，都将涉及大数据技术，本书能够帮助这些行业的大数据爱好者了解 Spark 生态圈和发展演进趋势。通过本书，读者不仅可以了解到 Spark 的特点和使用场景，而且如果希望继续深入学习 Spark 知识，那么本书也是很好的入门选择。

（2）Spark 开发人员

如果要进行 Spark 应用的开发，仅仅掌握 Spark 基本使用方法是不够的，还需深入了解 Spark 的设计原理、架构和运行机制。本书深入浅出地讲解了 Spark 的编程模型、作业运行机制、存储原理和运行架构等内容，通过对这些内容的学习，相信读者可以编写出更加高效的应用程序。

（3）Spark 运维人员

作为一名 Spark 运维人员，适当了解 Spark 的设计原理、架构和运行机制对于运维工作十分有帮助。通过对本书的学习，不仅能够更快地定位并排除故障，而且还能对 Spark 运行进行调优，让 Spark 运行得更加稳定和快速。

（4）数据科学家和算法研究

随着大数据技术的发展，实时流计算、机器学习、图计算等领域成为较热门的研究方向，而 Spark 有着较为成熟的生态圈，能够一站式解决类似场景的问题。这些研究人员可以通过本书加深对 Spark 的原理和应用场景的理解，从而能够更好地利用 Spark 各个组件进行数据计算和算法实现。

内容速览

本书分为三个部分，共计 12 章。

第一部分为基础篇（第 1~2 章），介绍了 Spark 诞生的背景、演进历程，以及 Spark 生态圈的组成，并详细介绍了如何搭建 Spark 实战环境。通过该环境不仅可以阅读 Spark 源代码，而且可以开发 Spark 应用程序。

第二部分为核心篇（第 3~6 章），讲解了 Spark 的编程模型、核心原理、存储原理和运行架构，在核心原理中对 Spark 通信机制、作业执行原理、调度算法、容错和监控管理等进行了深入分析，在分析原理和代码的同时结合实例进行演示。

第三部分为组件篇（第 7~12 章），介绍了 Spark 的各个组件，包括 Spark SQL 的即席查询、Spark Streaming 的实时流处理应用、MLbase/MLlib 的机器学习、GraphX 的图处理、SparkR 的数学计算和 Alluxio 的分布式内存文件系统等。

另外本书后面还包括 5 个附录：附录 A 为编译安装 Hadoop，附录 B 为安装 MySQL 数据库，附录 C 为编译安装 Hive，附录 D 为安装 ZooKeeper，附录 E 为安装 Kafka。由于本书篇幅有限，因此这些内容可到我的博客（http://www.cnblogs.com/shishanyuan）或博文视点网站（www.broadview.com.cn\30236）下载。

勘误和支持

由于笔者水平有限,加之编写时间跨度较长,同时 Spark 演进较快,因此在编写本书的过程中,难免会出现错误或者不准确的地方,恳请读者批评指正。如果本书存有错误,或者您有 Spark 的内容需要探讨,可以发送邮件到 jan98341@qq.com 与我联系,期待能够得到大家的反馈。

致谢

感谢中油瑞飞公司,让我接触到大数据的世界,并在工作的过程中深入了解 Spark。感谢吴建平、于鹏、李新宅、祝军、张文邃、马君博士、卢文君等领导同事,在本书编写过程中提供无私的帮助和宝贵的建议。

感谢京东商城的付彩宝、沈晓凯对我的工作和该书的支持,感谢付彩宝在繁忙的工作之余为本书写推荐,感谢京东数据挖掘架构师何云龙为本书作序,感谢大数据平台部的周龙波对该书提出了宝贵意见。

感谢 EMC 常雷博士为本书审稿并写推荐。

感谢 Alluxio 的 CEO 李浩源博士对本书的支持,感谢范斌在非常忙的工作中,抽出时间给 Alluxio 章节进行了审稿并提供了很好的建议。

非常感谢我的家人对我的理解和支持,特别是在写书过程中老婆又为我们家添了一位猴宝宝,让为我拥有一对健康可爱的儿女,这些都给了我莫大的动力,让我的努力更加有意义。

谨以此书先给我亲爱的家人,你们是我努力的源泉。

郭景瞻

2016 年 11 月

读者服务

轻松注册成为博文视点社区用户（www.broadview.com.cn），扫码直达本书页面。

- **提交勘误**：您对书中内容的修改意见可在 提交勘误 处提交，若被采纳，将获赠博文视点社区积分（在您购买电子书时，积分可用来抵扣相应金额）。
- **交流互动**：在页面下方 读者评论 处留下您的疑问或观点，与我们和其他读者一同学习交流。

页面入口：http://www.broadview.com.cn/30236

目录

第一篇 基础篇

第1章 Spark 及其生态圈概述 ... 1
1.1 Spark 简介 ... 1
1.1.1 什么是 Spark ... 1
1.1.2 Spark 与 MapReduce 比较 ... 3
1.1.3 Spark 的演进路线图 ... 4
1.2 Spark 生态系统 ... 5
1.2.1 Spark Core ... 6
1.2.2 Spark Streaming ... 7
1.2.3 Spark SQL ... 9
1.2.4 BlinkDB ... 11
1.2.5 MLBase/MLlib ... 12
1.2.6 GraphX ... 12
1.2.7 SparkR ... 13
1.2.8 Alluxio ... 14
1.3 小结 ... 15

第2章 搭建 Spark 实战环境 ... 16
2.1 基础环境搭建 ... 16
2.1.1 搭建集群样板机 ... 17
2.1.2 配置集群环境 ... 22
2.2 编译 Spark 源代码 ... 25
2.2.1 配置 Spark 编译环境 ... 26
2.2.2 使用 Maven 编译 Spark ... 27
2.2.3 使用 SBT 编译 Spark ... 29
2.2.4 生成 Spark 部署包 ... 30
2.3 搭建 Spark 运行集群 ... 31

    2.3.1 修改配置文件 31
    2.3.2 启动 Spark 33
    2.3.3 验证启动 33
    2.3.4 第一个实例 33
  2.4 搭建 Spark 实战开发环境 35
    2.4.1 CentOS 中部署 IDEA 36
    2.4.2 使用 IDEA 开发程序 37
    2.4.3 使用 IDEA 阅读源代码 42
  2.5 小结 47

第二篇 核心篇

第 3 章 Spark 编程模型ﾠ... 48
  3.1 RDD 概述 48
    3.1.1 背景 48
    3.1.2 RDD 简介 49
    3.1.3 RDD 的类型 50
  3.2 RDD 的实现 51
    3.2.1 作业调度 51
    3.2.2 解析器集成 52
    3.2.3 内存管理 53
    3.2.4 检查点支持 54
    3.2.5 多用户管理 54
  3.3 编程接口 55
    3.3.1 RDD 分区（Partitions）... 55
    3.3.2 RDD 首选位置（PreferredLocations）... 56
    3.3.3 RDD 依赖关系（Dependencies）.. 56
    3.3.4 RDD 分区计算（Iterator）.. 58
    3.3.5 RDD 分区函数（Partitioner）... 58
  3.4 创建操作 59
    3.4.1 并行化集合创建操作 59
    3.4.2 外部存储创建操作 61
  3.5 转换操作 63
    3.5.1 基础转换操作 63

	3.5.2	键值转换操作	70
3.6		控制操作	77
3.7		行动操作	80
	3.7.1	集合标量行动操作	80
	3.7.2	存储行动操作	84
3.8		小结	87

第4章 Spark 核心原理 89

- 4.1 消息通信原理 90
  - 4.1.1 Spark 消息通信架构 90
  - 4.1.2 Spark 启动消息通信 91
  - 4.1.3 Spark 运行时消息通信 94
- 4.2 作业执行原理 102
  - 4.2.1 概述 102
  - 4.2.2 提交作业 104
  - 4.2.3 划分调度阶段 106
  - 4.2.4 提交调度阶段 109
  - 4.2.5 提交任务 112
  - 4.2.6 执行任务 117
  - 4.2.7 获取执行结果 119
- 4.3 调度算法 122
  - 4.3.1 应用程序之间 122
  - 4.3.2 作业及调度阶段之间 126
  - 4.3.3 任务之间 130
- 4.4 容错及 HA 136
  - 4.4.1 Executor 异常 136
  - 4.4.2 Worker 异常 137
  - 4.4.3 Master 异常 138
- 4.5 监控管理 139
  - 4.5.1 UI 监控 139
  - 4.5.2 Metrics 150
  - 4.5.3 REST 152
- 4.6 实例演示 154
  - 4.6.1 计算年降水实例 154

第5章 Spark 存储原理 161

5.1 存储分析 161
5.1.1 整体架构 161
5.1.2 存储级别 167
5.1.3 RDD 存储调用 168
5.1.4 读数据过程 170
5.1.5 写数据过程 177

5.2 Shuffle 分析 186
5.2.1 Shuffle 简介 186
5.2.2 Shuffle 的写操作 186
5.2.3 Shuffle 的读操作 193

5.3 序列化和压缩 200
5.3.1 序列化 200
5.3.2 压缩 201

5.4 共享变量 202
5.4.1 广播变量 202
5.4.2 累加器 203

5.5 实例演示 204

5.6 小结 208

第6章 Spark 运行架构 209

6.1 运行架构总体介绍 209
6.1.1 总体介绍 209
6.1.2 重要类介绍 210

6.2 本地（Local）运行模式 211
6.2.1 运行模式介绍 211
6.2.2 实现原理 213

6.3 伪分布（Local-Cluster）运行模式 215
6.3.1 运行模式介绍 215
6.3.2 实现原理 216

6.4 独立（Standalone）运行模式 218

（前文）

4.6.2 HA 配置实例 157

4.7 小结 160

　　　　6.4.1 运行模式介绍 218
　　　　6.4.2 实现原理 219
　6.5 YARN 运行模式 220
　　　　6.5.1 YARN 运行框架 220
　　　　6.5.2 YARN-Client 运行模式介绍 221
　　　　6.5.3 YARN-Client 运行模式实现原理 223
　　　　6.5.4 YARN-Cluster 运行模式介绍 227
　　　　6.5.5 YARN-Cluster 运行模式实现原理 229
　　　　6.5.6 YARN-Client 与 YARN-Cluster 对比 232
　6.6 Mesos 运行模式 233
　　　　6.6.1 Mesos 介绍 233
　　　　6.6.2 粗粒度运行模式介绍 234
　　　　6.6.3 粗粒度实现原理 236
　　　　6.6.4 细粒度运行模式介绍 239
　　　　6.6.5 细粒度实现原理 240
　　　　6.6.6 Mesos 粗粒度和 Mesos 细粒度对比 243
　6.7 实例演示 243
　　　　6.7.1 独立运行模式实例 243
　　　　6.7.2 YARN-Client 实例 247
　　　　6.7.3 YARN-Cluster 实例 250
　6.8 小结 253

第三篇　组件篇

第 7 章 Spark SQL 255
　7.1 Spark SQL 简介 255
　　　　7.1.1 Spark SQL 发展历史 255
　　　　7.1.2 DataFrame/Dataset 介绍 258
　7.2 Spark SQL 运行原理 261
　　　　7.2.1 通用 SQL 执行原理 261
　　　　7.2.2 SparkSQL 运行架构 262
　　　　7.2.3 SQLContext 运行原理分析 265
　　　　7.2.4 HiveContext 介绍 276
　7.3 使用 Hive-Console 278

7.3.1 编译 Hive-Console 278
7.3.2 查看执行计划 280
7.3.3 应用 Hive-Console 281
7.4 使用 SQLConsole 284
7.4.1 启动 HDFS 和 Spark Shell 284
7.4.2 与 RDD 交互操作 284
7.4.3 读取 JSON 格式数据 287
7.4.4 读取 Parquet 格式数据 288
7.4.5 缓存演示 289
7.4.6 DSL 演示 290
7.5 使用 Spark SQL CLI 290
7.5.1 配置并启动 Spark SQL CLI 291
7.5.2 实战 Spark SQL CLI 292
7.6 使用 Thrift Server 293
7.6.1 配置并启动 Thrift Server 293
7.6.2 基本操作 295
7.6.3 交易数据实例 296
7.6.4 使用 IDEA 开发实例 298
7.7 实例演示 299
7.7.1 销售数据分类实例 299
7.7.2 网店销售数据统计 303
7.8 小结 306

第 8 章 Spark Streaming 308
8.1 Spark Streaming 简介 308
8.1.1 术语定义 309
8.1.2 Spark Streaming 特点 312
8.2 Spark Streaming 编程模型 314
8.2.1 DStream 的输入源 314
8.2.2 DStream 的操作 315
8.3 Spark Streaming 运行架构 319
8.3.1 运行架构 319
8.3.2 消息通信 320
8.3.3 Receiver 分发 323

　　　　8.3.4　容错性 329
　8.4　Spark Streaming 运行原理 331
　　　　8.4.1　启动流处理引擎 331
　　　　8.4.2　接收及存储流数据 334
　　　　8.4.3　数据处理 341
　8.5　实例演示 346
　　　　8.5.1　流数据模拟器 346
　　　　8.5.2　销售数据统计实例 348
　　　　8.5.3　Spark Streaming+Kafka 实例 351
　8.6　小结 356

第 9 章　Spark MLlib 358

　9.1　Spark MLlib 简介 358
　　　　9.1.1　Spark MLlib 介绍 358
　　　　9.1.2　Spark MLlib 数据类型 360
　　　　9.1.3　Spark MLlib 基本统计方法 365
　　　　9.1.4　预言模型标记语言 369
　9.2　线性模型 370
　　　　9.2.1　数学公式 370
　　　　9.2.2　线性回归 371
　　　　9.2.3　线性支持向量机 372
　　　　9.2.4　逻辑回归 373
　　　　9.2.5　线性最小二乘法、Lasso 和岭回归 373
　　　　9.2.6　流式线性回归 373
　9.3　决策树 374
　9.4　决策模型组合 375
　　　　9.4.1　随机森林 376
　　　　9.4.2　梯度提升决策树 377
　9.5　朴素贝叶斯 377
　9.6　协同过滤 378
　9.7　聚类 380
　　　　9.7.1　K-means 380
　　　　9.7.2　高斯混合 382
　　　　9.7.3　快速迭代聚类 384

9.7.4 LDA ... 384
9.7.5 二分 K-means ... 385
9.7.6 流式 K-means ... 386
9.8 降维 ... 386
9.8.1 奇异值分解降维 ... 386
9.8.2 主成分分析降维 ... 387
9.9 特征提取和变换 ... 388
9.9.1 词频—逆文档频率 ... 388
9.9.2 词向量化工具 ... 389
9.9.3 标准化 ... 390
9.9.4 范数化 ... 390
9.10 频繁模式挖掘 ... 391
9.10.1 频繁模式增长 ... 391
9.10.2 关联规则挖掘 ... 391
9.10.3 PrefixSpan ... 391
9.11 实例演示 ... 392
9.11.1 K-means 聚类算法实例 ... 392
9.11.2 手机短信分类实例 ... 396
9.12 小结 ... 401

第 10 章 Spark GraphX ... 402

10.1 GraphX 介绍 ... 402
10.1.1 图计算 ... 402
10.1.2 GraphX 介绍 ... 403
10.1.3 发展历程 ... 404
10.2 GraphX 实现分析 ... 405
10.2.1 GraphX 图数据模型 ... 406
10.2.2 GraphX 图数据存储 ... 408
10.2.3 GraphX 图切分策略 ... 410
10.2.4 GraphX 图操作 ... 412
10.3 实例演示 ... 418
10.3.1 图例演示 ... 418
10.3.2 社区发现演示 ... 425
10.4 小结 ... 429

第 11 章　SparkR 430

11.1　概述 430
- 11.1.1　R 语言介绍 430
- 11.1.2　SparkR 介绍 431

11.2　SparkR 与 DataFrame 432
- 11.2.1　DataFrames 介绍 432
- 11.2.2　与 DataFrame 的相关操作 434

11.3　编译安装 SparkR 435
- 11.3.1　编译安装 R 语言 435
- 11.3.2　安装 SparkR 运行环境 437
- 11.3.3　安装 SparkR 438
- 11.3.4　启动并验证安装 439

11.4　实例演示 440

11.5　小结 444

第 12 章　Alluxio 445

12.1　Alluxio 简介 445
- 12.1.1　Alluxio 介绍 445
- 12.1.2　Alluxio 系统架构 446
- 12.1.3　HDFS 与 Alluxio 450

12.2　Alluxio 编译部署 451
- 12.2.1　编译 Alluxio 451
- 12.2.2　单机部署 Alluxio 453
- 12.2.3　集群模式部署 Alluxio 455

12.3　Alluxio 命令行使用 457
- 12.3.1　接口说明 457
- 12.3.2　接口操作示例 459

12.4　实例演示 462
- 12.4.1　启动环境 462
- 12.4.2　Alluxio 上运行 Spark 462
- 12.4.3　Alluxio 上运行 MapReduce 465

12.5　小结 466

本书附录部分请到博文视点网站下载 www.broadview.com.cn/30236。

第一篇 基础篇

第 1 章

Spark 及其生态圈概述

1.1 Spark 简介

1.1.1 什么是 Spark

Spark 是加州大学伯克利分校 AMP 实验室（Algorithms、Machines and People Lab）开发的通用大数据处理框架。Spark 生态系统也称为 BDAS，是伯克利 APM 实验室所开发的，力图在算法（Algorithms）、机器（Machines）和人（People）三者之间通过大规模集成来展现大数据应用的一个开源平台。AMP 实验室运用大数据、云计算等各种资源以及各种灵活的技术方案，对海量的数据进行分析并转化为有用的信息，让人们更好地了解世界。

Spark 在 2013 年 6 月进入 Apache 成为孵化项目，8 个月后成为 Apache 顶级项目，速度之快足见过人之处。Spark 以其先进的设计理念，迅速成为社区的热门项目，围绕着 Spark 推出了 Spark SQL、Spark Streaming、MLlib、GraphX 和 SparkR 等组件，这些组件逐渐形成大数据处理一站式解决平台。Spark 并非池鱼，它的志向不是作为 Hadoop 的绿叶，而是期望替代 Hadoop 在大数据中的地位，成为大数据处理的主流标准。

Spark 使用 Scala 语言进行实现，它是一种面向对象、函数式编程语言，能够像操作本地集合对象一样轻松地操作分布式数据集。Spark 具有运行速度快、易用性好、通用性强和随处运行等特点。

1. 运行速度快

Spark 的中文意思是"电光火石"，Spark 确实如此！官方提供的数据表明，如果数据由磁盘读取，速度是 Hadoop MapReduce 的 10 倍以上；如果数据从内存中读取，速度可以高达 100

多倍。图1-1是在逻辑回归算法中Hadoop与Spark处理时间的比较，左边是Hadoop，耗时110秒，而Spark耗时0.9秒。

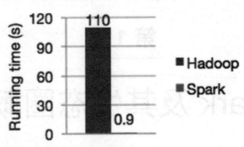

图1-1 逻辑回归算法在Hadoop和Spark上处理时间的比较

Spark相对于Hadoop有如此快的计算速度有数据本地性、调度优化和传输优化等原因，其中最主要的是基于内存计算和引入DAG执行引擎。

（1）Spark默认情况下迭代过程的数据保存到内存中，后续的运行作业利用这些结果进行计算，而Hadoop每次计算结果都直接存储到磁盘中，在随后的计算中需要从磁盘中读取上次计算的结果。由于从内存读取数据时间比磁盘读取时间低两个数量级，这就造成了Hadoop的运行速度较慢，这种情况在迭代计算中尤为明显。

（2）由于较复杂的数据计算任务需要多个步骤才能实现，且步骤之间具有依赖性。对于这些步骤之间，Hadoop需要借助Oozie等工具进行处理。而Spark在执行任务前，可以将这些步骤根据依赖关系形成DAG图（有向无环图），任务执行可以按图索骥，不需要人工干预，从而优化了计算路径，大大减少了I/O读取操作。

2. 易用性好

Spark不仅支持Scala编写应用程序，而且支持Java和Python等语言进行编写。Scala是一种高效、可拓展的语言，能够用简洁的代码处理较为复杂的处理工作。比如经典的WordCount例子，使用Scala编写，仅用简单两条语句就能够实现。具体代码如下：

```
scala> val textFile = sc.textFile("file:///home/hadoop/README.md")
scala> val counts = textFile.flatMap(line => line.split(" ")).map(word => (word, 1)).reduceByKey(_ + _)
```

3. 通用性强

Spark生态圈即BDAS（伯克利数据分析栈）所包含的组件：Spark Core提供内存计算框架、SparkStreaming的实时处理应用、Spark SQL的即席查询、MLlib的机器学习和GraphX的图处理，它们都是由AMP实验室提供，能够无缝地集成，并提供一站式解决平台，如图1-2所示。

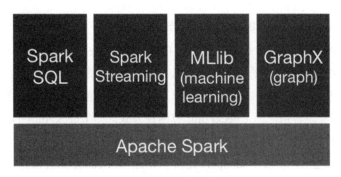

图 1-2　Spark 技术堆栈

4. 随处运行

Spark 具有很强的适应性，能够读取 HDFS、Cassandra、HBase、S3 和 Tachyon，为持久层读写原生数据，能够以 Mesos、YARN 和自身携带的 Standalone 作为资源管理器调度作业来完成 Spark 应用程序的计算。图 1-3 是 Spark 支持的技术框架。

图 1-3　Spark 支持的技术框架

1.1.2　Spark 与 MapReduce 比较

Spark 是通过借鉴 Hadoop MapReduce 发展而来的，继承了其分布式并行计算的优点，并改进了 MapReduce 明显的缺陷，具体体现在以下几个方面。

（1）Spark 把中间数据放在内存中，迭代运算效率高。MapReduce 中的计算结果是保存在磁盘上，这样势必会影响整体的运行速度，而 Spark 支持 DAG 图的分布式并行计算的编程框架，减少了迭代过程中数据的落地，提高了处理效率。

（2）Spark 的容错性高。Spark 引进了弹性分布式数据集（Resilient Distributed Dataset，RDD）的概念，它是分布在一组节点中的只读对象集合，这些集合是弹性的，如果数据集一部分丢失，则可以根据"血统"（即允许基于数据衍生过程）对它们进行重建。另外，在 RDD 计算时可以通过 CheckPoint 来实现容错，而 CheckPoint 有两种方式，即 CheckPoint Data 和 Logging The Updates，用户可以控制采用哪种方式来实现容错。

（3）Spark 更加通用。不像 Hadoop 只提供了 Map 和 Reduce 两种操作，Spark 提供的数据集操作类型有很多种，大致分为转换操作和行动操作两大类。转换操作包括 Map、Filter、FlatMap、Sample、GroupByKey、ReduceByKey、Union、Join、Cogroup、MapValues、Sort 和 PartionBy 等多种操作类型，行动操作包括 Collect、Reduce、Lookup 和 Save 等操作类型。另外，各个处理节点之间的通信模型不再像 Hadoop 只有 Shuffle 一种模式，用户可以命名、物化，控制中间结果的存储、分区等。

1.1.3　Spark 的演进路线图

Spark 由 Lester 和 Matei 在 2009 年算法比赛的思想碰撞中诞生，随后 4 年中，Spark 在 Berkeley's AMPLab 逐渐形成了现有的 Spark 雏形。Spark 在 2013 年 6 月进入 Apache 成为孵化项目，8 个月后成为 Apache 顶级项目，从此 Spark 的发展进入了快车道。2014 年 5 月底发布了第一个正式版本 Spark 1.0.0，在随后的时间里大致以 3 个月为周期发布一个小版本，并经过两年沉淀在 2016 年 7 月推出了 Spark 2.0 正式版本，其具体演进时间如下：

- 2009 年由 Berkeley's AMPLab 开始编写最初的源代码
- 2010 年开放源代码
- 2012 年 2 月发布 0.6.0 版本
- 2013 年 6 月进入 Apache 孵化器项目
- 2013 年年中 Spark 主要成员创立 Databricks 公司
- 2014 年 2 月成为 Apache 的顶级项目（8 个月的时间）
- 2014 年 5 月底 Spark 1.0.0 发布
- 2014 年 9 月 Spark 1.1.0 发布
- 2014 年 12 月 Spark 1.2.0 发布
- 2015 年 3 月 Spark 1.3.0 发布
- 2015 年 6 月 Spark 1.4.0 发布
- 2015 年 9 月 Spark 1.5.0 发布

- 2016 年 1 月 Spark 1.6.0 发布
- 2016 年 5 月 Spark 2.0.0 Preview 版本发布
- 2016 年 7 月 Spark 2.0.0 正式版本发布

图 1-4 为 Spark 的演进时间轴。

图 1-4 Spark 演进时间轴

Spark 进入 Apache 后以其代码开源、内存计算和一栈式解决方案风靡大数据生态圈，成为该生态圈和 Apache 基金会内最活跃的项目，得到了大数据研究人员、机构和众多厂商的支持。

- Spark 成为整个大数据生态圈和 Apache 基金会内最活跃的项目。
- Hadoop 最大的厂商 Cloudera 宣称加大 Spark 框架的投入来取代 MapReduce。
- Hortonworks 加大 Hadoop 与 Spark 整合。
- Hadoop 厂商 MapR 投入 Spark 阵营。
- Apache Mahout 放弃 MapReduce，将使用 Spark 作为后续算子的计算平台。

……

1.2 Spark 生态系统

Spark 生态系统以 Spark Core 为核心，能够读取传统文件（如文本文件）、HDFS、Amazon S3、Alluxio 和 NoSQL 等数据源，利用 Standalone、YARN 和 Mesos 等资源调度管理，完成应用程序分析与处理。这些应用程序来自 Spark 的不同组件，如 Spark Shell 或 Spark Submit 交互式批处理方式、Spark Streaming 的实时流处理应用、Spark SQL 的即席查询、采样近似查询引擎 BlinkDB 的权衡查询、MLbase/MLlib 的机器学习、GraphX 的图处理和 SparkR 的数学计算等，如图 1-5 所示，正是这个生态系统实现了"One Stack to Rule Them All"目标。

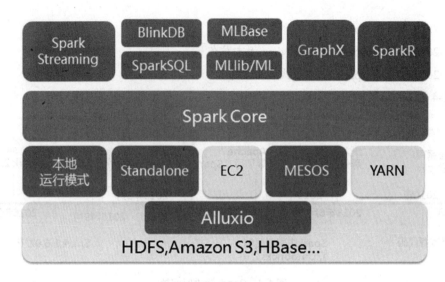

图 1-5　Spark 生态系统

1.2.1　Spark Core

Spark Core 是整个 BDAS 生态系统的核心组件，是一个分布式大数据处理框架。Spark Core 提供了多种资源调度管理，通过内存计算、有向无环图（DAG）等机制保证分布式计算的快速，并引入了 RDD 的抽象保证数据的高容错性，其重要特性描述如下。

- Spark Core 提供了多种运行模式，不仅可以使用自身运行模式处理任务，如本地模式、Standalone，而且可以使用第三方资源调度框架来处理任务，如 YARN、MESOS 等。相比较而言，第三方资源调度框架能够更细粒度管理资源。
- Spark Core 提供了有向无环图（DAG）的分布式并行计算框架，并提供内存机制来支持多次迭代计算或者数据共享，大大减少迭代计算之间读取数据的开销，这对于需要进行多次迭代的数据挖掘和分析性能有极大提升。另外，在任务处理过程中移动计算而非移动数据，RDD Partition 可以就近读取分布式文件系统中的数据块到各个节点内存中进行计算。
- 在 Spark 中引入了 RDD 的抽象，它是分布在一组节点中的只读对象集合，这些集合是弹性的，如果数据集一部分丢失，则可以根据"血统"对它们进行重建，保证了数据的高容错性。

1.2.2 Spark Streaming

Spark Streaming 是一个对实时数据流进行高吞吐、高容错的流式处理系统，可以对多种数据源（如 Kafka、Flume、Twitter 和 ZeroMQ 等）进行类似 Map、Reduce 和 Join 等复杂操作，并将结果保存到外部文件系统、数据库或应用到实时仪表盘，如图 1-6 所示。相比其他的处理引擎要么只专注于流处理，要么只负责批处理（仅提供需要外部实现的流处理 API 接口），而 Spark Streaming 最大的优势是提供的处理引擎和 RDD 编程模型可以同时进行批处理与流处理。

图 1-6　Spark Streaming 的输入/输出类型

对于传统流处理中一次处理一条记录的方式而言，Spark Streaming 使用的是将流数据离散化处理（Discretized Streams），通过该处理方式能够进行秒级以下的数据批处理。在 Spark Streaming 处理过程中，Receiver 并行接收数据，并将数据缓存至 Spark 工作节点的内存中。经过延迟优化后，Spark 引擎对短任务（几十毫秒）能够进行批处理，并且可将结果输出至其他系统中。与传统连续算子模型不同，其模型是静态分配给一个节点进行计算，而 Spark 可基于数据的来源以及可用资源情况动态分配给工作节点，如图 1-7 所示。

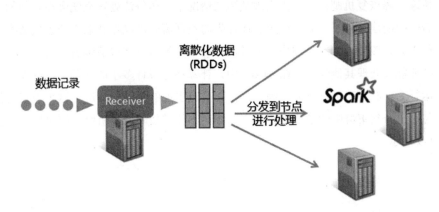

图 1-7　Spark Streaming 处理架构

使用离散化流数据（DStreaming），Spark Streaming 将具有如下特性。

- **动态负载均衡**：Spark Streaming 将数据划分为小批量，通过这种方式可以实现对资源更细粒度的分配。例如，传统实时流记录处理系统在输入数据流以键值进行分区处理情况下，如果一个节点计算压力较大超出了负荷，该节点将成为瓶颈，进而拖慢整个系统的处理速度。而在 Spark Streaming 中，作业任务将会动态地平衡分配给各个节点，如图 1-8 所示，即如果任务处理时间较长，分配的任务数量将少些；如果任务处理时间较短，则分配的任务数据将更多些。

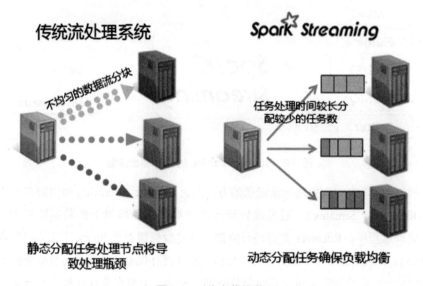

图 1-8　动态负载均衡

- **快速故障恢复机制**：在节点出现故障的情况下，传统流处理系统会在其他的节点上重启失败的连续算子，并可能重新运行先前数据流处理操作获取部分丢失数据。在此过程中只有该节点重新处理失败的过程，只有在新节点完成故障前所有计算后，整个系统才能够处理其他任务。在 Spark 中，计算将分成许多小的任务，保证能在任何节点运行后能够正确进行合并。因此，在某节点出现的故障的情况，这个节点的任务将均匀地分散到集群中的节点进行计算，相对于传递故障恢复机制能够更快地恢复，如图 1-9 所示。

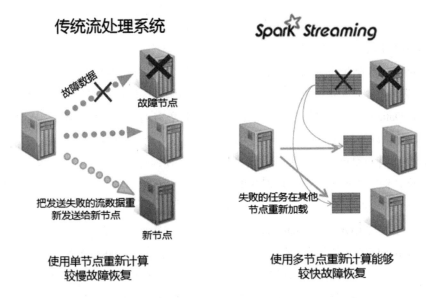

图 1-9　快速故障恢复机制

- 批处理、流处理与交互式分析的一体化：Spark Streaming 是将流式计算分解成一系列短小的批处理作业，也就是把 Spark Streaming 的输入数据按照批处理大小（如几秒）分成一段一段的离散数据流（DStream），每一段数据都转换成 Spark 中的 RDD，然后将 Spark Streaming 中对 DStream 流处理操作变为针对 Spark 中对 RDD 的批处理操作。另外，流数据都储存在 Spark 节点的内存里，用户便能根据所需进行交互查询。正是利用了 Spark 这种工作机制将批处理、流处理与交互式工作结合在一起。

1.2.3　Spark SQL

Spark SQL 的前身是 Shark，它发布时 Hive 可以说是 SQL on Hadoop 的唯一选择（Hive 负责将 SQL 编译成可扩展的 MapReduce 作业），鉴于 Hive 的性能以及与 Spark 的兼容，Shark 由此而生。

Shark 即 Hive on Spark，本质上是通过 Hive 的 HQL 进行解析，把 HQL 翻译成 Spark 上对应的 RDD 操作，然后通过 Hive 的 Metadata 获取数据库里的表信息，实际为 HDFS 上的数据和文件，最后由 Shark 获取并放到 Spark 上运算。Shark 的最大特性就是速度快，能与 Hive 的完全兼容，并且可以在 Shell 模式下使用 rdd2sql 这样的 API，把 HQL 得到的结果集继续在 Scala 环境下运算，支持用户编写简单的机器学习或简单分析处理函数，对 HQL 结果进一步分析计算。

在 2014 年 7 月 1 日的 Spark Summit 上，Databricks 宣布终止对 Shark 的开发，将重点放到

Spark SQL 上。在此次会议上，Databricks 表示，Shark 更多是对 Hive 的改造，替换了 Hive 的物理执行引擎，使之有一个较快的处理速度。然而，不容忽视的是，Shark 继承了大量的 Hive 代码，因此给优化和维护带来大量的麻烦。随着性能优化和先进分析整合的进一步加深，基于 MapReduce 设计的部分无疑成为了整个项目的瓶颈。因此，为了更好的发展，给用户提供一个更好的体验，Databricks 宣布终止 Shark 项目，从而将更多的精力放到 Spark SQL 上。

Spark SQL 允许开发人员直接处理 RDD，同时也可查询在 Hive 上存在的外部数据。Spark SQL 的一个重要特点是能够统一处理关系表和 RDD，使得开发人员可以轻松地使用 SQL 命令进行外部查询，同时进行更复杂的数据分析。

Spark SQL 的特点如下。

- 引入了新的 RDD 类型 SchemaRDD，可以像传统数据库定义表一样来定义 SchemaRDD。SchemaRDD 由定义了列数据类型的行对象构成。SchemaRDD 既可以从 RDD 转换过来，也可以从 Parquet 文件读入，还可以使用 HiveQL 从 Hive 中获取。
- 内嵌了 Catalyst 查询优化框架，在把 SQL 解析成逻辑执行计划之后，利用 Catalyst 包里的一些类和接口，执行了一些简单的执行计划优化，最后变成 RDD 的计算。
- 在应用程序中可以混合使用不同来源的数据，如可以将来自 HiveQL 的数据和来自 SQL 的数据进行 Join 操作。

Shark 的出现使得 SQL-on-Hadoop 的性能比 Hive 有了 10～100 倍的提高，那么，摆脱了 Hive 的限制，Spark SQL 的性能又有怎么样的表现呢？虽然没有 Shark 相对于 Hive 那样瞩目的性能提升，但也表现得优异，如图 1-10 所示（其中，右侧数据为 Spark SQL）。

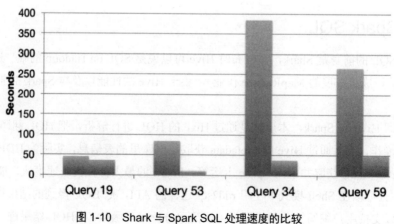

图 1-10　Shark 与 Spark SQL 处理速度的比较

为什么 Spark SQL 的性能会得到这么大的提升呢？主要是 Spark SQL 在以下几点做了优化。

- 内存列存储（In-Memory Columnar Storage）：Spark SQL 的表数据在内存中存储不是采用原生态的 JVM 对象存储方式，而是采用内存列存储。
- 字节码生成技术（Bytecode Generation）：Spark 1.1.0 在 Catalyst 模块的 Expressions 增加了 Codegen 模块，使用动态字节码生成技术，对匹配的表达式采用特定的代码动态编译。另外对 SQL 表达式都做了 CG 优化。CG 优化的实现主要还是依靠 Scala 2.10 运行时的反射机制（Runtime Reflection）。
- Scala 代码优化：Spark SQL 在使用 Scala 编写代码的时候，尽量避免低效的、容易 GC 的代码；尽管增加了编写代码的难度，但对于用户来说接口统一。

1.2.4 BlinkDB

BlinkDB 是一个用于在海量数据上运行交互式 SQL 查询的大规模并行查询引擎，它允许用户通过权衡数据精度来提升查询响应时间，其数据的精度被控制在允许的误差范围内。为了达到这个目标，BlinkDB 使用如下核心思想：

- 自适应优化框架，从原始数据随着时间的推移建立并维护一组多维样本。
- 动态样本选择策略，选择一个适当大小的示例，该示例基于查询的准确性和响应时间的紧迫性。

和传统关系型数据库不同，BlinkDB 是一个交互式查询系统，就像一个跷跷板，用户需要在查询精度和查询时间上做权衡；如果用户想要更快地获取查询结果，那么将牺牲查询结果的精度；反之，用户如果想获取更高精度的查询结果，就需要牺牲查询响应时间。图 1-11 为 BlinkDB 架构。

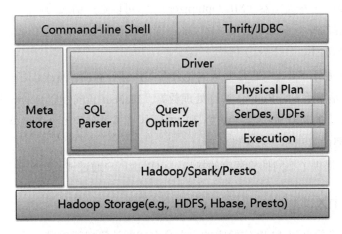

图 1-11　BlinkDB 架构

1.2.5 MLBase/MLlib

MLBase 是 Spark 生态系统中专注于机器学习的组件，它的目标是让机器学习的门槛更低，让一些可能并不了解机器学习的用户能够方便地使用 MLBase。MLBase 分为 4 个部分：MLRuntime、MLlib、MLI 和 ML Optimizer。

- MLRuntime：是由 Spark Core 提供的分布式内存计算框架，运行由 Optimizer 优化过的算法进行数据的计算并输出分析结果。
- MLlib：是 Spark 实现一些常见的机器学习算法和实用程序，包括分类、回归、聚类、协同过滤、降维以及底层优化。该算法可以进行可扩充。
- MLI：是一个进行特征抽取和高级 ML 编程抽象算法实现的 API 或平台。
- ML Optimizer：会选择它认为最适合的已经在内部实现好了的机器学习算法和相关参数，来处理用户输入的数据，并返回模型或其他帮助分析的结果。

图 1-12 为 MLBase/MLlib 结构。

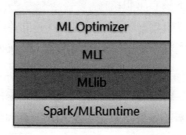

图 1-12 MLBase/MLlib 结构

MLBase 的核心是其优化器（ML Optimizer），它可以把声明式的任务转化成复杂的学习计划，最终产出最优的模型和计算结果。MLBase 与其他机器学习 Weka 和 Mahout 不同，三者各有特色，具体内容如下。

- MLBase 基于 Spark，它是使用的是分布式内存计算的；Weka 是一个单机的系统，而 Mahout 是使用 MapReduce 进行处理数据（Mahout 正向使用 Spark 处理数据转变）。
- MLBase 是自动化处理的；Weka 和 Mahout 都需要使用者具备机器学习技能，来选择自己想要的算法和参数来做处理。
- MLBase 提供了不同抽象程度的接口，可以由用户通过该接口实现算法的扩展。

1.2.6 GraphX

GraphX 最初是伯克利 AMP 实验室的一个分布式图计算框架项目，后来整合到 Spark 中成为一个核心组件。它是 Spark 中用于图和图并行计算的 API，可以认为是 GraphLab 和 Pregel 在 Spark 上的重写及优化。跟其他分布式图计算框架相比，GraphX 最大的优势是：在 Spark 基础上提供了一栈式数据解决方案，可以高效地完成图计算的完整的流水作业。

GraphX 的核心抽象是 Resilient Distributed Property Graph，一种点和边都带属性的有向多重图。GraphX 扩展了 Spark RDD 的抽象，它有 Table 和 Graph 两种视图，但只需要一份物理存储，两种视图都有自己独有的操作符，从而获得了灵活操作和执行效率。GraphX 的整体架构如图 10-4 所示，其中大部分的实现都是围绕 Partition 的优化进行的，这在某种程度上说明了，点分割的存储和相应的计算优化的确是图计算框架的重点和难点。

GraphX 的底层设计有以下几个关键点。

（1）对 Graph 视图的所有操作，最终都会转换成其关联的 Table 视图的 RDD 操作来完成。这样对一个图的计算，最终在逻辑上，等价于一系列 RDD 的转换过程。因此，Graph 最终具备了 RDD 的 3 个关键特性：Immutable、Distributed 和 Fault-Tolerant。其中最关键的是 Immutable（不变性）。逻辑上，所有图的转换和操作都产生了一个新图；物理上，GraphX 会有一定程度的不变顶点和边的复用优化，对用户透明。

（2）两种视图底层共用的物理数据，由 RDD[Vertex-Partition]和 RDD[EdgePartition]这两个 RDD 组成。点和边实际都不是以表 Collection[tuple] 的形式存储的，而是由 VertexPartition/EdgePartition 在内部存储一个带索引结构的分片数据块，以加速不同视图下的遍历速度。不变的索引结构在 RDD 转换过程中是共用的，降低了计算和存储开销。

（3）图的分布式存储采用点分割模式，而且使用 partitionBy 方法，由用户指定不同的划分策略（PartitionStrategy）。划分策略会将边分配到各个 EdgePartition，顶点 Master 分配到各个 VertexPartition，EdgePartition 也会缓存本地边关联点的 Ghost 副本。划分策略的不同会影响到所需要缓存的 Ghost 副本数量，以及每个 EdgePartition 分配的边的均衡程度，需要根据图的结构特征选取最佳策略。

1.2.7 SparkR

R 是遵循 GNU 协议的一款开源、免费的软件，广泛应用于统计计算和统计制图，但是它只能单机运行。为了能够使用 R 语言分析大规模分布式的数据，伯克利分校 AMP 实验室开发了 SparkR，并在 Spark 1.4 版本中加入了该组件。通过 SparkR 可以分析大规模的数据集，并通过 R Shell 交互式地在 SparkR 上运行作业。SparkR 特性如下：

- 提供了 Spark 中弹性分布式数据集（RDDs）的 API，用户可以在集群上通过 R Shell 交互性地运行 Spark 任务。

- 支持序化闭包功能，可以将用户定义函数中所引用到的变量自动序化发送到集群中其他的机器上。
- SparkR 还可以很容易地调用 R 开发包，只需要在集群上执行操作前用 includePackage 读取 R 开发包就可以了。

图 1-13 为 SparkR 的处理流程示意图。

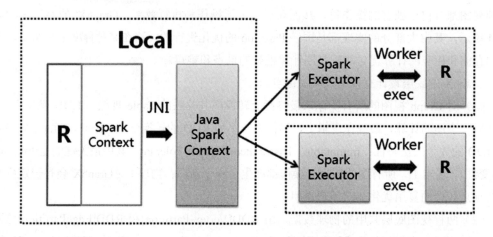

图 1-13　SparkR 处理流程

1.2.8　Alluxio

Alluxio 是一个分布式内存文件系统，它是一个高容错的分布式文件系统，允许文件以内存的速度在集群框架中进行可靠的共享，就像 Spark 和 MapReduce 那样。Alluxio 是架构在最底层的分布式文件存储和上层的各种计算框架之间的一种中间件。其主要职责是将那些不需要落地到 DFS 里的文件，落地到分布式内存文件系统中，来达到共享内存，从而提高效率。同时可以减少内存冗余、GC 时间等。

和 Hadoop 类似，Alluxio 的架构是传统的 Master-Slave 架构，所有的 Alluxio Worker 都被 Alluxio Master 所管理，Alluxio Master 通过 Alluxio Worker 定时发出的心跳来判断 Worker 是否已经崩溃以及每个 Worker 剩余的内存空间量，为了防止单点问题使用了 ZooKeeper 做了 HA，具体参见图 12-2 Alluxio 高可用架构图。

Alluxio 具有如下特性。

- JAVA-Like File API：Alluxio 提供类似 Java File 类的 API。
- 兼容性：Alluxio 实现了 HDFS 接口，所以 Spark 和 MapReduce 程序不需要任何修改即可运行。

- 可插拔的底层文件系统：Alluxio 是一个可插拔的底层文件系统，提供容错功能，它将内存数据记录在底层文件系统。它有一个通用的接口，可以很容易地插入到不同的底层文件系统。目前支持 HDFS、S3、GlusterFS 和单节点的本地文件系统，以后将支持更多的文件系统。Alluxio 所支持的应用如图 1-14 所示。

图 1-14　Alluxio 支持的应用

1.3　小结

本章先介绍了 Spark 诞生的背景，它继承了分布式并行计算的优点并改进了 MapReduce 明显的缺陷，然后介绍了 Spark 的演进路线，最后对 Spark 生态系统进行了介绍，由于这些组件的存在实现了"One Stack to Rule Them All"目标。

第 2 章 搭建 Spark 实战环境

本章不仅介绍基础环境搭建、编译部署 Spark，而且还介绍如何搭建 Spark 程序开发环境，以及使用 IDEA 开发 Spark 程序。如果为了快速掌握 Spark 相关知识，读者可跳过编译 Spark 源代码和编译 Spark 源代码步骤，直接从 Spark 官网下载安装包，搭建一个单机或伪分布式运行环境。

2.1 基础环境搭建

Spark 可以运行在 Windows 或 Linux 等操作系统中。如果使用 Windows 进行搭建，在 Windows 中需要安装 Cygwin。由于本书内容的限制，这里只讲解 Linux 环境下的安装。现在 Linux 操作系统版本较多，既有商业收费的，也有开源免费的，这里采用免费 CentOS 6.5 的 64 位版本，它是来自于 Red Hat Linux 企业版依照开放源代码规定释出的源代码所编译而成的。为什么使用 64 位版本呢？原因是在大多数生产环境使用该配置，更关键的是在该位数版本下才能使用更大的内存，这对 Spark 以内存计算而言，其重要性不言而喻。

为了更加真实地模拟 Spark 运行，在这里我们将搭建 Spark 集群运行模式（当然在该集群中将安装 Hadoop，实战中需要它提供 HDFS 文件存储和 YARN 资源调度管理）。该集群不少于 3 个节点，如果没有物理节点，可以利用虚拟机软件，如 VMware、VirtualBox 等工具进行搭建。本文使用 VMware 搭建虚拟机集群。

搭建集群之前需要规划集群环境，本书演示的集群网络环境配置见表 2-1。

表 2-1 集群网路环境配置

序号	IP 地址	机器名	运行进程	核数/内存	用户名	目录
1	192.168.1.10	master	NN/SNN/DN/RM Master/Worker	1 核/3G	spark	/app /app/soft
2	192.168.1.11	slave1	DN/NM/Worker	1 核/2G	spark	/app/compile
3	192.168.1.12	slave2	DN/NM/Worker	1 核/2G	spark	/app/spark /home/spark

- 所有节点均安装 CentOS 6.5 版本 64 位系统,防火墙/SElinux 均禁用,所有节点上均创建一个 spark 用户,该用户工作目录是/home/spark,上传文件存放在/home/spark/work 目录中。
- 所有节点上均创建一个目录/app 用于存放程序,并且拥有者是 spark 用户(一般做法是 root 用户在根目录下创建/app 目录,并使用 chown 命令修改该目录的拥有者为 spark,否则 spark 用户使用 SSH 命令往其他机器分发文件时会出现权限不足的提示)。在/app 目录下创建 soft、compile 和 spark 三个子目录:/app/soft 子目录用于存放支持 Spark 运行相关基础软件,/app/compile 子目录用于编译时存放源代码及编译结果,而/app/spark 子目录则存放 Spark 和 Hadoop 等。
- 在演示集群中一个节点为 master,其他两个节点为 slave。除在 master 节点中设置运行 Spark 的 Master、Worker 进程外,还运行 Hadoop 的 Name Node、Seconde NameNode、Data Node 和 Resource Manager 等进程。而在 slave 节点中除设置运行 Spark 的 Worker 进程外,还运行 Hadoop 的 Data Node 和 Node Manager 进程。

在保证集群环境目录的统一设置外,最好在每个节点中都采用统一的软件版本,以下为笔者使用软件版本,仅供参考,读者可根据实际情况进行选取。

- JDK:1.7.0_55 64 位
- Hadoop:2.7.2
- Scala:2.11.8
- Spark:2.0.0

2.1.1 搭建集群样板机

本节讲解搭建样板机环境。搭建分为安装操作系统、设置系统环境和配置运行环境 3 个步骤,其中,安装 CentOS 操作系统可以参考作者的博客进行搭建。需要注意的是,CentOS 操作系统选用模式为桌面(Desktop),方便后续实战中搭建开发环境。

1．设置系统环境

该部分对服务器的配置在服务器本地进行配置，配置完毕后需要重启服务器确认配置是否生效，特别是远程访问服务器需要设置固定 IP 地址。

（1）设置机器名：为了方便 Hadoop 和 Spark 设置配置的方便，将设置每个服务器的机器名。具体方法是以 root 用户登录，在命令行终端使用#vi /etc/sysconfig/network 打开配置文件，根据前面集群规划设置服务器的机器名，在 network 配置文件设置内容如下，新机器名在重启后生效。

```
NETWORKING=yes
HOSTNAME=master
```

（2）设置 IP 地址：在 CentOS 操作系统可以通过两种方法设置服务器的 IP 地址，一种是使用命令终端进行设置，另一种是在网络连接设置界面进行配置。

1）使用命令终端设置。以 root 用户登录，在配置文件#vi/etc/sysconfig/network-scripts/ifcfg-eth0 设置网卡信息，以下仅列出重要参数信息。

```
DEVICE=eth0                  #对应第一张网卡
TYPE=Ethernet                #以太网类型
ONBOOT=yes                   #是否启动时运行
BOOTPROTO= static            #使用静态 IP，而不是由 DHCP 分配 IP
NAME="System eth0"           #网络连接名称
HWADDR=00:50:56:94:04:3C     #eth0 的 MAC 地址，根据实际情况而定，对应配置
                             （/etc/udev/rules.d/70-persistent-net.rules）
IPADDR=192.168.1.10          #指定本机 IP 地址
NETMASK=255.255.255.0        #指定子网掩码
GATEWAY=192.168.1.1          #指定网关
DNS1=8.8.8.8                 #DNS 地址，配置后同步到/etc/resolv.conf 文件
```

通过以上方法设置网络配置以后，在命令终端使用#service network restart 重启网络或重启机器，并使用 ifconfig 命令查看设置 IP 地址是否生效（如果该机器在设置后需要通过远程访问，建议重启机器）。

2）使用网络连接设置界面。在系统菜单中单击 System→Preferences→Network Connections 或在导航栏单击网络连接图标，出现如图 2-1 所示的界面。

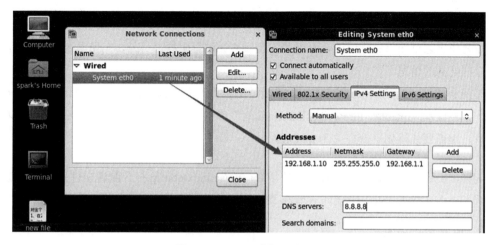

图 2-1　CentOS 进行网络设置

重建或修改网络连接，设置该连接为手工方式，网络信息参数如下：

IP 地址：192.168.1.10

子网掩码：255.255.255.0

网关：192.168.1.1

DNS：8.8.8.8

【注意】

- 以上参数信息仅做参考，IP 地址、网关和 DNS 等需要根据所在网络的实际情况进行设置。
- 设置连接方式为 Available to all users，否则通过远程连接时，会在服务器重启后无法连接服务器。
- 如果运行在 VMware 虚拟机上，网络模式使用桥接模式，保证能够连接到互联网中，以方便 Hadoop 和 Spark 源代码编译等实战，因为在编译过程中需要下载其依赖包。

（3）设置 Host 映射文件：以 root 用户身份，在命令行终端使用#vi /etc/hosts 打开配置文件，该文件保存了网址域名/机器名与其对应的 IP 地址建立一个关联的"数据库"。根据集群规划在该配置文件中加入如下内容：

```
192.168.1.10 master
192.168.1.11 slave1
192.168.1.12 slave2
```

设置完毕后,可以使用#ping master 命令直接检测 master 服务器是否连通以及检测服务器的响应速度。

（4）**关闭防火墙和 SELinux**：关闭防火墙和 SELinux 的原因在于 Hadoop 和 Spark 运行过程中需要使用通过端口进行通信，而这些安全设施会进行阻拦。另外，在使用 SSH 无密码访问时也存在同样的情况。

关闭 iptables 时，以 root 用户登录，在命令行终端使用#service iptables status 查看 iptables 状态，如果显示"iptables: Firewall is not running."表示 iptables 已经关闭；如果显示 iptables 设置信息，则需要使用如下命令关闭 iptables：

```
#chkconfig iptables off
```

同样，使用 root 用户在命令行终端使用#vi /etc/selinux/config 打开配置文件，在该配置文件中设置 SELINUX=disable，关闭防火墙和 SELinux 的设置需要重启机器才能够生效。

2．配置运行环境

样板机配置运行环境需要进行更新 OpenSSL、修改 OpenSSH 配置、创建 spark 组和用户、安装和配置 JDK、安装和配置 Scala 等。

（1）**更新 OpenSSL**：由于 CentOS 系统自带的 OpenSSL 存在 Bug，如果不更新 OpenSSL，在部署过程会出现无法通过 SSH 连接节点，使用如下命令进行更新 SSL：

```
#yum update openssl
```

（2）**修改 OpenSSH 配置**：在集群环境中需要进行 SSH 进行免密码登录，需要修改 OpenSSH 配置文件，确认使用 RSA 算法（非对称加密算法）进行公钥加密并确认生成私钥存放文件等。配置过程需以 root 用户身份，在命令行终端使用#vi /etc/ssh/sshd_config 打开配置文件，打开 3 个配置项，如图 2-2 所示：

```
RSAAuthentication yes                          #设置是否使用 RSA 算法进行加密
PubkeyAuthentication yes                       #设置是否使用口令验证
AuthorizedKeysFile .ssh/authorized_keys        #生成的密钥存放的文件
```

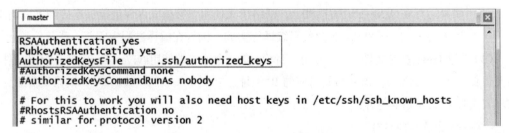

图 2-2　修改 sshd_config 配置文件

保存配置后，需要通过#service sshd restart 重启 SSH 服务，以便生效配置。

（3）增加 spark 组和用户：为了操作方便，将在节点中创建 spark 组和 spark 用户（如果在安装 CentOS 操作系统过程中，创建了该组和用户，该步骤可以省略）。创建命令如下：
```
#groupadd -g 1000 spark
#useradd -u 2000 -g spark spark
#passwd spark
```

在后面执行中需要使用到 sudo 命令，应把 spark 用户加入到 sudoers 文件中，先修改该配置文件的权限：
```
# chmod u+w /etc/sudoers
```

再使用# vi /etc/sudoers 命令打开该文件，找到这行 root ALL=(ALL) ALL，在它下面添加：
```
spark ALL=(ALL) ALL
```

根据前面集群安装配置规划，创建运行环境所需要的目录结构，创建 spark 用户临时上传文件目录，设置这些目录所属组和用户均为 spark：
```
#mkdir /app
#chown -R spark:spark /app
#mkdir /app/soft
#mkdir /app/compile
#mkdir /app/spark
#mkdir -p /home/spark/work
#chown -R spark:spark /home/spark/work
```

（4）安装和配置 JDK：运行环境使用的是 CentOS 64 位操作系统，对应在 Oracle JDK 页面下载 JDK 64 位安装包。另外，Spark 1.5 版本开始已经不支持 JDK 6，在这里我们需要选择 JDK 7 及以上版本的安装包，如 jdk-7u55-linux-x64.tar.gz。下载完毕后，把该安装包保存在 spark 用户临时上传文件目录/home/spark/work 中，解压该文件并移动到/app/soft 中。
```
$cd /home/spark/work
$tar -zxf jdk-7u55-linux-x64.tar.gz
$mv jdk1.7.0_55 /app/soft
$ll /app/soft
```

以 root 用户执行#vi /etc/profile 命令或者以 spark 用户执行$sudo vi /etc/profile 命令，配置/etc/profile（需要注意的是，/etc/profile 设置是全局配置，如防止用户间互相冲突也可以使用~/.bash_profile 进行设置）。设置的 JDK 相关配置内容如下：
```
export JAVA_HOME=/app/soft/jdk1.7.0_55
export PATH=$JAVA_HOME/bin:$PATH
export CLASSPATH=.:$JAVA_HOME/lib/dt.jar:$JAVA_HOME/lib/tools.jar
```

配置完毕后，需要编译该配置文件或重新登录以生效该配置，并验证 JDK 是否生效。
```
$source /etc/profile
```

```
$java -version
```
当显示如图 2-3 所示的信息时，表示 JDK 配置成功。

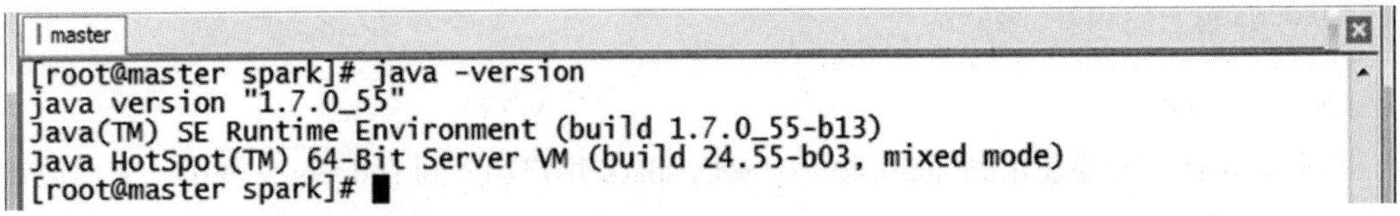

```
[root@master spark]# java -version
java version "1.7.0_55"
Java(TM) SE Runtime Environment (build 1.7.0_55-b13)
Java HotSpot(TM) 64-Bit Server VM (build 24.55-b03, mixed mode)
[root@master spark]#
```

图 2-3　验证 JDK 安装结果

（5）**安装和配置 Scala**：Spark 2.0 版本使用的 Scala 2.11.X，需要在 Scala 官方站点下载 Scala 2.11.6 及以上版本的安装包，如 scala-2.11.8.tgz，对应的操作系统为"Mac OS X, Unix, Cygwin"。下载完毕后，把该安装包保存在 spark 用户临时上传文件目录/home/spark/work 中，解压该文件并移动到/app/soft 中。

```
$cd /home/spark/work
$tar -zxf scala-2.11.8.tgz
$mv scala-2.11.8 /app/soft
$ll /app/soft
```

配置/etc/profile 文件，在该配置文件中加入相关 Scala 环境变量：

```
export SCALA_HOME=/app/soft/scala-2.11.8
export PATH=$PATH:${SCALA_HOME}/bin
```

配置完毕后，需要编译该配置文件并验证 Scala 是否生效：

```
$source /etc/profile
$scala -version
```

当显示如图 2-4 所示的信息时，表示 Scala 配置成功。

```
[spark@compiler ~]$ source /etc/profile
[spark@compiler ~]$ scala -version
Scala code runner version 2.11.6 -- Copyright 2002-2013, LAMP/EPFL
[spark@compiler ~]$
```

图 2-4　验证 Scala 安装结果

2.1.2　配置集群环境

通过前面的步骤搭建完毕集群样板机，以该机为模板复制出两份，按照前面规划集群配置，需要设置机器名和 IP 地址，同时设置 SSH 无密码登录。

1．复制样板机

用户可以通过复制样板机目录或使用 VMware 克隆功能进行样板机的复制,如图 2-5 所示。复制出两份副本,分别对应 slave1 和 slave2 节点。

图 2-5　复制样板机

2．设置机器名和 IP 地址

参见搭建样板机方法,以 root 用户登录,使用#vi /etc/sysconfig/network 打开配置文件,根据表 2-1 修改机器名和 IP 地址,修改后需要重新启动机器,新机器名在重启后生效。需要注意的是,复制的虚拟机第一次启动时选择"我已复制该虚拟机(P)"选项,如图 2-6 所示。这样启动后能够分配新的 MAC 地址,不会造成冲突。

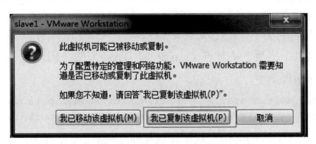

图 2-6　第一次启动复制样板机询问界面

3．配置 SSH 无密码登录

(1)生成私钥和公钥:使用 spark 用户登录,在 3 个节点中使用如下命令生成私钥和公钥,如图 2-7 所示。为简单起见,在生成私钥和公钥过程中提示问题均按回车键。

```
$ssh-keygen -t rsa
```

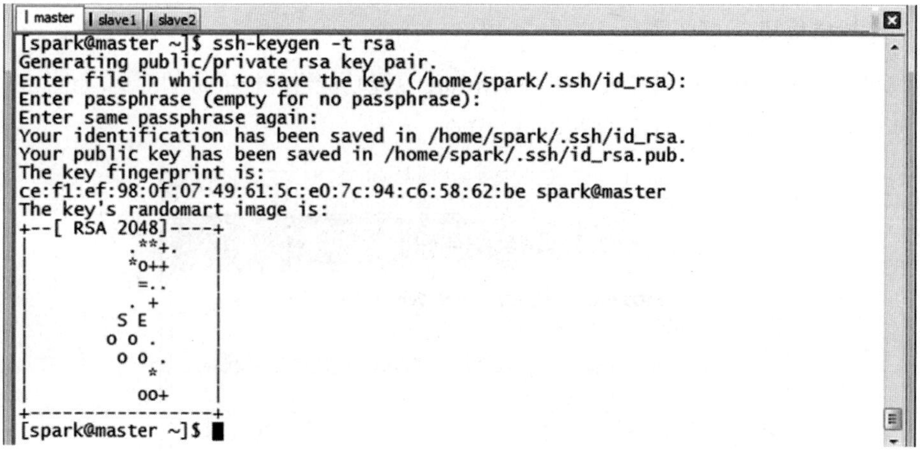

图 2-7　SSH 生成私钥和公钥

命令执行完毕后，可以在/home/spark/.ssh 目录中看见两个文件：id-rsa 和 id_rsa.pub，其中，以.pub 结尾的是公钥，把公钥命名为 authorized_keys_master.pub，使用的命令如下：

```
$cd /home/spark/.ssh
$mv id_rsa.pub authorized_keys_master.pub
```

同样地把其他两个节点的公钥命名为 authorized_keys_slave1.pub 和 authorized_keys_slave2.pub：

```
$mv id_rsa.pub authorized_keys_slave1.pub
$mv id_rsa.pub authorized_keys_slave2.pub
```

（2）合并公钥信息：把两个从节点（slave1 和 slave2）的公钥使用 scp 命令传送到 master 节点的/home/spark/.ssh 目录中。

```
$scp authorized_keys_slave1.pub spark@master:/home/spark/.ssh
$scp authorized_keys_slave2.pub spark@master:/home/spark/.ssh
```

使用 cat 命令把 3 个节点的公钥信息保存到 authorized_key 文件中：

```
$cat authorized_keys_master.pub>> authorized_keys
$cat authorized_keys_slave1.pub>> authorized_keys
$cat authorized_keys_slave2.pub>> authorized_keys
```

合并完毕后，查看 authorized_key 文件内容如图 2-8 所示。

图 2-8　合并 3 台节点公钥的内容

使用 scp 命令把密码文件分发到 slave1 和 slave2 节点，命令如下：

```
$scp authorized_keys spark@slave1:/home/spark/.ssh
$scp authorized_keys spark@slave2:/home/spark/.ssh
```

传输完毕以后，需要在 3 台节点中使用如下设置 authorized_keys 读写权限：

```
$cd /home/spark/.ssh
$chmod 400 authorized_keys
```

（3）验证免密码登录：在各节点中使用 ssh 命令，验证它们之间是否可以免密码登录。

```
$ssh master
$ssh slave1
$ssh slave2
```

2.2　编译 Spark 源代码

由于实际环境较为复杂，从 Spark 官方下载二进制安装包中可能不具有相关功能或不支持指定软件版本，这就需要我们根据实际情况编译 Spark 源代码，生成所需要的部署包。Spark 可以通过 Maven 和 SBT 两种方式进行编译，再通过 make-distribution.sh 脚本生成部署包。Maven 方式编译则需要 Maven 工具，而 SBT 方式编译需要安装 Git 工具，两种方式均需要在联网下下载依赖包。通过比较发现，SBT 编译速度较慢。其原因有可能是：①时间不一样，SBT 是白天进行编译，Maven 是深夜进行的，获取依赖包的速度不同；②Maven 下载大文件是多线程进行，而 SBT 是单进程。

2.2.1 配置 Spark 编译环境

1．安装 Git 并编译安装（SBT 编译使用）

（1）下载 Git 安装包：用户可以通过安装包或 yum 命令安装 Git，在 CentOS 系统上既可以使用#yum install Git 命令进行安装，也可以从以下地址下载安装包程序进行安装，这里采用下载安装包并编译。图 2-9 为下载 Git 安装包程序的页面。

```
http://www.onlinedown.net/softdown/169333_2.htm
https://www.kernel.org/pub/software/scm/git/
```

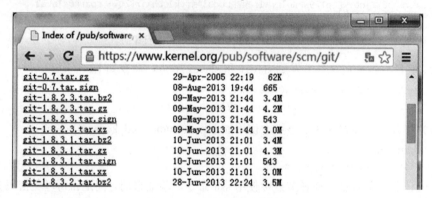

图 2-9　下载 git 安装包的页面

（2）上传 Git 并解压缩：把下载的安装包 git-1.7.6.tar.gz 上传到/home/spark/work 目录中，解压缩后移动到/app/soft 目录下，具体命令如下：

```
$cd /home/spark/work/
$tar -xzf git-1.7.6.tar.gz
$mv git-1.7.6 /app/soft
$ll /app/soft
```

（3）编译安装：编译安装 Git 前需要依赖安装程序，安装这些依赖程序后，以 root 用户在 Git 所在路径编译安装 Git，该过程时间比较长。

```
#yum install curl-devel
#yum install autoconf automake libtool cmake
#yum install ncurses-devel
#yum install openssl-devel
#yum install gcc*
#cd /app/soft/git-1.7.6
#./configure
#make
#make install
```

（4）把 Git 加入到 PATH 路径中：把 Git 路径加入到/etc/profile 配置文件的 PATH 中。

```
export GIT_HOME=/app/soft/git-1.7.6
export PATH=$PATH:$GIT_HOME/bin
```

配置完毕后，使用如下命令编译该配置文件，并验证 Git 是否生效。

```
$source /etc/profile
$git version
```

当显示如图 2-10 所示的信息，表示 Git 配置成功。

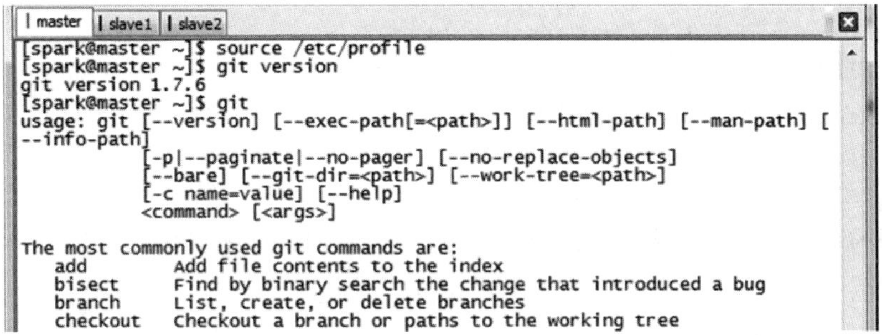

图 2-10　验证 Git 安装结果

2. 下载 Spark 源代码

（1）用户可以从 Apache 等网站下载 Spark 源代码，或在 Spark 源代码托管网站 Github 下载进行下载,在这里我们下载最新的版本 Spark 2.0.0，下载地址如下：

```
http://spark.apache.org/downloads.html
https://github.com/apache/spark
```

将下载好的源代码包 spark 2.0.0.tgz 上传到/home/spark/work 目录下。

（2）在主节点上解压缩，并把解压缩移动到 Spark 编译目录/app/compile 中，详细命令如下：

```
$cd /home/spark/work/
$tar -zxf spark-2.0.0.tgz
$mv spark-2.0.0 /app/compile/spark-2.0.0-src
$ll /app/compile
```

2.2.2　使用 Maven 编译 Spark

通常我们使用 Maven 来编译 Spark，它属于 Apache 基金会开源软件，需要注意的是，Spark 2.0.0 编译时强制要求 Maven 3.3.9 及以上版本，否则在编译开始检查会报错。

1. 下载 Maven 安装包

在 Maven 官网进行下载，具体地址如下：

```
http://maven.apache.org/download.cgi#
```
把下载的安装包 apache-maven-3.3.9-bin.tar.gz 移动到 spark 用户的文件临时存放目录 /home/spark/work 中。

2. 解压 Maven 并配置参数

解压 Maven 安装包 apache-maven-3.3.9-bin.tar.gz，并把解压后程序移动到/app/soft 目录下。

```
$cd /home/spark/work
$tar -zxf apache-maven-3.3.9-bin.tar.gz
$mv apache-maven-3.3.9 /app/soft/maven-3.3.9
$ll /app/soft
```

需要在/etc/profile 配置文件中加入如下设置：

```
export MAVEN_HOME=/app/soft/maven-3.3.9
export PATH=$PATH:$MAVEN_HOME/bin
```

配置完毕后，使用如下命令编译该配置文件，并验证 Maven 配置是否生效，当显示如图 2-11 所示的信息时，表示 Maven 配置成功。

```
$source /etc/profile
$mvn -version
```

图 2-11 验证 Maven 安装结果

3. 编译代码

编译过程需要保证编译机器联网状态，以保证 Maven 从网上下载其依赖包。另外，编译前需要设置 JVM 内存大小，否则在编译过程中，会由于默认内存小而出现内存溢出的错误。编译执行脚本如下，其中，参数-P 表示激活依赖的程序及版本，-Dskip Tests 表示编译时跳过测试环节。

```
$cp -r /app/compile/spark-2.0.0-src /app/compile/spark-2.0.0-mvn
$cd /app/compile/spark-2.0.0-mvn
$export MAVEN_OPTS="-Xmx2g -XX:MaxPermSize=512M -XX:ReservedCodeCacheSize=512m"
$mvn -Pyarn -Phadoop-2.7 -Pspark-ganglia-lgpl -Pkinesis-asl -Phive -DskipTests
   clean package
```

整个编译过程编译了约 31 个任务，如果是已经下载依赖包的情况，则编译耗时 1′5″，编译成功界面如图 2-12 所示。由于编译过程中需要下载较多的依赖包，因此整个编译时间取决于网速，如果出现异常或者假死的情况，可使用 Ctrl+C 组合键中断，然后再重新编译。编译时可以通过$du -sh /app/compile/spark-2.0.0-mvn 命令查看编译文件夹的大小，最终编译完成后的文件夹大小约为 899MB。

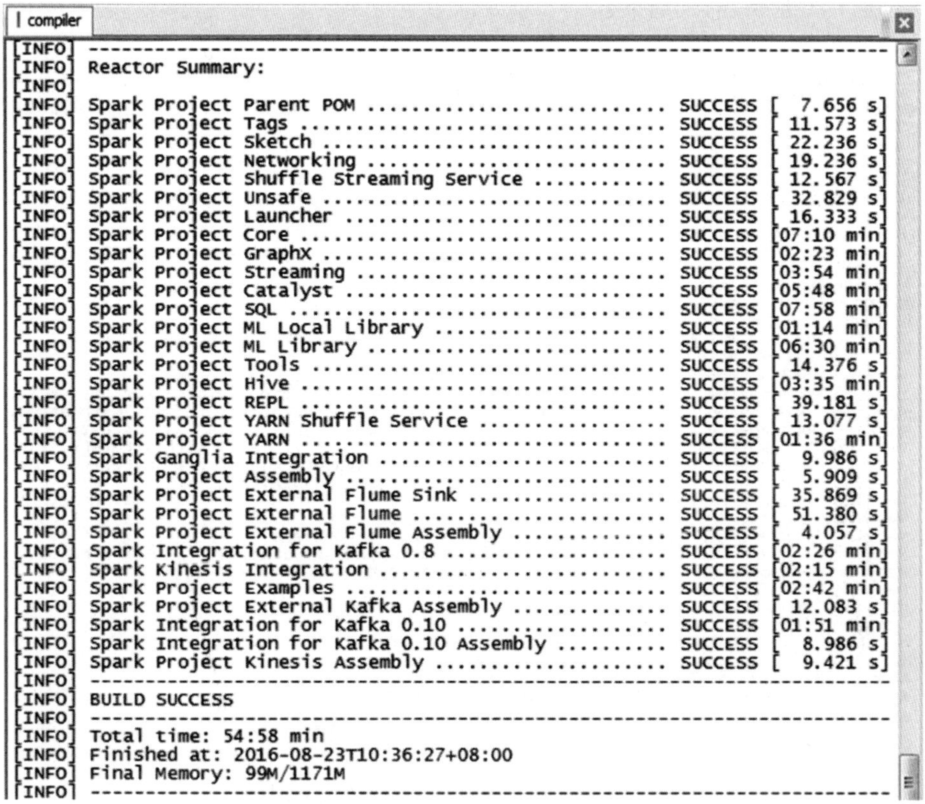

```
| compiler                                                                                    |X|
[INFO] ------------------------------------------------------------------------
[INFO] Reactor Summary:
[INFO]
[INFO] Spark Project Parent POM ........................... SUCCESS [  7.656 s]
[INFO] Spark Project Tags ................................. SUCCESS [ 11.573 s]
[INFO] Spark Project Sketch ............................... SUCCESS [ 22.236 s]
[INFO] Spark Project Networking ........................... SUCCESS [ 19.236 s]
[INFO] Spark Project Shuffle Streaming Service ............ SUCCESS [ 12.567 s]
[INFO] Spark Project Unsafe ............................... SUCCESS [ 32.829 s]
[INFO] Spark Project Launcher ............................. SUCCESS [ 16.333 s]
[INFO] Spark Project Core ................................. SUCCESS [07:10 min]
[INFO] Spark Project GraphX ............................... SUCCESS [02:23 min]
[INFO] Spark Project Streaming ............................ SUCCESS [03:54 min]
[INFO] Spark Project Catalyst ............................. SUCCESS [05:48 min]
[INFO] Spark Project SQL .................................. SUCCESS [07:58 min]
[INFO] Spark Project ML Local Library ..................... SUCCESS [01:14 min]
[INFO] Spark Project ML Library ........................... SUCCESS [06:30 min]
[INFO] Spark Project Tools ................................ SUCCESS [ 14.376 s]
[INFO] Spark Project Hive ................................. SUCCESS [03:35 min]
[INFO] Spark Project REPL ................................. SUCCESS [ 39.181 s]
[INFO] Spark Project YARN Shuffle Service ................. SUCCESS [ 13.077 s]
[INFO] Spark Project YARN ................................. SUCCESS [01:36 min]
[INFO] Spark Ganglia Integration .......................... SUCCESS [  9.986 s]
[INFO] Spark Project Assembly ............................. SUCCESS [  5.909 s]
[INFO] Spark Project External Flume Sink .................. SUCCESS [ 35.869 s]
[INFO] Spark Project External Flume ....................... SUCCESS [ 51.380 s]
[INFO] Spark Project External Flume Assembly .............. SUCCESS [  4.057 s]
[INFO] Spark Integration for Kafka 0.8 .................... SUCCESS [02:26 min]
[INFO] Spark Kinesis Integration .......................... SUCCESS [02:15 min]
[INFO] Spark Project Examples ............................. SUCCESS [02:42 min]
[INFO] Spark Project External Kafka Assembly .............. SUCCESS [ 12.083 s]
[INFO] Spark Integration for Kafka 0.10 ................... SUCCESS [01:51 min]
[INFO] Spark Integration for Kafka 0.10 Assembly .......... SUCCESS [  8.986 s]
[INFO] Spark Project Kinesis Assembly ..................... SUCCESS [  9.421 s]
[INFO] ------------------------------------------------------------------------
[INFO] BUILD SUCCESS
[INFO] ------------------------------------------------------------------------
[INFO] Total time: 54:58 min
[INFO] Finished at: 2016-08-23T10:36:27+08:00
[INFO] Final Memory: 99M/1171M
[INFO] ------------------------------------------------------------------------
```

图 2-12 使用 Maven 编译 Spark 成功的结果页面

2.2.3 使用 SBT 编译 Spark

SBT 是 Simple Build Tool 的缩写，可以使用它进行 Scala 或 Java 程序编译的工具，功能类似于 Maven 或 Ant。在 Spark 程序中自带了 SBT 编译的脚本和相关的程序（在 Spark 1.6 及以前版本，该脚本在 sbt 目录下），我们只需要直接调用 SBT 进行编译即可，编译脚本如下：

```
$cp -r /app/compile/spark-2.0.0-src /app/compile/spark-2.0.0-sbt
$cd /app/compile/spark-2.0.0-sbt
$build/sbt assembly -Pyarn -Phadoop-2.7 -Pspark-ganglia-lgpl -Pkinesis-asl
 -Phive
```

和 Maven 编译一样，整个编译过程机器必须保证在联网状态，从网上下载依赖包。在编译的过程中经常出现假死的情况，出现这种情况只需要重新执行编译脚本，整个过程需要几个小时才能编译完成（编译的快慢由下载依赖包的速度决定）。图 2-13 是使用 SBT 编译成功的界面（当然这也是多次重新运行的结果）。

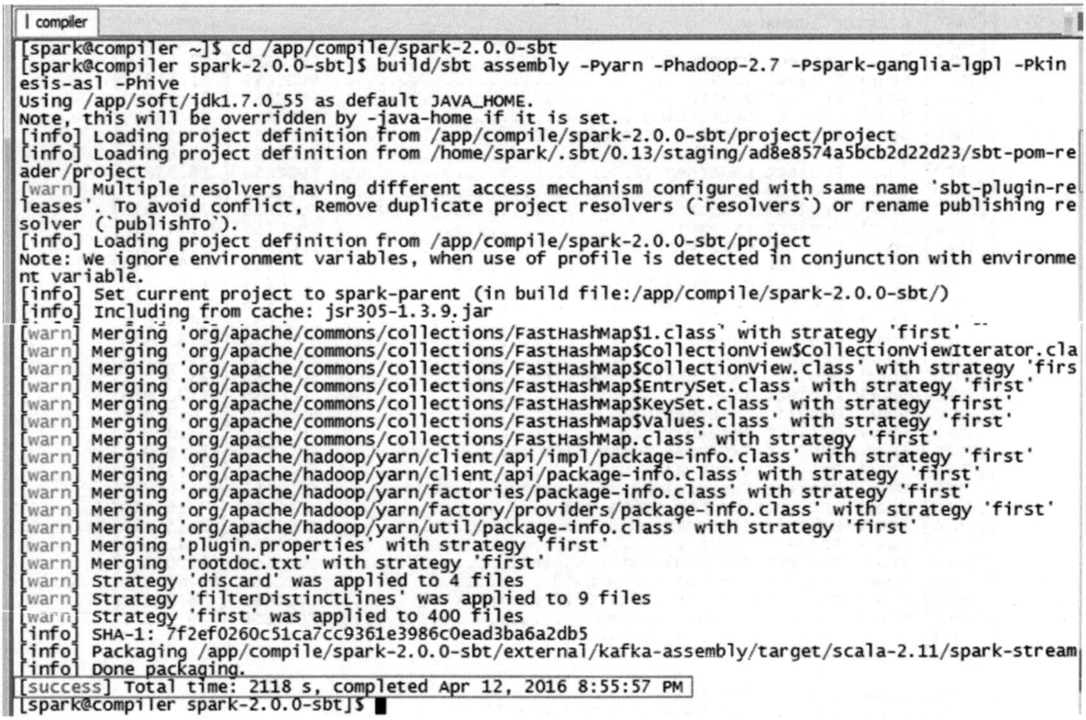

```
compiler
[spark@compiler ~]$ cd /app/compile/spark-2.0.0-sbt
[spark@compiler spark-2.0.0-sbt]$ build/sbt assembly -Pyarn -Phadoop-2.7 -Pspark-ganglia-lgpl -Pkin
esis-asl -Phive
Using /app/soft/jdk1.7.0_55 as default JAVA_HOME.
Note, this will be overridden by -java-home if it is set.
[info] Loading project definition from /app/compile/spark-2.0.0-sbt/project/project
[info] Loading project definition from /home/spark/.sbt/0.13/staging/ad8e8574a5bcb2d22d23/sbt-pom-re
ader/project
[warn] Multiple resolvers having different access mechanism configured with same name 'sbt-plugin-re
leases'. To avoid conflict, Remove duplicate project resolvers (`resolvers`) or rename publishing re
solver (`publishTo`).
[info] Loading project definition from /app/compile/spark-2.0.0-sbt/project
Note: We ignore environment variables, when use of profile is detected in conjunction with environme
nt variable.
[info] Set current project to spark-parent (in build file:/app/compile/spark-2.0.0-sbt/)
[info] Including from cache: jsr305-1.3.9.jar
[warn] Merging 'org/apache/commons/collections/FastHashMap$1.class' with strategy 'first'
[warn] Merging 'org/apache/commons/collections/FastHashMap$CollectionView$CollectionViewIterator.cla
[warn] Merging 'org/apache/commons/collections/FastHashMap$CollectionView.class' with strategy 'firs
[warn] Merging 'org/apache/commons/collections/FastHashMap$EntrySet.class' with strategy 'first'
[warn] Merging 'org/apache/commons/collections/FastHashMap$KeySet.class' with strategy 'first'
[warn] Merging 'org/apache/commons/collections/FastHashMap$Values.class' with strategy 'first'
[warn] Merging 'org/apache/commons/collections/FastHashMap.class' with strategy 'first'
[warn] Merging 'org/apache/hadoop/yarn/client/api/impl/package-info.class' with strategy 'first'
[warn] Merging 'org/apache/hadoop/yarn/client/api/package-info.class' with strategy 'first'
[warn] Merging 'org/apache/hadoop/yarn/factories/package-info.class' with strategy 'first'
[warn] Merging 'org/apache/hadoop/yarn/factory/providers/package-info.class' with strategy 'first'
[warn] Merging 'org/apache/hadoop/yarn/util/package-info.class' with strategy 'first'
[warn] Merging 'plugin.properties' with strategy 'first'
[warn] Merging 'rootdoc.txt' with strategy 'first'
[warn] Strategy 'discard' was applied to 4 files
[warn] Strategy 'filterDistinctLines' was applied to 9 files
[warn] Strategy 'first' was applied to 400 files
[info] SHA-1: 7f2ef0260c51ca7cc9361e3986c0ead3ba6a2db5
[info] Packaging /app/compile/spark-2.0.0-sbt/external/kafka-assembly/target/scala-2.11/spark-stream
[info] Done packaging.
[success] Total time: 2118 s, completed Apr 12, 2016 8:55:57 PM
[spark@compiler spark-2.0.0-sbt]$
```

图 2-13　使用 SBT 编译 Spark 成功的结果页面

2.2.4　生成 Spark 部署包

通过 Maven 或 SBT 编译 Spark 源代码后，可以使用 Spark 源代码 dev 目录下一个生成部署包的脚本 make-distribution.sh（在 Spark 1.6 及以前的版本，该脚本在根目录下）。

在 dev 目录下的 make-distribution.sh 命令脚本中，使用如下命令生成 Spark 部署包。

```
$cd /app/compile/spark-2.0.0-mvn/
$export MAVEN_OPTS="-Xmx2g -XX:MaxPermSize=512M -XX:ReservedCodeCacheSize=512m"
$./dev/make-distribution.sh -name custom-spark -tgz -Psparkr -Phadoop-2.7 -Phive
  -Phive-thriftserver -Pyarn
```

整个编译过程编译了约 31 个任务，耗时 1′38″，编译成功的界面如图 2-14 所示。

```
| compiler                                                                      □ ×
+ cp /app/compile/spark-2.0.0-mvn/LICENSE /app/compile/spark-2.0.0-mvn/dist
+ cp -r /app/compile/spark-2.0.0-mvn/licenses /app/compile/spark-2.0.0-mvn/dist
+ cp /app/compile/spark-2.0.0-mvn/NOTICE /app/compile/spark-2.0.0-mvn/dist
+ '[' -e /app/compile/spark-2.0.0-mvn/CHANGES.txt ']'
+ cp -r /app/compile/spark-2.0.0-mvn/data /app/compile/spark-2.0.0-mvn/dist
+ mkdir /app/compile/spark-2.0.0-mvn/dist/conf
+ cp /app/compile/spark-2.0.0-mvn/conf/docker.properties.template /app/compile/spark-2.0.
properties.template /app/compile/spark-2.0.0-mvn/conf/metrics.properties.template /app/co
park-defaults.conf.template /app/compile/spark-2.0.0-mvn/conf/spark-env.sh.template /app/
+ cp -r /app/compile/spark-2.0.0-mvn/README.md /app/compile/spark-2.0.0-mvn/dist
+ cp -r /app/compile/spark-2.0.0-mvn/bin /app/compile/spark-2.0.0-mvn/dist
+ cp -r /app/compile/spark-2.0.0-mvn/python /app/compile/spark-2.0.0-mvn/dist
+ cp -r /app/compile/spark-2.0.0-mvn/sbin /app/compile/spark-2.0.0-mvn/dist
+ '[' -d /app/compile/spark-2.0.0-mvn/R/lib/SparkR ']'
+ mkdir -p /app/compile/spark-2.0.0-mvn/dist/R/lib
+ cp -r /app/compile/spark-2.0.0-mvn/R/lib/SparkR /app/compile/spark-2.0.0-mvn/dist/R/lib
+ cp /app/compile/spark-2.0.0-mvn/R/lib/sparkr.zip /app/compile/spark-2.0.0-mvn/dist/R/li
+ '[' true == true ']'
+ TARDIR_NAME=spark-2.0.0-bin-custom-spark
+ TARDIR=/app/compile/spark-2.0.0-mvn/spark-2.0.0-SNAPSHOT-bin-custom-spark
+ rm -rf /app/compile/spark-2.0.0-mvn/spark-2.0.0-SNAPSHOT-bin-custom-spark
+ cp -r /app/compile/spark-2.0.0-mvn/dist /app/compile/spark-2.0.0-mvn/spark-2.0.0-SNAPSH
+ tar czf spark-2.0.0-SNAPSHOT-bin-custom-spark.tgz -C /app/compile/spark-2.0.0-mvn spark
+ rm -rf /app/compile/spark-2.0.0-mvn/spark-2.0.0-SNAPSHOT-bin-custom-spark
[spark@compiler spark-2.0.0-mvn]$
```

图 2-14　生成 Spark 部署包成功的界面

3．查看生成结果

生成的部署包位于根目录下，文件名为 spark-2.0.0-bin-custom-spark.tgz，如图 2-15 所示。

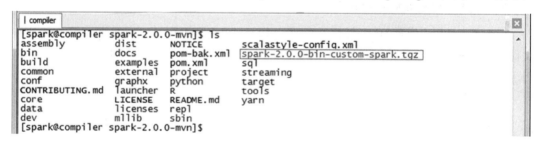

图 2-15　生成 Spark 部署包的结果页面

2.3　搭建 Spark 运行集群

2.3.1　修改配置文件

用户既可以使用上一步骤编译好的 spark-2.0.0-bin-custom-spark.tgz 文件作为安装包，也可以直接从官网下载 Spark 2.0.0 版本二进制安装包，上传到 master 节点，并把该安装包移动到 /home/spark/work 目录下，使用如下命令进行解压缩。

```
$cd /home/spark/work/
```

```
$tar -zxf spark-2.0.0-bin-2.7.2.tgz
$mv spark-2.0.0-bin-2.7.2 /app/spark/spark-2.0.0
$ll /app/spark
```

1. 配置/etc/profile

使用$sudo vi /etc/profile 命令打开配置文件，在 PATH 中加入 Spark 路径，加入后使用$source /etc/profile 使该配置生效。

```
export SPARK_HOME=/app/spark/spark-2.0.0
export PATH=$PATH:$SPARK_HOME/bin:$SPARK_HOME/sbin
```

2. 配置 conf/slaves

该配置文件用于设置集群中运行 Worker 节点信息，该文件在$SPARK_HOME/conf 目录下，使用$sudo vi slaves 打开，在该配置文件中加入如下内容。需要注意的是，在 master 节点不仅运行 Master 进程，也运行 Worker 进程。

```
master
slave1
slave2
```

3. 配置 conf/spark-env.sh

该配置文件用于设置 Spark 运行环境，默认在$SPARK_HOME/conf 目录中没有 spark-env.sh 文件，需要通过复制或修改 spark-env.sh.template 进行创建。使用的命令如下：

```
$cd /app/spark/spark-2.0.0/conf
$cp spark-env.sh.templatespark-env.sh
$sudo vi spark-env.sh
```

在文件中加入如下内容，设置 master 节点为运行 Master 进程节点，通信端口为 7077，在每个节点中运行 Worker 数量为 1，使用的核数为 1 个、内存为 1024MB。

```
export SPARK_MASTER_IP=master              #设置运行Master进程节点地址
export SPARK_MASTER_PORT=7077              #设置Mater通信端口
export SPARK_EXECUTOR_INSTANCES=1          #设置每个节点中运行Executor数量
export SPARK_WORKER_INSTANCES=1            #设置每个节点中运行Worker数量
export SPARK_WORKER_CORES=1                #每个Worker使用的核数
export SPARK_WORKER_MEMORY=1024M           #每个Worker使用的内存大小
export SPARK_MASTER_WEBUI_PORT=8080        #Master的监控界面WebUI端口
export SPARK_CONF_DIR=/app/spark/spark-2.0.0/conf
```

4. 分发 Spark 程序

配置完毕后进入 master 节点中的/app/spark 目录，使用 scp 命令把 spark 目录复制到 slave1 和 slave2 节点中。

```
$cd /app/spark
$scp -r spark-2.0.0 spark@slave1:/app/spark/
$scp -r spark-2.0.0 spark@slave2:/app/spark/
```

2.3.2 启动 Spark

用户可以使用两种方法启动 Spark 程序：一种是通过同时启动 Master 和所有的 Worker 进程，另一种是 Master 和 Worker 分别启动。

第一种方法使用 start-all.sh 脚本，通过读取 slaves 配置文件信息用来启动所有 Worker。

```
$cd /app/spark/spark-2.0.0/sbin
$./start-all.sh
```

第二种方法是先使用 start-master.sh 启动 Master 进程，然后使用 start-slaves.sh 启动 Worker，执行命令如下：

```
$cd /app/spark/spark-2.0.0/sbin
$./start-master.sh
$./start-slaves.sh spark://master:7077
```

2.3.3 验证启动

启动完毕以后可以通过 jps 命令查看启动情况，在 master 节点应该同时启动 Master 和 Worker 进程，在 slave1 和 slave2 节点中只启动 Worker 进程。用户还可以通过 $netstat -nlt 命令查看 master 节点的网络情况，或者使用带有条件的端口命令 $netstat -an | grep 8080 查看 8080 端口的使用情况。

另外，用户可以在 Spark 监控界面的 WebUI 进行查看，如图 4-18 所示。在浏览器中输入 http://master:8080，可以查看到 Spark 集群的信息，具体监控界面的使用将在下文详细介绍。

2.3.4 第一个实例

不能免俗，我们也从 WordCount 开始 Spark 实战之旅。该实例将用 Scala 语言编写并在 Spark Shell 执行。相对于 MapReduce 的 WordCount 过程分为 Map、Reduce 和 Job 处理三个部分，在 Scala 中利用其函数式特点能够简洁实现该过程，废话少说，让我们赶紧动手吧！

在实战前需要启动 Spark 集群和 Spark Shell，Spark Shell 启动脚本如下，启动界面如图 2-16 所示。

```
$cd /app/spark/spark-2.0.0/bin
$./spark-shell --master spark://master:7077 --executor-memory 1024m
  --driver-memory1024m
```

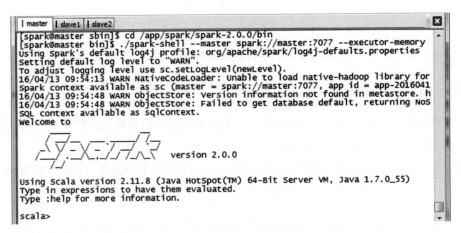

图 2-16　启动 Spark Shell 界面

在 WordCount 程序中，首先会读取 Spark 根目录下的 README.md 文件，然后把读取进来的内容进行分词，在这里分词的方法是使用空格进行分割，最后统计单词出现的次数，按照倒序打印显示出现次数最多的 10 个单词。下面就是使用 Scala 语言编写的 WordCount 执行脚本，用一行进行实现，这也可以看出来 Scala 是一种精练的语言。

```
scala> sc.textFile("/app/spark/spark-2.0.0/README.md").flatMap(_.split(" "))
  .map(x=>(x,1)).reduceByKey(_+_).map(x=>(x._2,x._1)).sortByKey(false).map(x
  =>(x._2,x._1)).take(10)
res0: Array[(String, Int)] = Array(("",67), (the,21), (to,14), (Spark,13),
  (for,11), (and,10), (##,8), (a,8), (run,7), (can,6))
```

为了更好看到实现过程，下面将逐行进行实现。

（1）在下面一行通过 SparkContext 的 textFile()方法读入根目录下的 README.md 文件，保存的形式为 MappedRDD。

```
scala> val rdd = sc.textFile("/app/spark/spark-2.0.0/README.md")
rdd: org.apache.spark.rdd.RDD[String] = /app/spark/spark-2.0.0/README.md
  MapPartitionsRDD[21] at textFile at <console>:25
```

（2）在下面一行处理中先使用 split()方法进行分词，分词的方法为按照空格进行分割，分割完毕数据格式为 RDD[(String)]，然后通过 flatMap()对分割的单词进行展平，展平完毕后使用 map(x=>(x,1))对每个单词计数 1，此时数据格式为 RDD[(String, Int)]，最后使用 reduceByKey(_+_)根据 Key 也就是单词进行计数，这个过程是一个 Shuffle 过程，数据格式为 ShuffledRDD。

```
scala> val wordcount = rdd.flatMap(_.split(" ")).map(x=>(x,1)).reduceByKey(_+_)
wordcount: org.apache.spark.rdd.RDD[(String,Int)]=ShuffledRDD[24] at reduceByKey
  at <console>:27
```

（3）在下面一行处理中先使用 map(x=>(x._2,x._1))对单词统计结果 RDD[(String, Int)]的键和值进行互换，其对应单词和词频，形成 RDD[(Int,String)]数据格式，然后通过 sortByKey(false) 根据键值也就是词频进行排序，false 要求按照倒序进行排列，最后再次通过 map(x=>(x._2,x._1)) 对 RDD 的键和值进行互换，最终形成排序后的 RDD[(String, Int)]。

```
scala> val wordsort = wordcount.map(x=>(x._2,x._1)).sortByKey(false).map(x=>
   (x._2,x._1))
wordsort: org.apache.spark.rdd.RDD[(String, Int)] = MapPartitionsRDD[29] at map
   at <console>:29
```

（4）在最后一行处理中，对前面排序的结果获取前面 10 个结果。

```
scala> wordsort.take(10)
res3: Array[(String, Int)] = Array(("",67), (the,21), (to,14), (Spark,13),
   (for,11), (and,10), (##,8), (a,8), (run,7), (can,6))
```

通过 http://master:8080 查看 Spark 运行情况，如图 2-17 所示，可以看到 Spark 为 3 个节点，每个节点各为 1 个内核/1024MB 内存，Spark Shell 分配 3 个核，每个核有 1024MB 内存。

Workers

Worker Id	Address	State	Cores	Memory
worker-20160328220829-192.168.1.11-47100	192.168.1.11:47100	ALIVE	1 (1 Used)	1024.0 MB (1024.0 MB Used)
worker-20160328220829-192.168.1.12-37456	192.168.1.12:37456	ALIVE	1 (1 Used)	1024.0 MB (1024.0 MB Used)
worker-20160328220835-192.168.1.10-33380	192.168.1.10:33380	ALIVE	1 (1 Used)	1024.0 MB (1024.0 MB Used)

Running Applications

Application ID		Name	Cores	Memory per Node	Submitted Time	User	State	Duration
app-20160328221022-0000	(kill)	Spark shell	3	1024.0 MB	2016/03/28 22:10:22	spark	RUNNING	9.8 min

图 2-17　查看 Spark 集群及应用程序使用资源情况

2.4　搭建 Spark 实战开发环境

Spark 应用程序可以使用 Scala、Java 和 Python 等语言进行编写，在本文我们将着重讲解 Scala 如何编写 Spark 应用程序。至于开发 Scala 开发环境可以选择个人比较熟悉的 IDE，比如 IDEA、Eclipse 等都是非常优秀的开发工具，笔者在 Spark 应用程序开发过程中还是比较喜欢 IDEA。

IDEA 全称 IntelliJ IDEA，是 Java 语言开发的集成环境，在业界它被公认为最好的 Java 开发工具之一，其功能包括智能代码助手、代码自动提示、重构、J2EE 支持、Ant、JUnit、代码审查等。IDEA 是 JetBrains 公司的产品，这家公司总部位于捷克共和国的首都布拉格，开发人员以严谨著称的东欧程序员为主。

IDEA 每个版本提供 Community 和 Ultimate 两个版本，其中 Community 是完全免费的，而 Ultimate 版本可以免费使用 30 天，之后需要进行注册。在 JetBrains 官网的下载页面 http://www.jetbrains.com/idea/download/选择需要的版本进行下载，在这里我们下载 Windows 或 Linux 版本 Community14 版本，下载界面如图 2-18 所示。

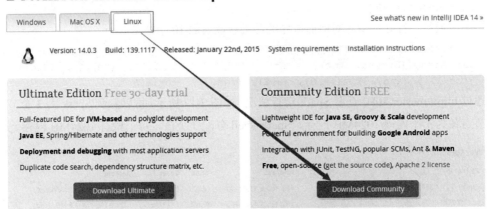

图 2-18　选择 IDEA 下载界面

2.4.1　CentOS 中部署 IDEA

把下载的安装包 ideaIC-14.0.2.tar.gz 移动到/home/spark/work 目录下，使用如下命令进行解压缩并移动到/app/soft 目录下。

```
$cd /home/spark/work
$tar -zxf ideaIC-14.0.2.tar.gz
$mv idea-IC-139.659.2 /app/soft/idea-IC
$ll /app/soft
```

用户可以通过下面几种方式启动 IntelliJ IDEA：

- 在 IntelliJ IDEA 安装所在目录下，进入 bin 目录中双击 idea.sh 启动 IntelliJ IDEA。
- 在命令行终端中，进入$IDEA_HOME/bin 目录，输入./idea.sh 启动 IntelliJ IDEA。
- 进入$IDEA_HOME/bin 目录创建 idea.sh 快捷方式，把该快捷方式移动到桌面上，通过单击该图标启动 IntelliJ IDEA（建议）。

IDEA 默认情况下并没有安装 Scala 插件，需要手动进行安装。安装方法是启动 IDEA，如果是第一次启动，可以在启动界面上选择 Configure→Plugins 选项；如果不是第一次启动，可以在菜单中选择 File→Settings→Plugins，弹出插件管理界面，在该界面上列出了所有安装好的插件。由于 Scala 插件没有安装，需要单击 Install JetBrains plugins 浏览插件。在该界面中待安装的插件很多，可以通过查询或者字母顺序找到 Scala 插件。选择插件后在界面的右侧出现该插件的详细信息，单击 Install plugin 按钮开始安装插件，如图 2-19 所示。

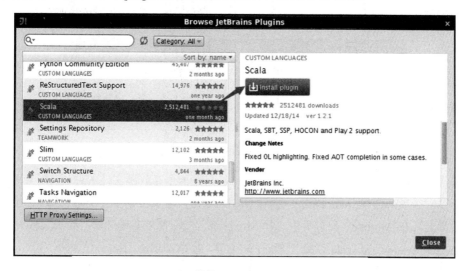

图 2-19　选择 IDEA Scala 插件界面

注意：用户也可以从 Jetbrains 公司官网下载 Scala 插件，然后通过本地的方式进行安装，需要特别注意的是，插件和 IDEA 版本是严格对应的，具体参见 http://plugins.jetbrains.com/plugin/?idea&id=1347 列表。笔者在提供的安装包资源中提供了 IDEA 编辑器对应的插件 scala-intellij-bin-1.2.1.zip。

安装插件后，在启动界面中选择创建新项目，在弹出的界面中将会出现 Scala 类型项目，选择后将出现提示创建的项目是 Scala 还是 SBT 代码项目。

从 IntelliJ IDEA 12 开始推出了 Darcula 主题的全新用户界面，该界面以黑色为主题风格得到很多开发人员的喜爱。具体方法是在主界面菜单中选择 File→Setting 子菜单，在弹出的界面选择 Appearance &Behavior→Appearance，在 Theme 中选择 Darcula 主题，保存该主题重新进入，可以看到非常酷的界面！

2.4.2　使用 IDEA 开发程序

启动 IDEA 后，在菜单栏中选择 File→New Project，选择创建 Scala 项目，然后出现如图 2-20 所示的界面。在该界面中输入项目名称、项目存放位置、选择 Project SDK 和 Scala SDK，具体内容如下。

- Project name：项目名称，命名为sparklearning，该项目将用来编写后续章节程序代码。
- Project location：项目存放位置，在这里设置为~/ideaProjects/sparklearning。
- Project SDK：项目使用SDK，选择JDK 1.7。
- Scala SDK：选择Scala 版本为2.11.8。

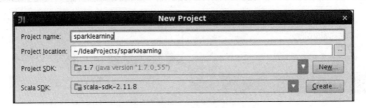

图 2-20　在 IDEA 中创建 Scala 项目

创建该项目后，可以看到现在还没有源文件，只有一个存放源文件的目录 src 以及存放工程其他信息的杂项。选择 sparklearing 项目，单击 F4 键或者单击菜单上的项目结构图标，打开项目配置界面，在 Modules 配置页中选择 src 目录，然后单击右键选择"新加文件夹"，添加 src/main/scala 目录，设置 scala 目录为 Sources 类型，如图 2-21 所示。

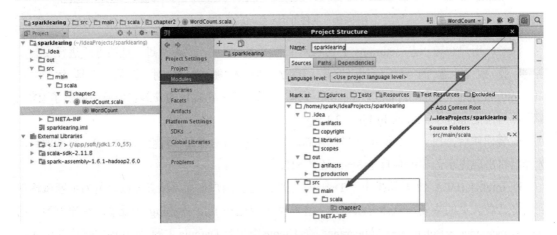

图 2-21　创建项目代码结构

设置完项目目录后，设置以下项目运行 Library，如图 2-22 所示。

- 添加 Scala SDK Library，选择 scala-2.11.8 版本。
- 添加 Java Library，选择$SPARK_HOME/jar/目录下所有 jar 包。

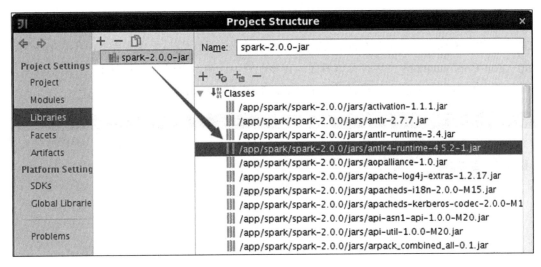

图 2-22　配置项目依赖 JAR

下面使用 IDEA 对第一个实例进行重写，具体方法是在 src/main/scala 目录下创建 chapter2 包，在包中创建 WordCount.scala 文件，程序代码如下所示：

```scala
import org.apache.spark.SparkContext._
import org.apache.spark.{SparkConf, SparkContext}
object WordCount
{
  def main(args: Array[String])
  {
    val conf = new SparkConf().setAppName("WordCount")
    val sc = new SparkContext(conf)
    val rdd = sc.textFile(args(0))
    val wordcount = rdd.flatMap(_.split(" ")).map(x=>(x,1)).reduceByKey(_+_)
    val wordsort = wordcount.map(x=>(x._2,x._1)).sortByKey(false).map(x=>
      (x._2,x._1))
    wordsort.saveAsTextFile(args(1))
    sc.stop()
  }
}
```

在 IDEA 里编写的代码既可以直接在 IDEA 中运行或调试，也可以使用 IDEA 打包后通过 Spark Submit 工具提交到集群中运行，以下演示不同的操作。

1. 直接运行

代码在运行之前需要进行编译，单击菜单 Build→Make Project 或者 Ctrl+F9 组合键对代码进行编译，编译结果会在 Event Log 进行提示，如果出现异常可以根据提示进行修改。编译通过以后在代码界面中单击鼠标右键，选择运行程序弹出"运行/调试配置"界面，或者单击菜单 Run→Edit Configurations 也可打开该界面。在该配置界面中，设置 WordCount 运行时需要输入单词计数的文件路径和输出结果路径两个参数，如图 2-23 所示。

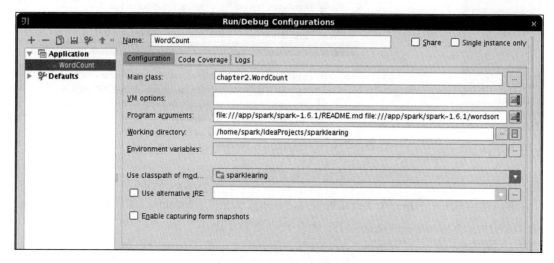

图 2-23　设置项目运行参数界面

- 单词计数的文件路径径: 设置 Spark 说明文件 file:///app/spark/spark-2.0.0/README.md。
- 输出结果路径：本地路径 file:///app/spark/spark-2.0.0/wordsort。

启动 Spark 集群，单击菜单 Run→Run 或者 Shift+F10 组合键运行 WordCount，在运行结果窗口中可以查看运行过程。当然，如果需要观察程序运行的详细过程，可以加入断点，使用调试模式根据程序运行过程。

2．打包运行

使用打包运行需重新打开项目配置界面，选择 Artifacts 配置页，在右边的操作界面中选择"+"号，选择添加 JAR 包的 From modules with dependencies 方式，在该界面中选择主函数入口为 WordCount，同时设置包输出路径，这里配置为 /home/spark/IdeaProjects/sparklearning/out/artifacts/，如图 2-24 所示。

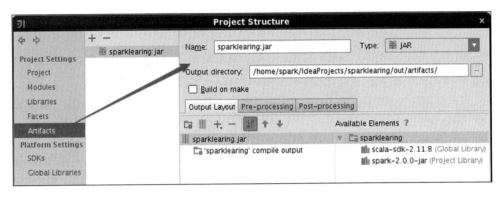

图 2-24　设置项目打包参数界面

默认情况下，Output Layout 会附带 Scala 相关的类包，但由于运行环境已经有 Scala 相关类包，可以去除这些包只保留项目的输出内容。设置完毕后单击菜单 Build→Build Artifacts 或在程序代码区域中单击右键，选择 Build 或者 Rebuild 动作生成项目的打包文件，将打包好的文件移动到 Spark 根目录下。

```
$cd /home/spark/IdeaProjects/sparklearning/out/artifacts/
$cp sparklearing.jar/app/spark/spark-2.0.0/
$ls /app/spark/spark-2.0.0/
```

先启动 Spark 集群，然后使用 Spark Submit 在集群中运行：

```
$cd /app/spark/spark-2.0.0
$bin/spark-submit --class chapter2.WordCount \
  --master spark://master:7077 \
  --executor-memory 1g \
  sparklearing.jar \
  file:///app/spark/spark-2.0.0/README.md \
  file:///app/spark/spark-2.0.0/ wordsort2
```

运行的结果如图 2-25 所示。

图 2-25　查看单词计数结果界面

2.4.3 使用 IDEA 阅读源代码

在后面的章节中，将剖析 Spark 的核心代码，进而解析其原理和运行机制，所以搭建 Spark 源码阅读环境是必不可少，可以利用前面安装的 IDEA 轻松搭建。其过程如下：

（1）可以使用 2.2.1 节下载的 Spark 源代码，复制一份专门作为 IDEA 源码阅读，命令如下：

```
$cd /app/compile
$cp spark-2.0.0-srcspark-2.0.0-idea
```

不过更建议使用 2.2.2 节编译过的 Spark 源代码，该源代码目录中包括编译文件、依赖包等，可以减少 IDEA 下载依赖包的时间。

```
$cd /app/compile
$cp -r spark-2.0.0-mvn spark-2.0.0-idea
```

（2）打开 IDEA 开发环境，选择导入项目，在弹出的界面中使用 Maven，如图 2-26 所示。

（3）选择源代码来源方式。在选择编译子系统界面中选中如下几个选项，如图 2-27 所示。

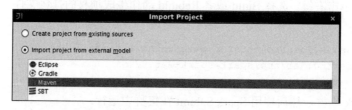

图 2-26　选择 Maven

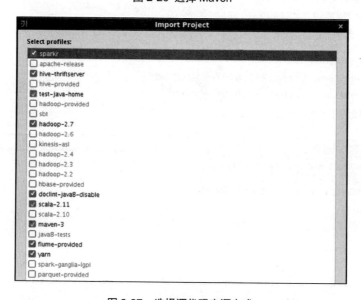

图 2-27　选择源代码来源方式

（4）选择导入 Spark 子项目界面。

选择完毕后，IDEA 将根据 Maven 的配置文件 pom.xml 文件解析项目的依赖关系，下载依赖包和初始化项目设置。

（5）编译并解决异常问题。

IDEA 同步 Spark 源代码后，可以选择 Run→rebuild 编译 Spark 源代码项目，在编译过程中会提示类似如图 2-28 所示的错误。

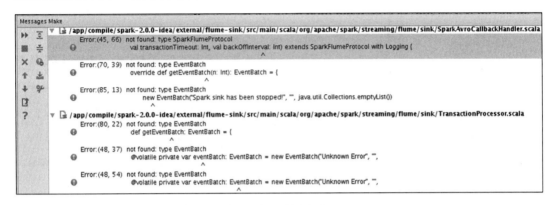

图 2-28　提示出错信息

（6）编译 Spark 源代码提示错误信息。

出现错误信息的原因是由于插件使用协议的限制，在 Spark 源代码未包含某些类的源码，解决办法是把 external/flume-sink/target 目录下的 spark-streaming-flume-sink_2.11-2.0.0-sources.jar 解压（需要注意的是，只有经过编译过的 Spark 才有，否则参见作者博客中提供资源），把 SparkFlumeProtocol.java、EventBatch.java 和 SparkSinkEvent.java 三个类复制到 org.apache.spark.streaming.flume.sink 包中再重新编译即可，如图 2-29 所示。

Spark 源代码编译通过以后，我们利用 IDEA 启动本地运行 Spark 环境。使用 Ctrl+N 打开 src/scala/main/org/apache/spark/deploy/master/Master.scala 代码，通过右键或者菜单 Run→Edit Configurations 打开运行配置界面，如图 2-30 所示，在该界面中设定应用名称为 Master，主程序入口为 org.apache.spark.deploy.master.Master。

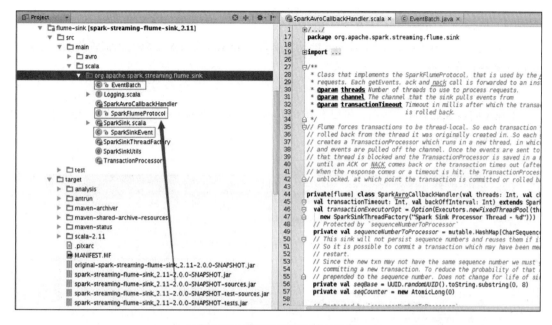

图 2-29　编译出错需新增 3 个类

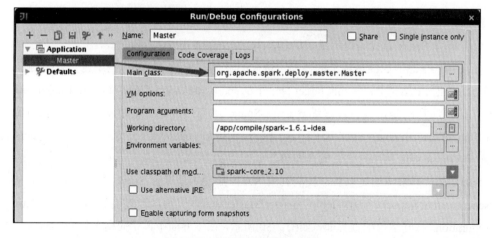

图 2-30　使用源代码启动 Master

【注意】启动时如果出现 java.lang.NoClassDefFoundError：com/google/common/collect/Maps 错误，具体信息如下：

```
Exception in thread "main" java.lang.NoClassDefFoundError: com/google/common/
  collect/Maps
  at org.apache.hadoop.metrics2.lib.MetricsRegistry.<init>(MetricsRegistry.java:42)
  at org.apache.hadoop.metrics2.impl.MetricsSystemImpl.<init>(MetricsSystemImpl.java:87)
```

```
at org.apache.hadoop.metrics2.impl.MetricsSystemImpl.<init>(MetricsSystemImpl.java:133)
………………
at org.apache.spark.deploy.master.Master$.startRpcEnvAndEndpoint(Master.scala:1148)
at org.apache.spark.deploy.master.Master$.main(Master.scala:1133)
at org.apache.spark.deploy.master.Master.main(Master.scala)
……………………..
at java.lang.reflect.Method.invoke(Method.java:606)
at com.intellij.rt.execution.application.AppMain.main(AppMain.java:134)
```

修改引用包 guava-14.0.1.jar 范围为编译就可以解决，如图 2-31 所示。

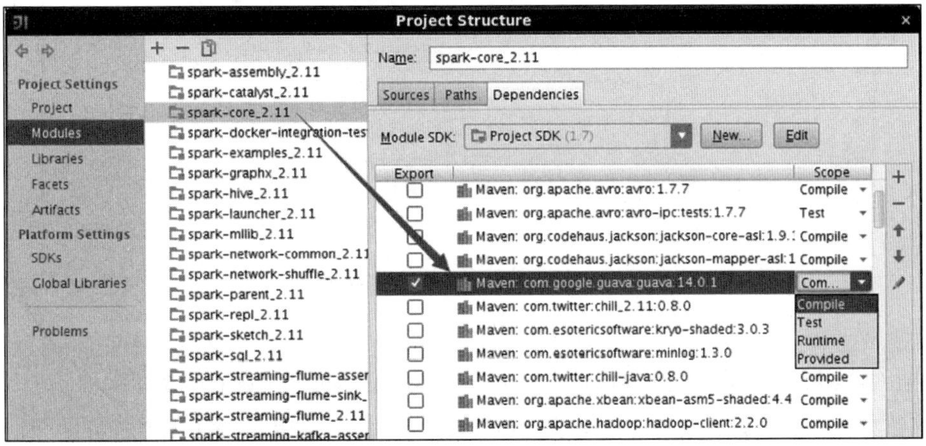

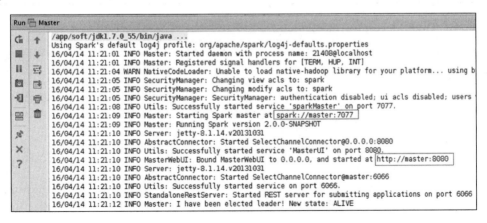

图 2-31　解决类无法找到错误方法

设置完毕后，启动 Master 应用程序，在 IDEA 的运行日志中看到以本地方式启动，其地址为 spark://master:7077，如图 2-32 所示。

图 2-32　使用源代码启动 Master 日志信息

打开 src/scala/main/org/apache/spark/deploy/worker/Worker.scala 代码，使用设置 Master 应用程序的方法打开 Worker 应用程序配置界面，区别 Master 配置，需要在程序参数中加入 --webui-port 8081 spark://master:7077，如图 2-33 所示。

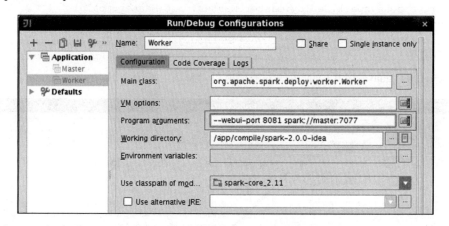

图 2-33　使用源代码启动 Worker

启动 Worker 应用程序，可以在运行日志中看到 Worker 连接到 spark://master:7077 的 Master 中，如图 2-34 所示。通过监控界面可以看到在该 Spark 运行环境中有一个本地启动的 Worker，CPU 核数和内存由于没有设置，使用了能够设置的最大值。如果看到上面监控界面的信息表示通过源码方式启动了 Spark 运行环境，可以通过 Spark Shell 等工具提交操作，更重要的是，利用该环境以 Debug 方式运行设置断点，跟踪 Spark 运行过程，通过源码阅读更加深入了解 Spark 核心技术。

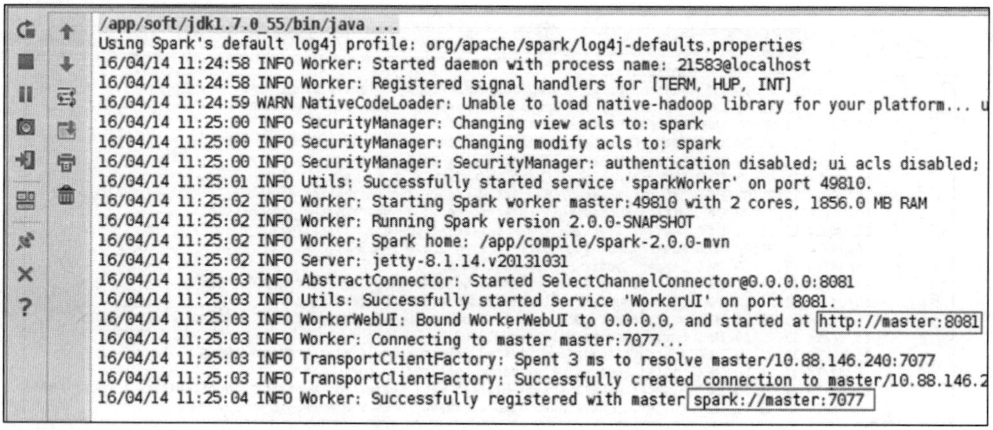

图 2-34　使用源代码启动 Worker 日志信息。

2.5 小结

为了更加深入地了解 Spark 运行的核心原理，需要通过阅读源代码并开发 Spark 实例加以实践。在本章中先介绍搭建集群基础环境，然后在此基础上配置了 Spark 编译环境，利用该环境分别使用 Maven 和 SBT 编译 Spark 源代码，并对编译好的代码进行打包，接着使用打包程序部署 Spark 集群，最后搭建 Spark 实战开发环境，利用该环境开发应用程序和阅读源代码。

第二篇 核心篇

第 3 章

Spark 编程模型

3.1 RDD 概述

3.1.1 背景

在过去的十几年时间里，计算机普遍应用和移动互联网大发展导致了数据量爆发式增长，单台机器处理能力和 I/O 性能远远满足不了这种增长，越来越多的企业不得把计算扩展到集群中。但是集群运行程序出现了一些挑战：首先是并行化处理，以前的应用程序需要以并行化的方式重写，并且这种编程模型能够处理范围广泛的计算；其次是集群的容错，在大规模的情况下节点故障和慢节点将成为新常态，这种情况可能极大地影响应用程序的性能；最后是多用户共享，在集群运行中需要动态的扩展和缩减计算资源，如 CPU、内存和磁盘等。

针对这些问题，不同的企业和研究机构设计了不同的编程模式。起初 Google 公司提出了 MapReduce 编程模式，这是一种简单通用而且能够自动处理故障的批处理计算模型。随着技术的发展，出现了 Storm 流处理系统、Impala 交互式 SQL 查询系统和 Bulk Synchronous Parallel(BSP) 并行迭代图计算模型。在这些模型中都需要高效的数据共享，例如，迭代算法（PageRank、K-means 聚类和逻辑回归），都需要进行多次访问相同的数据集；交互数据挖掘需要对同一数据子集进行多个特定的查询；而流式应用则需要随时间对状态进行维护和共享。但是不幸的是，尽管这些框架支持大量的计算操作运算，但是它们缺乏针对数据共享的高效元语。在这些框架中，实现计算之间（例如两个 MapReduce 作业之间）数据共享只有一个办法，就是把数据写到外部存储系统，如分布式文件系统 HDFS，这会引入数据备份、磁盘 I/O 以及序列化等开销，从而占据了大部分的执行时间。

而 Spark 和前面系统解决办法思路相反，它设计了统一的编程抽象——弹性分布式数据集（RDD），这种全新的模型可以令用户直接控制数据的共享，使得用户可以指定数据存储到硬盘还是内存、控制数据的分区方法和数据集上进行的操作。RDD 不仅增加了高效的数据共享元语，而且大大增加了其通用性。相比较前面的系统架构，Spark 有如下优势：

- 在相同的运行环境中，支持迭代、批处理、交互式和流处理，相对单一模式的系统能更好地发挥其性能。
- 以很小的代价在该计算模式提供节点故障和慢节点容错处理功能，特别是在流和交互式 SQL 查询中，基于 RDD 产生的新系统比现有的系统有更强的容错性。
- 相对其他计算模型有更为快速的计算能力，比 MapReduce 的性能要高 100 倍，能够媲美各个应用领域的专业系统。
- 适合多用户管理，允许应用程序弹性地扩展和缩减计算资源。

3.1.2 RDD 简介

Spark 编程模型是弹性分布式数据集（Resilient Distributed Dataset，RDD），它是 MapReduce 模型的扩展和延伸，但它解决了 MapReduce 的缺陷：在并行计算阶段高效地进行数据共享。运用高效的数据共享概念和类似于 MapReduce 的操作方式，使得并行计算能够高效地进行，并可以在特定的系统中得到关键的优化。

相比以前集群容错处理模型，如 MapReduce、Dryad，它们将计算转换为一个有向无环图（DAG）的任务集合。这使在这些模型中能够有效地恢复 DAG 中故障和慢节点执行的任务，但是这些模型中除了文件系统外没有提供其他的存储方式，这就导致了在网络上进行频繁的数据复制而造成 I/O 压力。由于 RDD 提供一种基于粗粒度变换（如 map、filter 和 join 等）的接口，该接口会将相同的操作应用到多个数据集上，这就使得它们可以记录创建数据集的"血统"（Lineage），而不需要存储真正的数据，从而达到高效的容错性。当某个 RDD 分区丢失的时候，RDD 记录有足够的信息来重新计算，而且只需要计算该分区，这样丢失的数据可以很快地恢复，不需要昂贵的复制代价。

基于 RDD 机制实现了多类模型计算，包括多个现有的集群编程模式。在这些模型中，RDD 不仅在性能方面达到了之前的系统水平，也带来了现有系统所缺少的新特性，如容错性、慢节点执行和弹性资源分配等。这些模型包括以下几方面的内容。

（1）**迭代计算**：目前最常见的工作方式，比如应用于图处理、数值优化以及机器学习中的算法。RDD 可以支持各类型模型，包括 Pregel、MapReduce、GraphLab 和 PowerGraph 模型。

（2）**交互式 SQL 查询**：在 MapReduce 集群中大部分需求是执行 SQL 查询，而 MapReduce

相对并行数据库在交互式查询有很大的不足。而 Spark 的 RDD 不仅拥有很多常见数据库引擎的特性，达到可观的性能，而且在 Spark SQL 中提供完善的容错机制，能够在短查询和长查询中很好地处理故障和慢节点。

（3）**MapReduceRDD**：通过提供 MapReduce 的超集，能够高效地执行 MapReduce 程序，同样也可以如 DryadLINQ 这样常见的 DAG 数据流的应用。

（4）**流式数据处理**：流式数据处理已经在数据库和系统领域进行了很长时间的研究，但是大规模流式数据处理仍是一项挑战。当前的模型没有解决在大规模集群中频繁出现的慢节点的问题，同时对故障解决办法有限，需要大量的复制或浪费很长的恢复时间。为了恢复一个丢失的节点，当前的系统需要保存每一个操作的两个副本，或通过一系列耗费大量开销的串行处理对故障点之前的数据进行重新处理。

Spark 提出了离散数据流（D-Streams）来解决这样的问题，D-Streams 把流式计算的执行当作一系列短而确定的批量计算的序列，并将状态保存在 RDD 中。D-Streams 根据相关 RDD 的依赖关系图进行并行化恢复，可以达到快速故障恢复，避免了数据复制。另外通过推测执行来支持对 Straggler 迁移执行，例如，对慢任务运行经过推测的备份副本。尽管 D-Streams 将计算转换为许多不相关联的作业来运行而增加延迟，但这种延迟在 D-Streams 集群处理中只耗费次秒级时间。

RDD 还能够支持一些现有系统不能表示的新应用。例如，许多数据流应用程序还需要加入历史数据的信息；通过使用 RDD 可以在同一程序中同时使用批处理和流式处理，这样来实现所有模型中数据共享和容错恢复；同样地，流式应用的操作者常常需要在数据流的状态上执行及时查询。一般来说，每一个批处理应用常常需要整合多个处理类型，比如，一个应用可能需要使用 SQL 提取数据，在数据集上训练一个机器学习模型，然后对这个模型进行查询。由于计算的大部分时间花在系统之间共享数据的分布式文件系统 I/O 的开销上，因此使用当前多个系统组合而成的工作流效率非常低下。而使用一个基于 RDD 机制的系统，这些计算可以在同一个引擎中紧接着执行，不需要额外的 I/O 操作，处理效率大大提高。

3.1.3　RDD 的类型

Spark 编程中开发者需要编写一个驱动程序（Driver Program）来连接到工作进程（Worker），如图 3-1 所示。驱动程序定义一个或多个 RDD 以及相关行动操作，驱动程序同时记录 RDD 的继承关系，即"血统"。而工作进程（Worker）是一直运行的进程，它将经过一系列操作后的 RDD 分区数据保存在内存中。

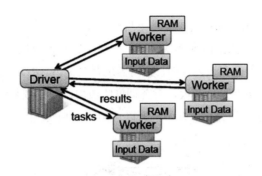

图 3-1　Spark 运行架构图

Spark 中的操作大致可以分为四类操作，分别为创建操作、转换操作、控制操作和行为操作。

- 创建操作（Creation Operation）：用于 RDD 创建工作。RDD 创建只有两种方法，一种是来自于内存集合和外部存储系统，另一种是通过转换操作生成的 RDD。
- 转换操作（Transformation Operation）：将 RDD 通过一定的操作变换成新的 RDD，比如 HadoopRDD 可以使用 map 操作变换为 MappedRDD，RDD 的转换操作是惰性操作，它只是定义了一个新的 RDDs，并没有立即执行。
- 控制操作（Control Operation）：进行 RDD 持久化，可以让 RDD 按不同的存储策略保存在磁盘或者内存中，比如 cache 接口默认将 RDD 缓存在内存中。
- 行动操作（Action Operation）：能够触发 Spark 运行的操作，例如，对 RDD 进行 collect 就是行动操作。Spark 中行动操作分为两类，一类的操作结果变成 Scala 集合或者变量，另一类将 RDD 保存到外部文件系统或者数据库中。

3.2　RDD 的实现

3.2.1　作业调度

当对 RDD 执行转换操作时，调度器会根据 RDD 的"血统"来构建由若干调度阶段（Stage）组成的有向无环图（DAG），每个调度阶段包含尽可能多的连续窄依赖转换。调度器按照有向无环图顺序进行计算，并最终得到目标 RDD。

调度器向各节点分配任务采用延时调度机制并根据数据存储位置（数据本地性）来确定。若一个任务需要处理的某个分区刚好存储在某个节点的内存中，则该任务会分配给该节点；如果在内存中不包含该分区，调度器会找到包含该 RDD 的较佳位置，并把任务分配给所在节点。

对应宽依赖的操作，在 Spark 将中间结果物化到父分区的节点上，这和 MapReduce 物化 map 的输出类似，可以简化数据的故障恢复过程。如图 3-2 所示，实线圆角方框标识的是 RDD。阴影背景的矩形是分区，若已存于内存中，则用黑色背景标识。RDD 上一个行动操作的执行将会以宽依赖为分区来构建各个调度阶段，对各调度阶段内部的窄依赖则前后连接构成流水线。在本例中，Stage 1 的输出已经存在内存中，所以直接执行 Stage 2，然后执行 Stage 3。

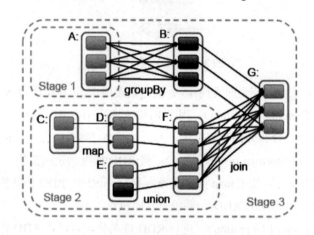

图 3-2　Spark 如何计算作业调度阶段

对于执行失败的任务，只要它对应调度阶段父类信息仍然可用，该任务会分散到其他节点重新执行。如果某些调度阶段不可用（例如，因为 Shuffle 在 map 节点输出丢失了），则重新提交相应的任务，并以并行方式计算丢失的分区。在作业中如果某个任务执行缓慢（即 Straggler），系统则会在其他节点上执行该任务的副本。该方法与 MapReduce 推测执行做法类似，并取最先得到的结果作为最终的结果。

3.2.2　解析器集成

与 Ruby 和 Python 类似，Scala 提供了一个交互式 Shell（解析器），借助内存数据带来的低延迟特性，可以让用户通过解析器对大数据进行交互式查询。

Spark 解析器将用户输入的多行命令解析为相应 Java 对象的示例如图 3-3 所示。

Scala 解析器处理过程一般为：①将用户输入的每一行编译成一个类；②将该类载入到 JVM 中；③调用该类的某个函数。在该类中包含一个单利对象，对象中包含当前行的变量或函数，在初始化方法中包含处理该行的代码。例如，如果用户输入 "var x=5"，在换行输入 println(x)，那解析器会定义一个叫 Line1 的类，该类包含 x，第二行编译成 println (Line1.getInstance().x)。

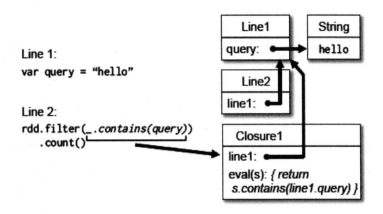

图 3-3 Scala 解析器处理过程

Spark 中做了以下两个改变。

（1）类传输：为了让工作节点能够从各行生成的类中获取到字节码，通过 HTTP 传输。

（2）代码生成器的改动：通常各种代码生成的单例对象是由类的静态方法来提供的。也就是说，当序列化一个引用上一行定义变量的闭包（例如上面例子的 Line1.x），Java 不会通过检索对象树的方式去传输包含 x 的 Line1 实例。因此工作节点不能够得到 x，在 Spark 中修改了代码生成器的逻辑，让各行对象的实例可以被字节应用。在图 3-3 中显示了 Spark 修改之后解析器是如何把用户输入的每一行变成 Java 对象的。

3.2.3 内存管理

Spark 提供了 3 种持久化 RDD 的存储策略：未序列化 Java 对象存在内存中、序列化的数据存于内存中以及存储在磁盘中。第一个选项的性能是最优的，因为可以直接访问在 Java 虚拟机内存里的 RDD 对象；在空间有限的情况下，第二种方式可以让用户采用比 Java 对象更有效的内存组织方式，但代价是降低了性能；第三种策略使用于 RDD 太大的情形，每次重新计算该 RDD 会带来额外的资源开销（如 I/O 等）。

对于内存使用 LRU 回收算法来进行管理，当计算得到一个新的 RDD 分区，但没有足够空间来存储时，系统会从最近最少使用的 RDD 回收其一个分区的空间。除非该 RDD 是新分区对应的 RDD，这种情况下 Spark 会将旧的分区继续保留在内存中，防止同一个 RDD 的分区被循环调入/调出。这点很关键，因为大部分的操作会在一个 RDD 的所有分区上进行，那么很有可能已经存在内存中的分区将再次被使用。

3.2.4 检查点支持

虽然"血统"可以用于错误后 RDD 的恢复,但是对于很长的"血统"的 RDD 来说,这样的恢复耗时比较长,因此需要通过检查点操作(Checkpoint)保存到外部存储中。

通常情况下,对于包含宽依赖的长"血统"的 RDD 设置检查点操作是非常有用的。在这种情况下,集群中某个节点出现故障时,会使得从各个父 RDD 计算出的数据丢失,造成需要重新计算。相反,对于那些窄依赖的 RDD,对其进行检查点操作就不是有必须。在这种情况下如果一个节点发生故障,RDD 在该节点中丢失的分区数据可以通过并行的方式从其他节点中重新计算出来,计算成本只是复制 RDD 的很小部分。

Spark 提供为 RDD 设置检查点操作的 API,可以让用户自行决定需要为那些数据设置检查点操作。另外由于 RDD 的只读特性,使得不需要关心数据一致性问题,比常用的共享内存更容易做检查点。

3.2.5 多用户管理

RDD 模型将计算分解为多个相互独立的细粒度任务,这使得它在多用户集群能够支持多种资源共享算法。特别地,每个 RDD 应用可以在执行过程中动态调整访问资源。

- 在每个应用程序中,Spark 运行多线程同时提交作业,并通过一种等级公平调度器来实现多个作业对集群资源的共享,这种调度器和 Hadoop Fair Scheduler 类似。该算法主要用于创建基于针对相同内存数据的多用户应用,例如:Spark SQL 引擎有一个服务模式支持多用户并行查询。公平调度算法确保短的作业能够在即使长作业占满集群资源的情况下尽早完成。
- Spark 的公平调度也使用延迟调度,通过轮询每台机器的数据,在保持公平的情况下给予作业高的本地性。Spark 支持多级本地化访问策略(本地化),包括内存、磁盘和机架。
- 由于任务相互独立,调度器还支持取消作业来为高优先级的作业腾出资源。
- Spark 中可以使用 Mesos 来实现细粒度的资源共享,这使得 Spark 应用能相互之间或在不同的计算框架之间实现资源的动态共享。
- Spark 使用 Sparrow 系统扩展支持分布式调度,该调度允许多个 Spark 应用以去中心化的方式在同一集群上排队工作,同时提供数据本地性、低延迟和公平性。

3.3 编程接口

Spark 中提供了通用接口来抽象每个 RDD，这些接口包括：①分区信息，它们是数据集的最小分片；②依赖关系，指向其父 RDD；③函数，基于父 RDD 计算方法；④划分策略和数据位置的元数据。例如：一个 HDFS 文件的 RDD 将文件的每个文件块表示为一个分区，并且知道每个文件块的位置信息，当对 RDD 进行 map 操作后分区将具有相同的划分。表 3-1 列出了 RDD 编程接口。

表 3-1　RDD 编程接口

序号	操作	含义
1	Partitions()	返回分片对象列表
2	PreferredLocations(p)	根据数据的本地特性，列出分片 p 的首选位置
3	Dependencies()	返回依赖列表
4	Iterator(p,parentIters)	给定 p 的父分片的迭代器，计算分片 p 的元素
5	Partitioner()	返回说明 RDD 是否是 Hash 或者是范围分片的元数据

3.3.1　RDD 分区（Partitions）

RDD 划分成很多的分区（Partition）分布到集群的节点中，分区的多少涉及对这个 RDD 进行并行计算的粒度。分区是一个逻辑概念，变换前后的新旧分区在物理上可能是同一块内存或存储，这种优化防止函数式不变性导致的内存需求无限扩张。在 RDD 操作中用户可以使用 Partitions 方法获取 RDD 划分的分区数，当然用户也可以设定分区数目。如果没有指定将使用默认值，而默认数值是该程序所分配到 CPU 核数，如果是从 HDFS 文件创建，默认为文件的数据块数。

以下以 scala>开头的程序运行在 Spark Shell 命令行中，具体启动可以参见搭建 Spark 集群 2.2.4 节的内容。

```
//使用textFile方法获取spark根目录的文件，未设置分区数
scala> val part = sc.textFile("/app/spark/spark-2.0.0/README.md")
part: org.apache.spark.rdd.RDD[String] = /app/spark/spark-2.0.0/README.md
  MapPartitionsRDD[1] at textFile at <console>:25

scala> part.partitions.size
res3: Int = 2

//显式地设置RDD为6个分区
scala> val part1 = sc.textFile("/app/spark/spark-2.0.0/README.md",6)
```

```
part1: org.apache.spark.rdd.RDD[String] = /app/spark/spark-2.0.0/README.md
  MapPartitionsRDD[2] at textFile at <console>:25

scala> part1.partitions.size
res4: Int = 6
```

3.3.2 RDD 首选位置（PreferredLocations）

在 Spark 形成任务有向无环图（DAG）时，会尽可能地把计算分配到靠近数据的位置，减少数据网络传输。当 RDD 产生的时候存在首选位置，如 HadoopRDD 分区的首选位置就是 HDFS 块所在的节点；当 RDD 分区被缓存，则计算应该发送到缓存分区所在的节点进行，再不然回溯 RDD 的"血统"一直找到具有首选位置属性的父 RDD，并据此决定子 RDD 的位置。

3.3.3 RDD 依赖关系（Dependencies）

在 RDD 中将依赖划分成了两种类型：窄依赖（Narrow Dependencies）和宽依赖（Wide Dependencies），如图 3-4 所示。窄依赖是指每个父 RDD 的分区都至多被一个子 RDD 的分区使用，而宽依赖是多个子 RDD 的分区依赖一个父 RDD 的分区。例如，map 操作是一种窄依赖，而 join 操作是一种宽依赖（除非父 RDD 已经基于 Hash 策略被划分过了）。

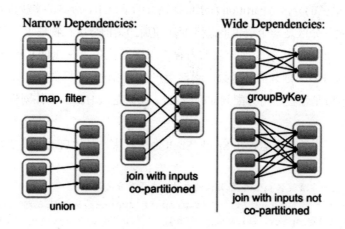

图 3-4　RDD 窄依赖和宽依赖

这两种依赖的区别从两个方面来说比较有用。第一，窄依赖允许在单个集群节点上流水线式执行，这个节点可以计算所有父级分区。例如，可以逐个元素地依次执行 filter 操作和 map 操作。相反，宽依赖需要所有的父 RDD 数据可用，并且数据已经通过类 MapReduce 的操作 Shuffle

完成。第二，在窄依赖中，节点失败后的恢复更加高效。因为只有丢失的父级分区需要重新计算，并且这些丢失的父级分区可以并行地在不同节点上重新计算。与此相反，在宽依赖的继承关系中，单个失败的节点可能导致一个 RDD 的所有先祖 RDD 中的一些分区丢失，导致计算的重新执行。

```scala
//读取Spark说明文件，先对分词，使用map操作生成wordmap RDD
scala> val rdd = sc.textFile("/app/spark/spark-2.0.0/README.md")
rdd: org.apache.spark.rdd.RDD[String] = /app/spark/spark-2.0.0/README.md
   MapPartitionsRDD[3] at textFile at <console>:25

scala> val wordmap = rdd.flatMap(_.split(" ")).map(x=>(x,1))
wordmap: org.apache.spark.rdd.RDD[(String, Int)] = MapPartitionsRDD[5] at map
   at <console>:27

scala> println(wordmap)
MapPartitionsRDD[5] at map at <console>:27

//wordmap的依赖关系为OneToOneDependency，属于窄依赖
scala> wordmap.dependencies.foreach { dep =>
     |   println("dependency type:" + dep.getClass)
     |   println("dependency RDD:" + dep.rdd)
     |   println("dependency partitions:" + dep.rdd.partitions)
     |   println("dependency partitions size:" + dep.rdd.partitions.length)
     | }
dependency type:class org.apache.spark.OneToOneDependency
dependency RDD:MapPartitionsRDD[4] at flatMap at <console>:27
dependency partitions:[Lorg.apache.spark.Partition;@593e0a4c
dependency partitions size:2

//使用reduceByKey操作对单词进行计数
scala> val wordreduce = wordmap.reduceByKey(_+_)
wordreduce: org.apache.spark.rdd.RDD[(String, Int)] = ShuffledRDD[6] at
   reduceByKey at <console>:29

scala> println(wordreduce)
ShuffledRDD[6] at reduceByKey at <console>:29

//wordreduce的依赖类型为ShuffleDependency，属于宽依赖
scala> wordreduce.dependencies.foreach { dep =>
     |   println("dependency type:" + dep.getClass)
     |   println("dependency RDD:" + dep.rdd)
     |   println("dependency partitions:" + dep.rdd.partitions)
     |   println("dependency partitions size:" + dep.rdd.partitions.length)
```

```
        | }
dependency type:class org.apache.spark.ShuffleDependency
dependency RDD:MapPartitionsRDD[5] at map at <console>:27
dependency partitions:[Lorg.apache.spark.Partition;@593e0a4c
dependency partitions size:2
```

3.3.4　RDD 分区计算（Iterator）

Spark 中 RDD 计算是以分区为单位的，而且计算函数都是在对迭代器复合，不需要保存每次计算的结果。分区计算一般使用 mapPartitions 等操作进行，mapPartitions 的输入函数是应用于每个分区，也就是把每个分区中的内容作为整体来处理的：

```
def     mapPartitions[U:    ClassTag](f:    Iterator[T]    =>    Iterator[U],
    preservesPartitioning: Boolean = false): RDD[U]
```

f 即为输入函数，它处理每个分区里面的内容。每个分区中的内容将以 Iterator[T] 传递给输入函数 f，f 的输出结果是 Iterator[U]。最终的 RDD 由所有分区经过输入函数处理后的结果合并起来的。在下面的例子中，函数 iterfunc 是把分区中一个元素和它的下一个元素组成一个 Tuple，因为分区中最后一个元素没有下一个元素了，所以(3,4)和(6,7)不在结果中。

```
scala> val a = sc.parallelize(1 to 9, 3)
a: org.apache.spark.rdd.RDD[Int] = ParallelCollectionRDD[7] at parallelize at
    <console>:25
scala> def iterfunc [T](iter: Iterator[T]) : Iterator[(T, T)] = {
    |     var res = List[(T, T)]()
    |     var pre = iter.next
    |     while(iter.hasNext) {
    |         val cur = iter.next
    |         res ::= (pre, cur)
    |         pre = cur
    |     }
    |     res.iterator
    | }
iterfunc: [T](iter: Iterator[T])Iterator[(T, T)]
scala> a.mapPartitions(iterfunc).collect
res7: Array[(Int, Int)] = Array((2,3), (1,2), (5,6), (4,5), (8,9), (7,8))
```

3.3.5　RDD 分区函数（Partitioner）

分区划分对于 Shuffle 类操作很关键，它决定了该操作的父 RDD 和子 RDD 之间的依赖类型。例如 Join 操作，如果协同划分的话，两个父 RDD 之间、父 RDD 与子 RDD 之间能形成一致的分区安排，即同一个 Key 保证被映射到同一个分区，这样就能形成窄依赖。反之，如果没有协同划分,导致宽依赖。这里所说的协同划分是指定分区划分器以产生前后一致的分区安排。

在 Spark 默认提供两种划分器:哈希分区划分器(HashPartitioner)和范围分区划分器(RangePartitioner),且 Partitioner 只存在于(K, V)类型的 RDD 中,对于非(K, V)类型的 Partitioner 值为 None。

在以下程序中,首先构造一个 MappedRDD,其 Partitioner 的值为 None,然后对 RDD 进行 groupByKey 操作得出 group_rdd 变量,对于 groupByKey 操作而言,这里创建了新的 HashPartitioner 对象,参数"4"代表 group_rdd 最终会拥有 4 个分区。

```
scala> val part = sc.textFile("/app/spark/spark-2.0.0/README.md")
part: org.apache.spark.rdd.RDD[String] = /app/spark/spark-2.0.0/README.md
  MapPartitionsRDD[10] at textFile at <console>:25

scala> part.partitioner
res8: Option[org.apache.spark.Partitioner] = None

scala> val group_rdd = part.map(x=>(x,x)).groupByKey(new org.apache.spark.
  HashPartitioner(4))
group_rdd: org.apache.spark.rdd.RDD[(String, Iterable[String])] =
  ShuffledRDD[12] at groupByKey at <console>:27

scala>group_rdd.partitioner
res9: Option[org.apache.spark.Partitioner] = Some(org.apache.spark.HashPartitioner@4)
```

3.4 创建操作

目前有两种类型的基础 RDD:一种是并行集合(Parallelized Collections),接收一个已经存在的 Scala 集合,然后进行各种并行计算;另一种是从外部存储创建 RDD,外部存储可以是文本文件或 Hadoop 文件系统 HDFS,还可以是从 Hadoop 接口 API 创建。

在一个文件的每条记录上运行函数。只要文件系统是 HDFS,或者 Hadoop 支持的任意存储系统即可。这两种类型的 RDD 都可以通过相同的方式进行操作,从而获得子 RDD 等一系列拓展,形成"血统"关系图。

3.4.1 并行化集合创建操作

并行化集合是通过调用 SparkContext 的 parallelize 方法,在一个已经存在的 Scala 集合上创建的(一个 Seq 对象)。集合的对象将会被复制,创建出一个可以被并行操作的分布式数据集。

例如，下面的解释器输出，演示了如何从一个数组创建一个并行集合。在 SparkContext 类中实现了 parallelize 和 makeRDD 两个并行化集合创建操作：

- parallelize[T](seq: Seq[T], numSlices: Int = defaultParallelism):RDD[T]

例如：val rdd = sc.parallelize(Array(1 to 10)) 根据能启动的 Executor 的数量来进行切分多个分区，每一个分区启动一个任务来进行处理。当然可以在该方法指定分区数量，如 val rdd = sc.parallelize(Array(1 to 10), 5) 中指定了 partition 的数量为 5。

```
//使用parallelize操作并行化1到10数据集，由于运行的Spark集群有3个Worker节点，
//存在3个Executor,故该数据集有3个分区
scala> var rdd = sc.parallelize(1 to 10)
rdd: org.apache.spark.rdd.RDD[Int] = ParallelCollectionRDD[13] at parallelize
    at <console>:25

scala> rdd.collect
res10: Array[Int] = Array(1, 2, 3, 4, 5, 6, 7, 8, 9, 10)

scala> rdd.partitions.size
res11: Int = 3

//显式地设置RDD为4个分区
scala> var rdd2 = sc.parallelize(1 to 10,4)
rdd2: org.apache.spark.rdd.RDD[Int] = ParallelCollectionRDD[14] at parallelize
    at <console>:25

scala> rdd2.collect
res12: Array[Int] = Array(1, 2, 3, 4, 5, 6, 7, 8, 9, 10)

scala> rdd2.partitions.size
res13: Int = 4
```

- makeRDD[T](seq: Seq[(T, Seq[String])]):RDD[T]
- makeRDD[T](seq: Seq[T], numSlices: Int = defaultParallelism):RDD[T]

用法与 parallelize 类似，不过该方法可以指定每一个分区的首选位置。在下面的例子中，指定了 1 到 10 的首选位置节点名为"master""slave1"，而 11 到 15 的首选位置节点名为"slave2""slave3"。

```
//指定1到10的首选位置为"master""slave1",11到15的首选位置节点名为"slave2""slave3"
scala> var collect = Seq((1 to 10,Seq("master","slave1")),(11 to 15,
    Seq("slave2"," slave3")))
collect: Seq[(scala.collection.immutable.Range.Inclusive, Seq[String])] =
    List((Range(1, 2, 3, 4, 5, 6, 7, 8, 9, 10),List(master, slave1)), (Range(11,
    12, 13, 14, 15),List(slave2, " slave3")))
```

```
scala> var rdd = sc.makeRDD(collect)
rdd: org.apache.spark.rdd.RDD[scala.collection.immutable.Range.Inclusive] =
  ParallelCollectionRDD[15] at makeRDD at <console>:27

//该 RDD 分区数为 2，分区和首选位置分布一致
scala> rdd.partitions.size
res14: Int = 2

scala> rdd.preferredLocations(rdd.partitions(0))
res15: Seq[String] = List(master, slave1)

scala> rdd.preferredLocations(rdd.partitions(1))
res16: Seq[String] = List(slave2, slave3)
```

3.4.2 外部存储创建操作

Spark 可以将任何 Hadoop 所支持的存储资源转化成 RDD，如本地文件（需要网络文件系统，所有的节点都必须能访问到）、HDFS、Cassandra、HBase、Amazon S3 等。Spark 支持文本文件、SequenceFiles 和任何 Hadoop InputFormat 格式。

- textFile(path: String, minPartitions: Int = defaultMinPartitions): RDD[String]

使用 textFile 操作可以将本地文件或 HDFS 文件转换成 RDD，该操作支持整个文件目录读取，文件可以是文本或者压缩文件（如 gzip 等，自动执行解压缩并加载数据）。

在演示之前，先在本地目录中创建 sparklearning.txt 文件，然后把该文件上传到 HDFS 中：

```
$touch /home/spark/work/chapter3/sparklearning.txt
$vi /home/spark/work/chapter3/sparklearning.txt
```

加入如下内容：

```
soft kitty
warm kitty
little ball of fur
happy kitty
sleepy kitty
pur pur pur
```

复制到 slave1 和 slave2，并上传到 HDFS 中：

```
$scp /home/spark/work/chapter3/sparklearning.txt spark@slave1:/home/spark/
  work/chapter3/
$scp /home/spark/work/chapter3/sparklearning.txt spark@slave2:/home/spark/
  work/chapter3/
$hadoop fs -copyFromLocal /home/spark/work/chapter3/sparklearning.txt/
```

chapter3/sparklearning.txt

该 textFile 操作分别读取本地和 HDFS 文件：

```
//从本地文件创建，这里需要注意的是，所有节点的目录下均要有该文件，否则运行中会
//报 "FileNotFoundException" 的错误，这也是前面使用 scp 复制到 slave1 和 slave2 的原因
scala> var rdd = sc.textFile("/home/spark/work/chapter3/sparklearning.txt")
rdd: org.apache.spark.rdd.RDD[String] = /home/spark/work/chapter3/sparklearning.txt
    MapPartitionsRDD[17] at textFile at <console>:25

scala> rdd.count
res8: Long = 7

//从 HDFS 读取文件创建 RDD
scala> var rdd = sc.textFile("hdfs://master:9000/chapter3/sparklearning.txt")
rdd: org.apache.spark.rdd.RDD[String] = hdfs://master:9000/chapter3/
    sparklearning.txt MapPartitionsRDD[2] at textFile at <console>:27

scala> rdd.count
res9: Long =7
```

textFile 可选第二个参数 slice，默认情况下为每一个数据块分配一个 slice。用户也可以通过 slice 指定更多的分片，但不能使用少于 HDFS 数据块的分片数。

- wholeTextFiles(path: String, minPartitions: Int = defaultMinPartitions): RDD[(String, String)]
- 使用 wholeTextFiles()读取目录里面的小文件，返回（用户名、内容）对
- sequenceFile[K,V](path:String,minPartitions:Int=defaultMinPartitions):RDD[(K, V)]
- sequenceFile[K,V](path:String,keyClass:Class[K], valueClass: Class[V]):RDD[(K, V)]
- sequenceFile[K, V](path: String, keyClass: Class[K], valueClass: Class[V], minPartitions: Int): RDD[(K, V)]
- 使用 sequenceFile[K,V]()操作可以将 SequenceFile 转换成 RDD。SequenceFile 文件是 Hadoop 用来存储二进制形式的 key-value 对而设计的一种文本文件（Flat File）。
- hadoopFile[K,V, F<: InputFormat[K, V]](path: String): RDD[(K, V)]
- hadoopFile[K,V,F<: InputFormat[K, V]](path:String, minPartitions: Int): RDD[(K, V)]
- hadoopFile[K, V](path: String, inputFormatClass: Class[_ <: InputFormat[K, V]], keyClass: Class[K], valueClass: Class[V], minPartitions: Int = defaultMinPartitions): RDD[(K, V)]
- newAPIHadoopFile[K, V, F <: InputFormat[K, V]](path: String, fClass: Class[F], kClass: Class[K], vClass: Class[V], conf: Configuration = hadoopConfiguration): RDD[(K, V)]
- newAPIHadoopFile[K, V, F <: InputFormat[K, V]](path: String)(implicit km: ClassTag[K],

vm: ClassTag[V], fm: ClassTag[F]): RDD[(K, V)]
- 由于 Hadoop 的接口有新旧两个版本，所以 Spark 为了能够兼容 Hadoop 的版本，也提供了两套创建操作接口。对于外部存储创建操作而言，hadoopRDD 和 newHadoopRDD 是最为抽象的两个函数接口。
- hadoopRDD[K, V](conf: JobConf, inputFormatClass: Class[_ <: InputFormat[K, V]], keyClass: Class[K], valueClass: Class[V], minPartitions: Int = defaultMinPartitions): RDD[(K, V)]
- newAPIHadoopRDD[K, V, F <: InputFormat[K, V]](conf: Configuration = hadoopConfiguration, fClass: Class[F], kClass: Class[K], vClass: Class[V]): RDD[(K, V)]

使用 SparkContext 的 hadoopRDD 操作可以将其他任何 Hadoop 输入类型转化成 RDD 使用操作。一般来说，HadoopRDD 中每一个 HDFS 数据块都成为一个 RDD 分区。此外，通过转换操作可以将 HadoopRDD 等转换成 FilterRDD（依赖一个父 RDD 产生）和 JoinedRDD（依赖所有父 RDD）等。

3.5 转换操作

3.5.1 基础转换操作

- map[U](f: (T) ⇒ U): RDD[U]
- distinct(): RDD[(T)]
- distinct(numPartitions: Int): RDD[T]
- flatMap[U](f: (T) ⇒ TraversableOnce[U]): RDD[U]

map 操作是对 RDD 中的每个元素都执行一个指定的函数来产生一个新的 RDD，任何原 RDD 中的元素在新 RDD 中都有且只有一个元素与之对应。distinct 操作是去除 RDD 重复的元素，返回所有元素不重复的 RDD。而 flatMap 操作与 map 类似，区别是原 RDD 中的每个元素经过 map 处理后只能生成一个元素，而在 flatMap 操作中原 RDD 中的每个元素可生成一个或多个元素来构建新 RDD。

```
//读取 HDFS 文件到 RDD，使用 map 操作
scala> var data = sc.textFile("/home/spark/work/chapter3/sparklearning.txt")
data: org.apache.spark.rdd.RDD[String] = /home/spark/work/chapter3/
  sparklearning.txt MapPartitionsRDD[23] at textFile at <console>:25
```

```
scala> data.map(line => line.split("\\s+")).collect
res21: Array[Array[String]] = Array(Array(oft, kitty), Array(warm, kitty),
    Array(little, ball, of, fur), Array(happy, kitty), Array(sleepy, kitty),
    Array(pur, pur, pur), Array(""))

//使用 flatMap 操作
scala> data.flatMap(line => line.split("\\s+")).collect
res22: Array[String] = Array(oft, kitty, warm, kitty, little, ball, of, fur, happy,
    kitty, sleepy, kitty, pur, pur, pur, "")

//使用 distinct 操作，去除重复的元素
scala> data.flatMap(line => line.split("\\s+")).distinct.collect
res23: Array[String] = Array(sleepy, "", little, happy, pur, fur, oft, warm, of,
    ball, kitty)
```

- coalesce(numPartitions: Int, shuffle: Boolean = false): RDD[T]
- repartition(numPartitions: Int): RDD[T]

coalesce 和 repartition 都是对 RDD 进行重新分区。coalesce 操作使用 HashPartitioner 进行重分区，第一个参数为重分区的数目，第二个为是否进行 shuffle，默认情况为 false。repartition 操作是 coalesce 函数第二个参数为 true 的实现。

```
//读取本地文件生成RDD，该RDD默认有两个分区
scala> var data = sc.textFile("/home/spark/work/chapter3/sparklearning.txt")
data: org.apache.spark.rdd.RDD[String] = /home/spark/work/chapter3/
    sparklearning.txt MapPartitionsRDD[31] at textFile at <console>:25
scala> data.partitions.size
res24: Int = 2

//重新分区的数目小于原分区数目，可以正常进行
scala> var rdd1 = data.coalesce(1)
rdd1: org.apache.spark.rdd.RDD[String] = CoalescedRDD[32] at coalesce at
    <console>:27
scala> rdd1.partitions.size
res25: Int = 1

//如果重分区的数目大于原来的分区数，那么必须指定shuffle参数为true，否则分区数不变
scala> var rdd1 = data.coalesce(4)
rdd1: org.apache.spark.rdd.RDD[String] = CoalescedRDD[33] at coalesce at
    <console>:27
scala> rdd1.partitions.size
res26: Int = 2
```

- randomSplit(weights: Array[Double], seed: Long = Utils.random.nextLong): Array[RDD[T]]

- glom(): RDD[Array[T]]

randomSplit 操作是根据 weights 权重将一个 RDD 分隔为多个 RDD,而 glom 操作则是 RDD 中每一个分区所有类型为 T 的数据转变成元素类型为 T 的数组[Array[T]]。

```
//使用 randomSplit 操作根据权重对 RDD 进行拆分多个 RDD
scala> var rdd = sc.makeRDD(1 to 10,10)
rdd: org.apache.spark.rdd.RDD[Int] = ParallelCollectionRDD[0] at makeRDD at
  <console>:25
scala> var splitRDD = rdd.randomSplit(Array(1.0,2.0,3.0,4.0))
splitRDD: Array[org.apache.spark.rdd.RDD[Int]] = Array(
MapPartitionsRDD[47] at randomSplit at <console>:27,
MapPartitionsRDD[48] at randomSplit at <console>:27,
MapPartitionsRDD[49] at randomSplit at <console>:27,
MapPartitionsRDD[50] at randomSplit at <console>:27)

//这里要注意: randomSplit 的结果是一个 RDD 数组
scala> splitRDD.size
res30: Int = 4

//由于 randomSplit 的第一个参数 weights 中传入的值有 4 个,因此,就会切分成 4 个 RDD,
//把原来的 RDD 按照权重 1.0,2.0,3.0,4.0,随机划分到这 4 个 RDD 中,权重高的 RDD,
//划分到的几率就大一些。注意,权重的总和加起来为 1,否则会不正常
scala> splitRDD(0).collect
res31: Array[Int] = Array(1, 7)
scala> splitRDD(1).collect
res32: Array[Int] = Array(4)
scala> splitRDD(2).collect
res33: Array[Int] = Array(2, 3, 6, 9)
scala> splitRDD(3).collect
res34: Array[Int] = Array(5, 8, 10)

//定义一个 3 个分区 RDD,使用 glom 操作将每个分区中的元素放到一个数组中,结果就变成了 3 个数组
scala> var rdd = sc.makeRDD(1 to 10,3)
rdd: org.apache.spark.rdd.RDD[Int] = ParallelCollectionRDD[51] at makeRDD at
  <console>:25
scala> rdd.glom().collect
res36: Array[Array[Int]] = Array(Array(1, 2, 3), Array(4, 5, 6), Array(7, 8, 9,
  10)
```

- union(other: RDD[T]): RDD[T]

- intersection(other: RDD[T]): RDD[T]

- intersection(other: RDD[T], numPartitions: Int): RDD[T]

- intersection(other: RDD[T], partitioner: Partitioner): RDD[T]

- subtract(other: RDD[T]): RDD[T]
- subtract(other: RDD[T], numPartitions: Int): RDD[T]
- subtract(other: RDD[T], p: Partitioner): RDD[T]

union 操作是将两个 RDD 合并，返回两个 RDD 的并集，返回元素不去重。intersection 操作类似 SQL 中的 inner join 操作，返回两个 RDD 交集，返回元素去重。subtract 返回在 RDD 中出现，且不在 otherRDD 中出现的元素，返回元素不去重。在 intersection 和 subtract 操作中参数 numPartitions 指定返回的 RDD 的分区数，参数 partitioner 用于指定分区函数。

```
//进行 union 操作，结果不去重
scala> var rdd1 = sc.makeRDD(1 to 2,1)
rdd1: org.apache.spark.rdd.RDD[Int] = ParallelCollectionRDD[53] at makeRDD at
   <console>:25

scala> var rdd2 = sc.makeRDD(2 to 3,1)
rdd2: org.apache.spark.rdd.RDD[Int] = ParallelCollectionRDD[54] at makeRDD at
   <console>:25

scala> rdd1.union(rdd2).collect
res37: Array[Int] = Array(1, 2, 2, 3)

//进行 intersection 操作，交集结果去重
scala> rdd1.intersection(rdd2).collect
res38: Array[Int] = Array(2)

scala> var rdd3 = rdd1.intersection(rdd2)
rdd3: org.apache.spark.rdd.RDD[Int] = MapPartitionsRDD[67] at intersection at
   <console>:29
```

- mapPartitions[U](f: (Iterator[T]) ⇒ Iterator[U], preservesPartitioning: Boolean = false): RDD[U]
- mapPartitionsWithIndex[U](f: (Int, Iterator[T]) ⇒ Iterator[U], preservesPartitioning: Boolean = false): RDD[U]

mapPartitions 操作和 map 操作类似，只不过映射的参数由 RDD 中的每一个元素变成了 RDD 中每一个分区的迭代器，其中参数 preservesPartitioning 表示是否保留父 RDD 的 partitioner 分区信息。如果在映射的过程中需要频繁创建额外的对象，使用 mapPartitions 操作要比 map 操作高效得多。比如，将 RDD 中的所有数据通过 JDBC 连接写入数据库，如果使用 map 函数，可能要为每一个元素都创建一个 connection，这样开销很大。如果使用 mapPartitions，那么只需要针对每一个分区建立一个 connection。mapPartitionsWithIndex 操作作用类似于 mapPartitions，只是

输入参数多了一个分区索引。

```
//创建rdd1，该RDD有两个分区
scala>var rdd1 = sc.makeRDD(1 to 5,2)
rdd1: org.apache.spark.rdd.RDD[Int] = ParallelCollectionRDD[80] at makeRDD at
  <console>:25

//使用mapPartitions对rdd1进行重新分区
scala> var rdd3 = rdd1.mapPartitions{ x => {
    | var result = List[Int]()
    |    var i = 0
    |    while(x.hasNext){
    |      i += x.next()
    |    }
    |    result.::(i).iterator
    | }}
rdd3: org.apache.spark.rdd.RDD[Int] = MapPartitionsRDD[81] at mapPartitions at
  <console>:27

//rdd3将rdd1中每个分区中的数值累加
scala> rdd3.collect
res42: Array[Int] = Array(3, 12)
scala> rdd3.partitions.size
res43: Int = 2

//使用mapPartitionsWithIndex对rdd1进行重新分区，带有分区参数
scala> var rdd2 = rdd1.mapPartitionsWithIndex{
    |   (x,iter) => {
    |     var result = List[String]()
    |       var i = 0
    |       while(iter.hasNext){
    |         i += iter.next()
    |       }
    |       result.::(x + "|" + i).iterator
    |     }
    | }
rdd2: org.apache.spark.rdd.RDD[String] = MapPartitionsRDD[82] at
  mapPartitionsWithIndex at <console>:27

//rdd2将rdd1中每个分区的数字累加，并在每个分区的累加结果前面加了分区索引
scala> rdd2.collect
res44: Array[String] = Array(0|3, 1|12)
```

- zip[U](other: RDD[U]): RDD[(T, U)]

- zipPartitions[B, V](rdd2: RDD[B])(f: (Iterator[T], Iterator[B]) ⇒ Iterator[V]): RDD[V]
- zipPartitions[B, V](rdd2: RDD[B], preservesPartitioning: Boolean)(f: (Iterator[T], Iterator[B]) ⇒ Iterator[V]): RDD[V]
- zipPartitions[B, C, V](rdd2: RDD[B], rdd3: RDD[C])(f: (Iterator[T], Iterator[B], Iterator[C]) ⇒ Iterator[V]): RDD[V]
- zipPartitions[B, C, V](rdd2: RDD[B], rdd3: RDD[C], preservesPartitioning: Boolean)(f: (Iterator[T], Iterator[B], Iterator[C]) ⇒ Iterator[V]): RDD[V]
- zipPartitions[B, C, D, V](rdd2: RDD[B], rdd3: RDD[C], rdd4: RDD[D])(f: (Iterator[T], Iterator[B], Iterator[C], Iterator[D]) ⇒ Iterator[V]): RDD[V]
- zipPartitions[B, C, D, V](rdd2: RDD[B], rdd3: RDD[C], rdd4: RDD[D], preservesPartitioning: Boolean)(f: (Iterator[T], Iterator[B], Iterator[C], Iterator[D]) ⇒ Iterator[V]): RDD[V]

zip 操作用于将两个 RDD 组合成 Key/Value 形式的 RDD,这里默认两个 RDD 的 partition 数量以及元素数量都相同,否则会抛出异常。zipPartitions 操作将多个 RDD 按照 partition 组合成为新的 RDD,该操作需要组合的 RDD 具有相同的分区数,但对于每个分区内的元素数量没有要求。

```
//zip 操作将两个同样分区 RDD 进行合并,键值分别对照组合
scala> var rdd1 = sc.makeRDD(1 to 5,2)
rdd1: org.apache.spark.rdd.RDD[Int] = ParallelCollectionRDD[83] at makeRDD at
    <console>:25

scala> var rdd2 = sc.makeRDD(Seq("A","B","C","D","E"),2)
rdd2: org.apache.spark.rdd.RDD[String] = ParallelCollectionRDD[84] at makeRDD
    at <console>:25

scala> rdd1.zip(rdd2).collect
res45: Array[(Int, String)] = Array((1,A), (2,B), (3,C), (4,D), (5,E))

//如果两个 RDD 分区数不同,则会出现异常
scala> var rdd3 = sc.makeRDD(Seq("A","B","C","D","E"),3)
rdd3: org.apache.spark.rdd.RDD[String] = ParallelCollectionRDD[87] at makeRDD
    at <console>:25

scala> rdd1.zip(rdd3).collect
java.lang.IllegalArgumentException: Can't zip RDDs with unequal numbers of
    partitions: List(2, 3)
    at org.apache.spark.rdd.ZippedPartitionsBaseRDD.getPartitions(
```

```
        ZippedPartitionsRDD.scala:57)
    at org.apache.spark.rdd.RDD$$anonfun$partitions$2.apply(RDD.scala:247)
    ……

//使用 mapPartitionsWithIndex 操作，按照 partition 组合成为新的 RDD
scala> var rdd3 = sc.makeRDD(Seq("a","b","c","d","e"),2)
rdd3: org.apache.spark.rdd.RDD[String] = ParallelCollectionRDD[89] at makeRDD
    at <console>:25

scala> rdd3.mapPartitionsWithIndex{
    |         (x,iter) => {
    |             var result = List[String]()
    |             while(iter.hasNext){
    |               result ::= ("part_" + x + "|" + iter.next())
    |             }
    |             result.iterator
    |
    |         }
    |      }.collect
res48: Array[String] = Array(part_0|b, part_0|a, part_1|e, part_1|d, part_1|c)
```

- zipWithIndex(): RDD[(T, Long)]
- zipWithUniqueId(): RDD[(T, Long)]

zipWithIndex 操作将 RDD 中的元素和这个元素在 RDD 中的 ID（索引号）组合成键/值对。
zipWithUniqueId 操作将 RDD 中的元素和一个唯一 ID 组合成键/值对，该唯一 ID 生成算法如下：

- 每个分区中第一个元素的唯一 ID 值为：该分区索引号；
- 每个分区中第 N 个元素的唯一 ID 值为：（前一个元素的唯一 ID 值）+（该 RDD 总的分区数）。

zipWithIndex 需要启动一个 Spark 作业来计算每个分区的开始索引号，而 zipWithUniqueId 则不需要。

```
//分别定义 rdd1 和 rdd2，这两个 RDD 都有两个分区
scala> var rdd1 = sc.makeRDD(Seq("A","B","C","D","E","F"),2)
rdd1: org.apache.spark.rdd.RDD[String] = ParallelCollectionRDD[92] at makeRDD
    at <console>:25

scala> var rdd2 = sc.makeRDD(Seq("A","B","R","D","F"),2)
rdd2: org.apache.spark.rdd.RDD[String] = ParallelCollectionRDD[93] at makeRDD
    at <console>:25

//使用 zipWithIndex 操作，RDD 元素和索引号组成键值对
scala> rdd2.zipWithIndex().collect
```

```
res49: Array[(String, Long)] = Array((A,0), (B,1), (R,2), (D,3), (F,4))

//总分区数为2：第一个分区第一个元素ID为0，第二个分区第一个元素ID为1
//   第一个分区第二个元素ID为0+2=2，第一个分区第三个元素ID为2+2=4
//   第二个分区第二个元素ID为1+2=3，第二个分区第三个元素ID为3+2=5
scala> rdd1.zipWithUniqueId().collect
res50: Array[(String, Long)] = Array((A,0), (B,2), (C,4), (D,1), (E,3), (F,5))
```

3.5.2　键值转换操作

- partitionBy(partitioner: Partitioner): RDD[(K, V)]
- mapValues[U](f: (V) ⇒ U): RDD[(K, U)]
- flatMapValues[U](f: (V) ⇒ TraversableOnce[U]): RDD[(K, U)]

partitionBy 操作根据 partitioner 函数生成新的 ShuffleRDD，将原 RDD 重新分区。mapValues 类似于 map，只不过 mapValues 是针对[K,V]中的 V 值进行 map 操作，同样地，flatMapValues 相对 flatMap，flatMapValues 是针对[K,V]中的 V 值进行 flatMap 操作。

```
scala> var rdd1 = sc.makeRDD(Array((1,"A"),(2,"B"),(3,"C"),(4,"D")),2)
rdd1: org.apache.spark.rdd.RDD[(Int, String)] = ParallelCollectionRDD[0] at
    makeRDD at <console>:25

//查看rdd1每个分区的元素
scala> rdd1.mapPartitionsWithIndex{
     |           (partIdx,iter) => {
     |             var part_map = scala.collection.mutable.Map[String,List[(Int,
    String)]]()
     |             while(iter.hasNext){
     |               var part_name = "part_" + partIdx;
     |               var elem = iter.next()
     |               if(part_map.contains(part_name)) {
     |                 var elems = part_map(part_name)
     |                 elems ::= elem
     |                 part_map(part_name) = elems
     |               } else {
     |                 part_map(part_name) = List[(Int,String)]{elem}
     |               }
     |             }
     |             part_map.iterator
     |           }
     |         }.collect
//在分区part_0中为(2,B),(1,A),分区part_1为(4,D),(3,C)
res0: Array[(String, List[(Int, String)])] = Array((part_0,List((2,B), (1,A))),
```

```
(part_1,List((4,D), (3,C))))
```

//使用partitionBy 重分区
```
scala> var rdd2 = rdd1.partitionBy(new org.apache.spark.HashPartitioner(2))
rdd2: org.apache.spark.rdd.RDD[(Int, String)] = ShuffledRDD[2] at partitionBy
   at <console>:27
```

//查看rdd2 中每个分区的元素
```
scala> rdd2.mapPartitionsWithIndex{
     |       (partIdx,iter) => {
     |          var part_map = scala.collection.mutable.Map[String,List[(Int,
  String)]]()
     |          while(iter.hasNext){
     |            var part_name = "part_" + partIdx;
     |            var elem = iter.next()
     |            if(part_map.contains(part_name)) {
     |              var elems = part_map(part_name)
     |              elems ::= elem
     |              part_map(part_name) = elems
     |            } else {
     |              part_map(part_name) = List[(Int,String)]{elem}
     |            }
     |          }
     |          part_map.iterator
     |       }
     |     }.collect
```
//重新分区后，在分区part_0 中为(2,B)，(4,D)，分区part_1 为(3,C)，(1,A)
```
res2: Array[(String, List[(Int, String)])] = Array((part_0,List((4,D), (2,B))),
   (part_1,List((1,A), (3,C))))
```

//进行mapValues 操作
```
scala> var rdd1 = sc.makeRDD(Array((1,"A"),(2,"B"),(3,"C"),(4,"D")),2)
rdd1: org.apache.spark.rdd.RDD[(Int, String)] = ParallelCollectionRDD[4] at
   makeRDD at <console>:25
scala> rdd1.mapValues(x => x + "_").collect
res3: Array[(Int, String)] = Array((1,A_), (2,B_), (3,C_), (4,D_))
```

- combineByKey[C](createCombiner: (V) ⇒ C, mergeValue: (C, V) ⇒ C, mergeCombiners: (C, C) ⇒ C): RDD[(K, C)]

- combineByKey[C](createCombiner: (V) ⇒ C, mergeValue: (C, V) ⇒ C, mergeCombiners: (C, C) ⇒ C, numPartitions: Int): RDD[(K, C)]

- combineByKey[C](createCombiner: (V) ⇒ C, mergeValue: (C, V) ⇒ C, mergeCombiners:

(C, C) ⇒ C, partitioner: Partitioner, mapSideCombine: Boolean = true, serializer: Serializer = null): RDD[(K, C)]
- foldByKey(zeroValue: V)(func: (V, V) ⇒ V): RDD[(K, V)]
- foldByKey(zeroValue: V, numPartitions: Int)(func: (V, V) ⇒ V): RDD[(K, V)]
- foldByKey(zeroValue: V, partitioner: Partitioner)(func: (V, V) ⇒ V): RDD[(K, V)]

combineByKey 操作用于将 RDD[K,V] 转换成 RDD[K,C]，这里的 V 类型和 C 类型可以相同也可以不同。foldByKey 操作用于 RDD[K,V] 根据 K 将 V 做折叠、合并处理，其中的参数 zeroValue 表示先根据映射函数将 zeroValue 应用于 V，进行初始化 V，再将映射函数应用于初始化后的 V。其中 combineByKey 中的参数含义如下。

- createCombiner：组合器函数，用于将 V 类型转换成 C 类型，输入参数为 RDD[K,V] 中的 V，输出为 C。
- mergeValue：合并值函数，将一个 C 类型和一个 V 类型值合并成一个 C 类型，输入参数为 (C,V)，输出为 C。
- mergeCombiners：合并组合器函数，用于将两个 C 类型值合并成一个 C 类型，输入参数为 (C,C)，输出为 C。
- numPartitions：结果 RDD 分区数，默认保持原有的分区数。
- partitioner：分区函数,默认为 HashPartitioner。
- mapSideCombine：是否需要在 Map 端进行 combine 操作，类似于 MapReduce 中的 combine，默认为 true。

```
scala> var rdd1 = sc.makeRDD(Array(("A",1),("A",2),("B",1),("B",2),("C",1)))
rdd1: org.apache.spark.rdd.RDD[(String, Int)] = ParallelCollectionRDD[7] at
   makeRDD at <console>:25

// combineByKey 操作用于将 RDD[K,V] 转换成 RDD[K,C]
scala> rdd1.combineByKey(
     |     (v : Int) => v + "_",
     |     (c : String, v : Int) => c + "@" + v,
     |     (c1 : String, c2 : String) => c1 + "$" + c2
     |    ).collect
res5: Array[(String, String)] = Array((B,2_$1_), (C,1_), (A,1_$2_))

//将 rdd1 中每个 key 对应的 V 进行累加，注意 zeroValue=0,需要先初始化 V,映射函数为+操作，
//比如("A",0),("A",2)，先将 zeroValue 应用于每个 V，得到("A",0+0), ("A",2+0)，
//即("A",0),("A",2)，再将映射函数应用于初始化后的 V，最后得到(A,0+2),即(A,2)
scala> var rdd1 = sc.makeRDD(Array(("A",0),("A",2),("B",1),("B",2),("C",1)))
rdd1: org.apache.spark.rdd.RDD[(String, Int)] = ParallelCollectionRDD[9] at
```

```
                          makeRDD at <console>:25
scala> rdd1.foldByKey(0)(_+_).collect
res6: Array[(String, Int)] = Array((B,3), (C,1), (A,2))

//先将zeroValue = 2应用于每个V,得到("A",0+2),("A",2+2),即("A",2),("A",4),
//再将映射函数应用于初始化后的V,最后得到(A,2+4),即(A,6)
scala> rdd1.foldByKey(2)(_+_).collect
res7: Array[(String, Int)] = Array((B,7), (C,3), (A,6))

//先将zeroValue=0应用于每个V,注意,这次映射函数为乘法,得到("A",0*0), ("A",2*0),
//即 ("A",0),("A",0),再将映射函数应用于初始化后的V,最后得到(A,0*0),即(A,0)。
//其他K也一样,最终都得到了V=0
scala> rdd1.foldByKey(0)(_*_).collect
res8: Array[(String, Int)] = Array((B,0), (C,0), (A,0))

//映射函数为乘法时,需要将zeroValue设为1,才能得到我们想要的结果。
scala> rdd1.foldByKey(1)(_*_).collect
res9: Array[(String, Int)] = Array((B,2), (C,1), (A,0))
```

- reduceByKey(func: (V, V) ⇒ V): RDD[(K, V)]
- reduceByKey(func: (V, V) ⇒ V, numPartitions: Int): RDD[(K, V)]
- reduceByKey(partitioner: Partitioner, func: (V, V) ⇒ V): RDD[(K, V)]
- reduceByKeyLocally(func: (V, V) ⇒ V): Map[K, V]
- groupByKey(): RDD[(K, Iterable[V])]
- groupByKey(numPartitions: Int): RDD[(K, Iterable[V])]
- groupByKey(partitioner: Partitioner): RDD[(K, Iterable[V])]

groupByKey 操作用于将 RDD[K,V]中每个 K 对应的 V 值合并到一个集合 Iterable[V]中,reduceByKey 操作用于将 RDD[K,V]中每个 K 对应的 V 值根据映射函数来运算,其中参数 numPartitions 用于指定分区数,参数 partitioner 用于指定分区函数。reduceByKeyLocally 和 reduceByKey 功能类似,不同的是,reduceByKeyLocally 运算结果映射到一个 Map[K,V]中,而不是 RDD[K,V]。

```
scala> var rdd1 = sc.makeRDD(Array(("A",0),("A",2),("B",1),("B",2),("C",1)))
rdd1: org.apache.spark.rdd.RDD[(String, Int)] = ParallelCollectionRDD[14] at
    makeRDD at <console>:29

//使用groupByKey操作将RDD中每个K对应的V值合并到集合Iterable[V]中
scala> rdd1.groupByKey().collect
res15: Array[(String, Iterable[Int])] = Array((B,CompactBuffer(2, 1)),
    (C,CompactBuffer(1)), (A,CompactBuffer(0, 2)))
```

```
//使用reduceByKey操作将RDD[K,V]中每个K对应的V值根据映射函数来运算
scala> var rdd2 = rdd1.reduceByKey((x,y) => x + y)
rdd2: org.apache.spark.rdd.RDD[(String, Int)] = ShuffledRDD[16] at reduceByKey
    at <console>:27
scala> rdd2.collect
res17: Array[(String, Int)] = Array((B,3), (C,1), (A,2))

//对rdd1使用reduceByKey操作进行重新分区
scala> var rdd2 = rdd1.reduceByKey(new org.apache.spark.HashPartitioner(2),(x,y)
    => x + y)
rdd2: org.apache.spark.rdd.RDD[(String, Int)] = ShuffledRDD[17] at reduceByKey
    at <console>:27

scala> rdd2.collect
res19: Array[(String, Int)] = Array((B,3), (A,2), (C,1))

//使用reduceByKeyLocally操作，操作结果映射到一个Map[K,V]中
scala> rdd1.reduceByKeyLocally((x,y) => x + y)
res21: scala.collection.Map[String,Int] = Map(A -> 2, B -> 3, C -> 1)
```

- cogroup[W](other: RDD[(K, W)]): RDD[(K, (Iterable[V], Iterable[W]))]
- cogroup[W](other: RDD[(K, W)], numPartitions: Int): RDD[(K, (Iterable[V], Iterable[W]))]
- cogroup[W](other: RDD[(K, W)], partitioner: Partitioner): RDD[(K, (Iterable[V], Iterable[W]))]
- cogroup[W1, W2](other1: RDD[(K, W1)], other2: RDD[(K, W2)]): RDD[(K, (Iterable[V], Iterable[W1], Iterable[W2]))]
- cogroup[W1, W2](other1: RDD[(K, W1)], other2: RDD[(K, W2)], numPartitions: Int): RDD[(K, (Iterable[V], Iterable[W1], Iterable[W2]))]
- cogroup[W1, W2](other1: RDD[(K, W1)], other2: RDD[(K, W2)], partitioner: Partitioner): RDD[(K, (Iterable[V], Iterable[W1], Iterable[W2]))]
- cogroup[W1, W2, W3](other1: RDD[(K, W1)], other2: RDD[(K, W2)], other3: RDD[(K, W3)]): RDD[(K, (Iterable[V], Iterable[W1], Iterable[W2], Iterable[W3]))]
- cogroup[W1, W2, W3](other1: RDD[(K, W1)], other2: RDD[(K, W2)], other3: RDD[(K, W3)], numPartitions: Int): RDD[(K, (Iterable[V], Iterable[W1], Iterable[W2], Iterable[W3]))]
- cogroup[W1, W2, W3](other1: RDD[(K, W1)], other2: RDD[(K, W2)], other3: RDD[(K, W3)], partitioner: Partitioner): RDD[(K, (Iterable[V], Iterable[W1], Iterable[W2],

Iterable[W3]))]

cogroup 相当于 SQL 中的全外关联，返回左右 RDD 中的记录，关联不上的为空。可传入的参数有 1~3 个 RDD，参数 numPartitions 用于指定分区数，参数 partitioner 用于指定分区函数。

```
scala> var rdd1 = sc.makeRDD(Array(("A","1"),("B","2"),("C","3")),2)
rdd1: org.apache.spark.rdd.RDD[(String, String)] = ParallelCollectionRDD[0] at
    makeRDD at <console>:25

scala> var rdd2 = sc.makeRDD(Array(("A","a"),("C","c"),("D","d")),2)
rdd2: org.apache.spark.rdd.RDD[(String, String)] = ParallelCollectionRDD[1] at
    makeRDD at <console>:25

scala> var rdd3 = sc.makeRDD(Array(("A","A"),("E","E")),2)
rdd3: org.apache.spark.rdd.RDD[(String, String)] = ParallelCollectionRDD[2] at
    makeRDD at <console>:25

//rdd1 和 rdd2、rdd3 进行全外关联，返回左右 RDD 中的记录，关联不上的为空
scala> var rdd4 = rdd1.cogroup(rdd2,rdd3)
rdd4: org.apache.spark.rdd.RDD[(String, (Iterable[String], Iterable[String],
    Iterable[String]))] = MapPartitionsRDD[4] at cogroup at <console>:31

scala> rdd4.partitions.size
res0: Int = 2

scala> rdd4.collect
res1: Array[(String, (Iterable[String], Iterable[String], Iterable[String]))]
    = Array((B,(CompactBuffer(2),CompactBuffer(),CompactBuffer())),
    (D,(CompactBuffer(),CompactBuffer(d),CompactBuffer())),
    (A,(CompactBuffer(1),CompactBuffer(a),CompactBuffer(A))),
    (C,(CompactBuffer(3),CompactBuffer(c),CompactBuffer())),
    (E,(CompactBuffer(),CompactBuffer(),CompactBuffer(E))))
```

- join[W](other: RDD[(K, W)]): RDD[(K, (V, W))]
- join[W](other: RDD[(K, W)], numPartitions: Int): RDD[(K, (V, W))]
- join[W](other: RDD[(K, W)], partitioner: Partitioner): RDD[(K, (V, W))]
- fullOuterJoin[W](other: RDD[(K, W)]): RDD[(K, (Option[V], Option[W]))]
- fullOuterJoin[W](other: RDD[(K, W)], numPartitions: Int): RDD[(K, (Option[V], Option[W]))]
- fullOuterJoin[W](other: RDD[(K, W)], partitioner: Partitioner): RDD[(K, (Option[V], Option[W]))]
- leftOuterJoin[W](other: RDD[(K, W)]): RDD[(K, (V, Option[W]))]

- leftOuterJoin[W](other: RDD[(K, W)], numPartitions: Int): RDD[(K, (V, Option[W]))]
- leftOuterJoin[W](other: RDD[(K, W)], partitioner: Partitioner): RDD[(K, (V, Option[W]))]
- rightOuterJoin[W](other: RDD[(K, W)]): RDD[(K, (Option[V], W))]
- rightOuterJoin[W](other: RDD[(K, W)], numPartitions: Int): RDD[(K, (Option[V], W))]
- rightOuterJoin[W](other: RDD[(K, W)], partitioner: Partitioner): RDD[(K, (Option[V], W))]
- subtractByKey[W](other: RDD[(K, W)]): RDD[(K, V)]
- subtractByKey[W](other: RDD[(K, W)], p: Partitioner): RDD[(K, V)]
- subtractByKey[W](other: RDD[(K, W)], numPartitions: Int): RDD[(K, V)]

join、fullOuterJoin、leftOuterJoin 和 rightOuterJoin 都是针对 RDD[K,V]中 K 值相等的连接操作，分别对应内连接、全连接、左连接和右连接，这些操作都调用 cogroup 进行实现，subtractByKey 和基本操作 subtract，只是 subtractByKey 针对的是键值操作。其中参数 numPartitions 用于指定分区数，参数 partitioner 用于指定分区函数。

```
scala> var rdd1 = sc.makeRDD(Array(("A","1"),("B","2"),("C","3")),2)
rdd1: org.apache.spark.rdd.RDD[(String, String)] = ParallelCollectionRDD[5] at
    makeRDD at <console>:25

scala> var rdd2 = sc.makeRDD(Array(("A","a"),("C","c"),("D","d")),2)
rdd2: org.apache.spark.rdd.RDD[(String, String)] = ParallelCollectionRDD[6] at
    makeRDD at <console>:25

//进行内连接操作
scala> rdd1.join(rdd2).collect
res2: Array[(String, (String, String))] = Array((A,(1,a)), (C,(3,c)))

//进行左连接操作
scala> rdd1.leftOuterJoin(rdd2).collect
res3: Array[(String, (String, Option[String]))] = Array((B,(2,None)),
    (A,(1,Some(a))), (C,(3,Some(c))))

//进行右连接操作
scala> rdd1.rightOuterJoin(rdd2).collect
res4: Array[(String, (Option[String], String))] = Array((D,(None,d)),
    (A,(Some(1),a)), (C,(Some(3),c)))
```

3.6 控制操作

Spark 可以将 RDD 持久化到内存或磁盘文件系统中，把 RDD 持久化到内存中可以极大地提高迭代计算以及各计算模型之间的数据共享，一般情况下执行节点 60%内存用于缓存数据，剩下的 40%用于运行任务，具体内存管理可以参见 RDD 的内存管理章节。Spark 中使用 persist 和 cache 操作进行持久化，其中 cache 是 persist()的特例，操作详细信息如下：

- cache():RDD[T]
- persist():RDD[]
- persist(level:StorageLevel):RDD[T]

这里我们对搜狗的日志数据计算行数，搜狗日志数据可以从 http://www.sogou.com/labs/resource/q.php 下载，其中完整版大概 1.9GB 左右，文件中字段分别为：访问时间\t 用户 ID\t[查询词]\t 该 URL 在返回结果中的排名\t 用户点击的顺序号\t 用户点击的 URL。其中 SogouQ1.txt、SogouQ2.txt、SogouQ3.txt 分别是用 head -n 或者 tail -n 从 SogouQ 数据日志文件中截取，分别包含 100 万、200 万和 1000 万笔数据，这些测试数据通过笔者的博客可以找到下载链接。

把这些搜狗日志数据 SogouQ1.txt、SogouQ2.txt 和 SogouQ3.txt 存放到~/work 目录下，如图 3-5 所示，然后通过下面的命令上传到 HDFS 的/sogou 目录中

```
$cd /home/spark/work
$tar -zxf SogouQ1.txt.tar.gz
$tar -zxf SogouQ2.txt.tar.gz
$tar -zxf SogouQ3.txt.tar.gz
$ll .
$hadoop fs -mkdir /sogou
$hadoop fs -put SogouQ1.txt /sogou
$hadoop fs -put SogouQ2.txt /sogou
$hadoop fs -put SogouQ3.txt /sogou
$hadoop fs -ls /sogou
```

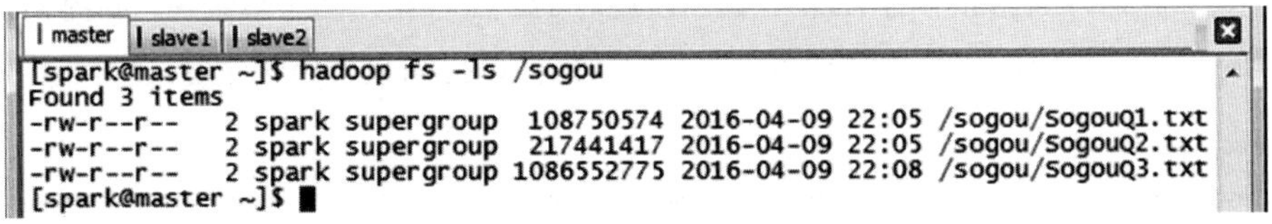

图 3-5 上传搜狗数据结果

【注意】如果需要使用 hadoop 命令查看 HDFS 中的内容，需要把 hadoop 工具加入到 PATH 目录中，在这里我们在/etc/profile 加入如下配置：

```
export HADOOP_HOME=/app/spark/hadoop-2.7.2
export PATH=$PATH:$HADOOP_HOME/bin
```

在第一次计数之前设置缓存这些数据，在计数过程中会从 HDFS 读取数据，然后将这些数据缓存在内存中；在第二次计数时，不会从 HDFS 读取数据，而是从内存中读取数据。

```
//从 HDFS 中读取 SogouQ2.txt 文件
scala>val sogou =sc.textFile("hdfs://master:9000/sogou/SogouQ2.txt")
sogou: org.apache.spark.rdd.RDD[String] = hdfs://master:9000/sogou/SogouQ2.txt
  MapPartitionsRDD[1] at textFile at <console>:25

//设置对处理的日志文件的 RDD 进行缓存
scala> sogou.cache()
res0: sogou.type = hdfs://master:9000/sogou/SogouQ2.txt MapPartitionsRDD[1] at
  textFile at <console>:27

//第一次计算行数，需要注意的是，在这里只能从 HDFS 读取，不是从内存中读取
scala> sogou.count()
res1: Long = 2000000

//第二次计算行数，使用的是从内存中读取的数据
scala> sogou.count()
res2: Long = 2000000
```

通过对监控页面可以观察到，第一次处理完毕后，所有文件都已经缓存在内存，缓存大小为 479.2MB，分为 2 个分区，如图 3-6 所示。

Storage

RDDs

RDD Name	Storage Level	Cached Partitions	Fraction Cached	Size in Memory	Size in ExternalBlockStore	Size on Disk
hdfs://master:9000/sogou/SogouQ2.txt	Memory Deserialized 1x Replicated	2	100%	479.2 MB	0.0 B	0.0 B

图 3-6　所有文件缓存到内存中

第一次计算耗时 2.1min（约 126s），第二次耗时 1s，两者相差两个数量级，如图 3-7 所示，所以在实际使用过程中，尽可能地利用内存缓存数据，以提高数据处理速度。

Completed Jobs (5)

Job Id	Description	Submitted	Duration	Stages: Succeeded/Total	Tasks (for all stages): Succeeded/Total
6	count at <console>:56	2016/04/08 17:10:41	1 s	1/1	2/2
5	count at <console>:56	2016/04/08 17:06:20	2.1 min	1/1	2/2

图 3-7　查看处理数据缓存后处理情况

在 persist()可以指定一个 StorageLevel，当 StorageLevel 为 MEMORY_ONLY 时就是 cache。StorageLevel 的列表可以在 StorageLevel 伴生单例对象中找到。

```
object StorageLevel {
  val NONE = new StorageLevel(false, false, false, false)
  val DISK_ONLY = new StorageLevel(true, false, false, false)
  val DISK_ONLY_2 = new StorageLevel(true, false, false, false, 2)
  val MEMORY_ONLY = new StorageLevel(false, true, false, true)
  val MEMORY_ONLY_2 = new StorageLevel(false, true, false, true, 2)
  val MEMORY_ONLY_SER = new StorageLevel(false, true, false, false)
  val MEMORY_ONLY_SER_2 = new StorageLevel(false, true, false, false, 2)
  val MEMORY_AND_DISK = new StorageLevel(true, true, false, true)
  val MEMORY_AND_DISK_2 = new StorageLevel(true, true, false, true, 2)
  val MEMORY_AND_DISK_SER = new StorageLevel(true, true, false, false)
  val MEMORY_AND_DISK_SER_2 = new StorageLevel(true, true, false, false, 2)
  val OFF_HEAP = new StorageLevel(false, false, true, false) // Alluxio
}
```

在 Spark 中可以使用 checkpoint 操作设置检查点，相对持久化操作 persist()操作，checkpoint 将切断与该 RDD 之前的依赖关系（"血统"）。设置检查点对包含宽依赖的长血统的 RDD 是非常有用的，可以避免占用过多的系统资源和节点失败情况下重新计算成本过高的问题，具体检查点更详细内容可以参见 3.3.4 节中 RDD 的检查点支持。

```
scala> val rdd=sc.makeRDD(1 to 4, 1)
rdd: org.apache.spark.rdd.RDD[Int] = ParallelCollectionRDD[19] at makeRDD at
   <console>:25
scala> val flatMapRDD=rdd.flatMap(x=>Seq(x, x))
flatMapRDD: org.apache.spark.rdd.RDD[Int] = MapPartitionsRDD[20] at flatMap at
   <console>:27
//设置检查点的存储位置在 HDFS，并使用 checkpoint 设置检查点，该操作属于懒加载
scala> sc.setCheckpointDir("hdfs://master:9000/chapter3/checkpoint/")
scala> flatMapRDD.checkpoint()

//在遇到行动操作时，进行检查点操作，检查点前为 ParallelCollectionRDD[0]，而检查点后
//为 ParallelCollectionRDD[1]
scala> flatMapRDD.dependencies.head.rdd
res12: org.apache.spark.rdd.RDD[_] = ParallelCollectionRDD[19] at makeRDD at
   <console>:25

scala> flatMapRDD.collect
res13: Array[Int] = Array(1, 1, 2, 2, 3, 3, 4, 4)

scala> flatMapRDD.dependencies.head.rdd
res14: org.apache.spark.rdd.RDD[_] = ReliableCheckpointRDD[21] at collect at
```

```
<console>:30
```

使用如下命令可以查看该检查点的保存目录:

```
$hadoop fs -ls /chapter3/checkpoint/
[spark@master ~]$ hadoop fs -ls /chapter3/checkpoint/
Found 1 items
drwxr-xr-x- spark supergroup0 2016-04-13 12:40 /chapter3/checkpoint/1926e95e-
   48a3-4e91-b10f-36047eb89b74
```

3.7 行动操作

3.7.1 集合标量行动操作

- first(): T 表示返回 RDD 中的第一个元素，不排序。
- count(): Long 表示返回 RDD 中的元素个数。
- reduce(f: (T, T) ⇒ T): T 根据映射函数 f，对 RDD 中的元素进行二元计算。
- collect(): Array[T]表示将 RDD 转换成数组。

```
//初始化 rdd1，进行 first 操作，获取 RDD 的第一个元素
scala> var rdd1 = sc.makeRDD(Array(("A","1"),("B","2"),("C","3")),2)
rdd1: org.apache.spark.rdd.RDD[(String, String)] = ParallelCollectionRDD[22] at
   makeRDD at <console>:25

scala> rdd1.first
res15: (String, String) = (A,1)

//对 RDD 进行二元计算，在这里进行求和
scala> var rdd1 = sc.makeRDD(1 to 10,2)
rdd1: org.apache.spark.rdd.RDD[Int] = ParallelCollectionRDD[23] at makeRDD at
   <console>:25

scala> rdd1.reduce(_ + _)
res17: Int = 55

//使用 reduce 操作，分别对二维数组元素进行字符串合并和求和操作
scala> var rdd2 = sc.makeRDD(Array(("A",0),("A",2),("B",1),("B",2),("C",1)))
rdd2: org.apache.spark.rdd.RDD[(String, Int)] = ParallelCollectionRDD[24] at
   makeRDD at <console>:25

scala> rdd2.reduce((x,y) => {
   |   (x._1 + y._1,x._2 + y._2)
```

```
    | })
res18: (String, Int) = (BCABA,6)
```

- take(num: Int): Array[T]表示获取 RDD 中从 0 到 num-1 下标的元素，不排序。
- top(num: Int): Array[T]表示从 RDD 中，按照默认（降序）或者指定的排序规则，返回前 num 个元素。
- takeOrdered(num: Int): Array[T]和 top 类似，只不过以和 top 相反的顺序返回元素。

```
//初始化 rdd1，使用 take 操作获取下标为 1 的元素
scala> var rdd1 = sc.makeRDD(Seq(10, 4, 2, 12, 3))
rdd1: org.apache.spark.rdd.RDD[Int] = ParallelCollectionRDD[25] at makeRDD at
  <console>:25

scala> rdd1.take(1)
res20: Array[Int] = Array(10)

//使用 takeOrdered 操作，按照升序获取前 2 个元素
scala> rdd1.takeOrdered(2)
res22: Array[Int] = Array(2, 3)
```

- aggregate[U](zeroValue: U)(seqOp: (U, T) ⇒ U, combOp: (U, U) ⇒ U)(implicit arg0: ClassTag[U]): U
- fold(zeroValue: T)(op: (T, T) ⇒ T): Taggregate 用户聚合 RDD 中的元素，先使用 seqOp 将 RDD 中每个分区中的 T 类型元素聚合成 U 类型，再使用 combOp 将之前每个分区聚合后的 U 类型聚合成 U 类型,特别注意 seqOp 和 combOp 都会使用 zeroValue 的值，zeroValue 的类型为 U。fold 是 aggregate 的简化，将 aggregate 中的 seqOp 和 combOp 使用同一个函数 op。

```
//定义 rdd1，设置第一个分区中包含 5,4,3,2,1，第二个分区中包含 10,9,8,7,6
scala> var rdd1 = sc.makeRDD(1 to 10,2)
rdd1: org.apache.spark.rdd.RDD[Int] = ParallelCollectionRDD[30] at makeRDD at
  <console>:25

scala> rdd1.mapPartitionsWithIndex{
     |         (partIdx,iter) => {
     |           var part_map = scala.collection.mutable.Map[String,List[Int]]()
     |             while(iter.hasNext){
     |               var part_name = "part_" + partIdx;
     |               var elem = iter.next()
     |               if(part_map.contains(part_name)) {
     |                 var elems = part_map(part_name)
     |                 elems ::= elem
     |                 part_map(part_name) = elems
```

```
    |               } else {
    |                 part_map(part_name) = List[Int]{elem}
    |               }
    |             }
    |             part_map.iterator
    |          }
    |        }.collect
res26: Array[(String, List[Int])] = Array((part_0,List(5, 4, 3, 2, 1)),
  (part_1,List(10, 9, 8, 7, 6)))
```

```
//进行 aggregate 操作，结果为什么是 58，看下面的计算过程：
//  先在每个分区中迭代执行 (x : Int,y : Int) => x + y，并且使用 zeroValue 的值 1
//    即 part_0 中 zeroValue+5+4+3+2+1 = 1+5+4+3+2+1 = 16
//       part_1 中 zeroValue+10+9+8+7+6 = 1+10+9+8+7+6 = 41
//  再将两个分区的结果合并(a : Int,b : Int) => a + b，并且使用 zeroValue 的值 1
//    即 zeroValue+part_0+part_1 = 1 + 16 + 41 = 58
scala> rdd1.aggregate(1)(
     |             {(x : Int,y : Int) => x + y},
     |             {(a : Int,b : Int) => a + b}
     |         )
res27: Int = 58
```

```
//结果同上面使用 aggregate 的第一个例子一样，即
scala> rdd1.fold(1)(
     |         (x,y) => x + y
     |     )
res28: Int = 58
```

```
scala> rdd1.aggregate(1)(
     |             {(x,y) => x + y},
     |             {(a,b) => a + b}
     |         )
res29: Int = 58
```

- **lookup(key: K): Seq[V]**

lookup 用于（K,V）类型的 RDD，指定 K 值，返回 RDD 中该 K 对应的所有 V 值。

```
//定义 rdd1 为键值对数组的 RDD，根据键值找出所有数值
scala> var rdd1 = sc.makeRDD(Array(("A",0),("A",2),("B",1),("B",2),("C",1)))
rdd1: org.apache.spark.rdd.RDD[(String, Int)] = ParallelCollectionRDD[32] at
makeRDD at <console>:25

scala> rdd1.lookup("A")
res30: Seq[Int] = WrappedArray(0, 2)
```

- countByKey(): Map[K, Long]
- foreach(f: (T) ⇒ Unit): Unit
- foreachPartition(f: (Iterator[T]) ⇒ Unit): Unit
- sortBy[K](f: (T) ⇒ K, ascending: Boolean = true, numPartitions: Int = this.partitions.length): RDD[T]

countByKey 统计 RDD[K,V]中每个 K 的数量。foreach 遍历 RDD，将函数 f 应用于每一个元素。要注意如果对 RDD 执行 foreach，只会在 Executor 端有效，而并不是 Driver 端，比如：rdd.foreach(println)，只会在 Executor 的 stdout 中打印出来，Driver 端是看不到的。foreachPartition 和 foreach 类似，只不过是对每一个分区使用 f。sortBy 根据给定的排序 k 函数将 RDD 中的元素进行排序。

```
//使用foreach操作,遍历打印RDD每一个元素
scala> var cnt = sc.accumulator(0)
cnt: org.apache.spark.Accumulator[Int] = 0

scala> var rdd1 = sc.makeRDD(1 to 10,2)
rdd1: org.apache.spark.rdd.RDD[Int] = ParallelCollectionRDD[38] at makeRDD at
  <console>:25
scala> rdd1.foreach(x => cnt += x)

scala> cnt.value
res39: Int = 55

//使用foreach操作,foreachPartition和foreach类似,只不过是对每一个分区使用函数f
scala> var rdd1 = sc.makeRDD(1 to 10,2)
rdd1: org.apache.spark.rdd.RDD[Int] = ParallelCollectionRDD[40] at makeRDD at
  <console>:25

scala> var allsize = sc.accumulator(0)
allsize: org.apache.spark.Accumulator[Int] = 0

scala> rdd1.foreachPartition { x => {
     |    allsize += x.size
     | }}
scala> println(allsize.value)
10


//定义RDD为键值类型,分别按照键升序排列和值降序排列
scala> var rdd1 = sc.makeRDD(Array(("A",2),("A",1),("B",6),("B",3),("B",7)))
```

```
rdd1: org.apache.spark.rdd.RDD[(String, Int)] = ParallelCollectionRDD[52] at
  makeRDD at <console>:25

scala> rdd1.sortBy(x => x).collect
res47: Array[(String, Int)] = Array((A,1), (A,2), (B,3), (B,6), (B,7))

scala> rdd1.sortBy(x => x._2,false).collect
res48: Array[(String, Int)] = Array((B,7), (B,6), (B,3), (A,2), (A,1))
```

3.7.2 存储行动操作

- saveAsTextFile(path: String): Unit
- saveAsTextFile(path: String, codec: Class[_ <: CompressionCodec]): Unit
- saveAsSequenceFile(path: String, codec: Option[Class[_ <: CompressionCodec]] = None): Unit
- saveAsObjectFile(path: String): Unit

saveAsTextFile 用于将 RDD 以文本文件的格式存储到文件系统中，codec 参数可以指定压缩的类名。saveAsSequenceFile 用于将 RDD 以 SequenceFile 的文件格式保存到 HDFS 上。saveAsObjectFile 用于将 RDD 中的元素序列化成对象，存储到文件中，对于 HDFS，默认采用 SequenceFile 保存。

```
//使用 saveAsTextFile 操作把 RDD 内容保存到 HDFS 中
scala> var rdd1 = sc.makeRDD(1 to 10,2)
rdd1: org.apache.spark.rdd.RDD[Int] = ParallelCollectionRDD[0] at makeRDD at
  <console>:25

scala>rdd1.saveAsTextFile("hdfs://master:9000/chapter3/TextFile1/")

//查看 HDFS 中的文件内容
$hadoop fs -ls /spark/TextFile1
Found 3 items
-rw-r--r--3 spark supergroup0  2016-04-06 23:03 /chapter3/TextFile1/_SUCCESS
-rw-r--r--3 spark supergroup10 2016-04-06 23:03 /chapter3/TextFile1/part-00000
-rw-r--r--3 spark supergroup11 2016-04-06 23:03 /chapter3/TextFile1/part-00001

//使用 saveAsObjectFile 操作把 RDD 内容保存到 HDFS 中
scala> var rdd1 = sc.makeRDD(1 to 10,2)
rdd1: org.apache.spark.rdd.RDD[Int] = ParallelCollectionRDD[2] at makeRDD at
  <console>:25

scala> rdd1.saveAsObjectFile("hdfs://master:9000/chapter3/TextFile2")
```

```
//查看HDFS目录
$hadoop fs -ls /chapter3/TextFile2
Found 3 items
-rw-r--r--3 spark supergroup0  2016-04-06 23:03 /chapter3/TextFile2/_SUCCESS
-rw-r--r--3 spark supergroup154 2016-04-06 23:03 /chapter3/TextFile2/part-00000
-rw-r--r--3 spark supergroup154 2016-04-06 23:03 /chapter3/TextFile2/part-00001

//对保存在HDFS文件并进行查看，文件内容为对象
$hadoop fs -cat/chapter3/sparkOfSaveAsTextFile2/part-00000
SEQ !org.apache.hadoop.io.NullWritable"org.apache.hadoop.io.BytesWritable
  ……
```

- saveAsHadoopFile[F <: OutputFormat[K, V]](path: String)(implicit fm: ClassTag[F]): Unit
- saveAsHadoopFile[F <: OutputFormat[K, V]](path: String, codec: Class[_ <: CompressionCodec]): Unit
- saveAsHadoopFile(path: String, keyClass: Class[_], valueClass: Class[_], outputFormatClass: Class[_ <: OutputFormat[_, _]], codec: Class[_ <: CompressionCodec]): Unit
- saveAsHadoopFile(path: String, keyClass: Class[_], valueClass: Class[_], outputFormatClass: Class[_ <: OutputFormat[_, _]], conf: JobConf = ……, codec: Option[Class[_ <: CompressionCodec]] = None): Unit
- saveAsHadoopDataset(conf: JobConf): Unit

saveAsHadoopFile是将RDD存储在HDFS上的文件中，支持老版本Hadoop API，可以指定outputKeyClass、outputValueClass以及压缩格式。

```
scala> import org.apache.hadoop.mapred.TextOutputFormat
scala> import org.apache.hadoop.io.Text
scala> import org.apache.hadoop.io.IntWritable
scala> var rdd1 = sc.makeRDD(Array(("A",2),("A",1),("B",6),("B",3),("B",7)))
rdd1: org.apache.spark.rdd.RDD[(String, Int)] = ParallelCollectionRDD[7] at
   makeRDD at <console>:34

scala> rdd1.saveAsHadoopFile("hdfs://master:9000/chapter3/HadoopFile/",
   classOf[Text],classOf[IntWritable],classOf[TextOutputFormat[Text,IntWritab
   le]])

//查看HDFS中的目录内容
$hadoop fs -ls /chapter3/HadoopFile
Found 4 items
```

```
-rw-r--r--3 spark supergroup0 2016-04-08 16:52 /chapter3/HadoopFile/_SUCCESS
-rw-r--r--3 spark supergroup4 2016-04-08 16:51 /chapter3/HadoopFile/part-00000
-rw-r--r--3 spark supergroup8 2016-04-08 16:51 /chapter3/HadoopFile/part-00001
-rw-r--r--3 spark supergroup8 2016-04-08 16:51 /chapter3/HadoopFile/part-00002

//查看HDFS中的文件内容
$hadoop fs -cat /chapter3/HadoopFile/part-00000
A    2
```

saveAsHadoopDataset用于将RDD保存到除了HDFS的其他存储中，比如HBase。在JobConf中，通常需要关注或者设置5个参数：文件的保存路径、key值的class类型、value值的class类型、RDD的输出格式（OutputFormat）以及压缩相关的参数。

```
scala> import org.apache.spark.SparkConf
scala> import org.apache.spark.SparkContext
scala> import SparkContext._
scala> import org.apache.hadoop.mapred.TextOutputFormat
scala> import org.apache.hadoop.io.Text
scala> import org.apache.hadoop.io.IntWritable
scala> import org.apache.hadoop.mapred.JobConf

scala> var rdd1 = sc.makeRDD(Array(("A",2),("A",1),("B",6),("B",3),("B",7)))
rdd1: org.apache.spark.rdd.RDD[(String, Int)] = ParallelCollectionRDD[14] at
    makeRDD at <console>:54
scala> var jobConf = new JobConf()
jobConf: org.apache.hadoop.mapred.JobConf = Configuration: core-default.xml,
    core-site.xml, mapred-default.xml, mapred-site.xml, yarn-default.xml,
    yarn-site.xml, hdfs-default.xml, hdfs-site.xml
scala> jobConf.setOutputFormat(classOf[TextOutputFormat[Text,IntWritable]])
scala> jobConf.setOutputKeyClass(classOf[Text])
scala> jobConf.setOutputValueClass(classOf[IntWritable])
scala> jobConf.set("mapred.output.dir","hdfs://master:9000/chapter3/Dataset/")
scala> rdd1.saveAsHadoopDataset(jobConf)

//查看HDFS中的目录内容
$hadoop fs -ls /chapter3/Dataset
Found 4 items
-rw-r--r--3 spark supergroup0 2016-04-08 17:01 /chapter3/Dataset/_SUCCESS
-rw-r--r--3 spark supergroup4 2016-04-08 17:01 /chapter3/Dataset/part-00000
-rw-r--r--3 spark supergroup8 2016-04-08 17:01 /chapter3/Dataset/part-00001
-rw-r--r--3 spark supergroup8 2016-04-08 17:01 /chapter3/Dataset/part-00002

//查看HDFS中的文件内容
$hadoop fs -cat /chapter3/Dataset/part-00001
```

```
A    1
B    6
```

- saveAsNewAPIHadoopFile[F <: OutputFormat[K, V]](path: String): Unit
- saveAsNewAPIHadoopFile(path: String, keyClass: Class[_], valueClass: Class[_], outputFormatClass: Class[_ <: OutputFormat[_, _]], conf: Configuration = self.context.hadoopConfiguration): Unit
- saveAsNewAPIHadoopDataset(conf: Configuration): Unit

saveAsNewAPIHadoopFile 用于将 RDD 数据保存到 HDFS 上，使用新版本 Hadoop API，用法基本同 saveAsHadoopFile。该方法作用同 saveAsHadoopDataset，只不过采用新版本 Hadoop API。

```
//使用 saveAsNewAPIHadoopFile 操作把 RDD 内容保存到 HDFS 中
scala> import org.apache.spark.SparkConf
scala> import org.apache.spark.SparkContext
scala> import SparkContext._
scala> import org.apache.hadoop.mapreduce.lib.output.TextOutputFormat
scala> import org.apache.hadoop.io.Text
scala> import org.apache.hadoop.io.IntWritable
scala> var rdd1 = sc.makeRDD(Array(("A",2),("A",1),("B",6),("B",3),("B",7)))
rdd1: org.apache.spark.rdd.RDD[(String, Int)] = ParallelCollectionRDD[48] at
  makeRDD at <console>:54
scala> rdd1.saveAsNewAPIHadoopFile("hdfs://master:9000/chapter3/NewAPI/",
  classOf[Text],classOf[IntWritable],classOf[TextOutputFormat[Text,IntWritab
  le]])

//对保存在 HDFS 的文件进行查看
$hadoop fs -ls /chapter3/NewAPI
Found 4 items
-rw-r--r--3 spark supergroup0 2016-04-06 23:17 /chapter3/NewAPI/_SUCCESS
-rw-r--r--3 spark supergroup4 2016-04-06 23:17 /chapter3/NewAPI/part-r-00000
-rw-r--r--3 spark supergroup8 2016-04-06 23:17 /chapter3/NewAPI/part-r-00001
-rw-r--r--3 spark supergroup8 2016-04-06 23:17 /chapter3/NewAPI/part-r-00002
$hadoop fs -cat /chapter3/NewAPI/part-r-00000
A    2
```

3.8 小结

在本章中首先介绍了 RDD 出现的背景和定义。在早期并行计算的编程模式，缺乏针对数据共享的高效元语，会造成磁盘 I/O 以及序列化等开销，Spark 提出了统一的编程抽象—弹性分布式数据集（RDD），该模型可以令并行计算阶段间高效地进行数据共享。Spark 处理数据时，会将计算转换为一个有向无环图（DAG）的任务集合，RDD 能够有效地恢复 DAG 中故障和慢节点执行的任务，并且 RDD 提供一种基于粗粒度变换的接口，记录创建数据集的"血统"，能够实现高效的容错性。

然后介绍 RDD 实现的五个方面，分别从作业调度、解析器集成、内存管理、检查点支持和多用户管理进行描述，这些内容是 Spark 底层原理和基础。接着，介绍了 Spark 提供通用接口，包括 RDD 分区、首选位置、依赖关系、分区计算和分区函数等接口。最后，介绍了 Spark 操作的四类操作，分别为创建操作、控制操作、转换操作和行为操作，并通过实例加以说明。

第 4 章

Spark 核心原理

在描述 Spark 运行基本流程前，我们先介绍一下 Spark 基本概念，参见图 4-1。

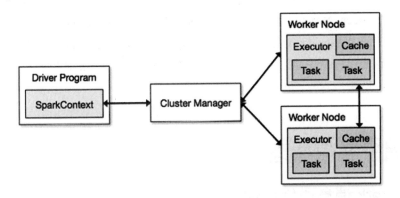

图 4-1 Spark 运行结构图

- **Application**（应用程序）：是指用户编写的 Spark 应用程序，包含驱动程序（Driver）和分布在集群中多个节点上运行的 Executor 代码，在执行过程中由一个或多个作业组成。
- **Driver**（驱动程序）：Spark 中的 Driver 即运行上述 Application 的 main 函数并且创建 SparkContext，其中创建 SparkContext 的目的是为了准备 Spark 应用程序的运行环境。在 Spark 中由 SparkContext 负责与 ClusterManager 通信，进行资源的申请、任务的分配和监控等；当 Executor 部分运行完毕后，Driver 负责将 SparkContext 关闭。通常用 SparkContext 代表 Driver。
- **Cluster Manager**（集群资源管理器）：是指在集群上获取资源的外部服务，目前有以下几种。
- **Standalone**：Spark 原生的资源管理，由 Master 负责资源的管理。

- **Hadoop Yarn**：由 YARN 中的 ResourceManager 负责资源的管理。
- **Mesos**：由 Mesos 中的 Mesos Master 负责资源的管理。
- **Worker**（工作节点）：集群中任何可以运行 Application 代码的节点，类似于 YARN 中的 NodeManager 节点。在 Standalone 模式中指的就是通过 Slave 文件配置的 Worker 节点，在 Spark on Yarn 模式中指的就是 NodeManager 节点。
- **Master**（总控进程）：Spark Standalone 运行模式下的主节点，负责管理和分配集群资源来运行 Spark Appliation。
- **Executor**（执行进程）：Application 运行在 Worker 节点上的一个进程，该进程负责运行 Task，并负责将数据存在内存或者磁盘上，每个 Application 都有各自独立的一批 Executor。在 Spark on Yarn 模式下，其进程名称为 CoarseGrainedExecutorBackend，类似于 Hadoop MapReduce 中的 YarnChild。一个 CoarseGrainedExecutorBackend 进程有且仅有一个 executor 对象，它负责将 Task 包装成 taskRunner，并从线程池中抽取出一个空闲线程运行 Task。每个 CoarseGrainedExecutorBackend 能并行运行 Task 的数量就取决于分配给它的 CPU 的个数了。

4.1 消息通信原理

4.1.1 Spark 消息通信架构

在 Spark 定义了通信框架接口，这些接口实现中调用 Netty 的具体方法（在 Spark 2.0 版本之前使用的是 Akka）。在框架中以 RpcEndPoint 和 RpcEndpointRef 实现了 Actor 和 ActorRef 相关动作，其中 RpcEndpointRef 是 RpcEndPoint 的引用，在消息通信中消息发送方持有引用 RpcEndpointRef，它们之间的关系如图 4-2 所示。

通信框架使用了工厂设计模式实现，这种设计方式实现了对 Netty 的解耦，能够根据需要引入其他的消息通信工具。通信框架创建参见上图左边四个类，具体实现步骤：首先定义了 RpcEnv 和 RpcEnvFactory 两个抽象类，在 RpcEnv 定义了 RPC 通信框架启动、停止和关闭等抽象方法，在 RpcEnvFactory 中定义了创建抽象方法；然后在 NettyRpcEnv 和 NettyRpcEnvFactory 类中使用 Netty 对继承的方法进行了实现，需要注意的是在 NettyRpcEnv 中启动终端点方法 setupEndpoint，在这个方法中会把 RpcEndPoint 和 RpcEndpointRef 相互以键值方式存放在线程安全的 ConcurrentHashMap 中；最后在 RpcEnv 的 object 类中通过反射方式实现了创建 RpcEnv

的实例的静态方法。

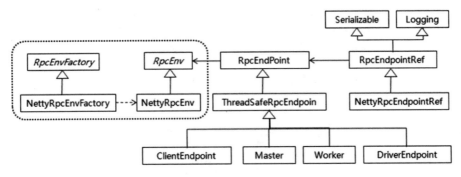

图 4-2　Spark 消息通信类图

在各模块使用中，如 Master、Worker 等，会先使用 RpcEnv 的静态方法创建 RpcEnv 实例，然后实例化 Master，由于 Master 继承于 ThreadSafeRpcEndpoin，创建的 Master 实例是一个线程安全的终端点，接着调用 RpcEnv 启动终端点方法，把 Master 的终端点和其对应的引用注册到 RpcEnv 中。在消息通信中，其他对象只要获取了 Master 终端点的引用，就能够发送消息给 Master 进行通信。下面为 Master.scala 类的 startRpcEnvAndEndPoint 方法中，启动消息通信框架的代码：

```
def startRpcEnvAndEndpoint(host: String, port: Int, webUiPort: Int, conf:
  SparkConf) : (RpcEnv, Int, Option[Int]) = {
  val securityMgr = new SecurityManager(conf)
  val rpcEnv = RpcEnv.create(SYSTEM_NAME, host, port, conf, securityMgr)
  val masterEndpoint = rpcEnv.setupEndpoint(ENDPOINT_NAME,
    new Master(rpcEnv, rpcEnv.address, webUiPort, securityMgr, conf))
  val portsResponse = masterEndpoint.askWithRetry[BoundPortsResponse]
    (BoundPortsRequest)
  (rpcEnv, portsResponse.webUIPort, portsResponse.restPort)
}
```

Spark 运行过程中 Master、Driver、Worker 以及 Executor 等模块之间由事件驱动消息的发送，下面以 Standalone 运行架构为例，分析 Spark 启动过程和应用程序运行过程中是如何进行通信的？

4.1.2　Spark 启动消息通信

Spark 启动过程中主要是进行 Master 与 Worker 之间的通信，其消息发送关系如图 4-3 所示。首先由 Worker 节点向 Master 发送注册消息，然后 Master 处理完毕后，返回注册成功消息或失败消息，如果成功注册，则 Worker 定时发送心跳消息给 Master。

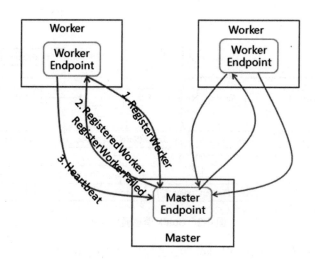

图 4-3 Spark 启动消息通信交互过程

其详细过程如下。

（1）当 Master 启动后，随之启动各 Worker，Worker 启动时会创建通信环境 RpcEnv 和终端点 EndPoint，并向 Master 发送注册 Woker 的消息 RegisterWorker。

由于 Woker 可能需要注册到多个 Master 中（如 HA 环境），在 Worker 的 tryRegisterAllMasters 方法中创建注册线程池 registerMasterThreadPool，把需要申请注册的请求放在该线程池中，然后通过该线程池启动注册线程。在该注册过程中，获取 Master 终端点引用，接着调用 registerWithMaster 方法，根据 Master 终端点引用的 send 方法发送注册 RegisterWorker 消息。Worker.tryRegisterAllMasters 方法代码如下所示：

```
private def tryRegisterAllMasters(): Array[JFuture[_]] = {
  masterRpcAddresses.map { masterAddress =>
    registerMasterThreadPool.submit(new Runnable {
      override def run(): Unit = {
        try {
          logInfo("Connecting to master " + masterAddress + "……")
          //获取Master 终端点的引用
          val masterEndpoint = rpcEnv.setupEndpointRef(masterAddress,
            Master.ENDPOINT_NAME)

          //调用 registerWithMaster 方法注册消息
          registerWithMaster(masterEndpoint)
        } catch {……}
……
}
```

其中 registerWithMaster 方法代码如下：

```
private def registerWithMaster(masterEndpoint: RpcEndpointRef): Unit = {
  //根据 Master 终端点引用，发送注册信息
  masterEndpoint.ask[RegisterWorkerResponse](
    RegisterWorker(workerId,host,port,self,cores,memory,workerWebUiUrl)).onCo
      mplete{
      //返回注册成功或失败的结果
      case Success(msg) =>Utils.tryLogNonFatalError {handleRegisterResponse(msg) }
      case Failure(e) =>
        logError(s"Cannot register with master: ${masterEndpoint.address}", e)
        System.exit(1)
    }(ThreadUtils.sameThread)
}
```

（2）Master 收到消息后，需要对 Worker 发送的信息进行验证、记录。如果注册成功，则发送 RegisteredWorker 消息给对应的 Woker，告诉 Woker 已经完成注册，随之进行步骤 3，即 Worker 定期发送心跳信息给 Master；如果在注册过程中失败，则会发送 RegisterWorkerFailed 消息，Woker 打印出错日志并结束 Worker 启动。

在 Master 中，Master 接收到 Worker 注册消息后，先判断 Master 当前状态是否处于 STANDBY 状态，如果是则忽略该消息，如果在注册列表中发现了该 Worker 的编号，则发送注册失败的消息。判断完毕后使用 registerWorker 方法把该 Worker 加入到列表中，用于集群进行处理任务时进行调度。Master.receiveAndReply 方法中注册 Worker 代码实现如下所示：

```
case RegisterWorker(id, workerHost, workerPort, workerRef, cores, memory,
  workerWebUiUrl) => {
  //Master 处于 STANDBY 状态，返回"Master 处于 STANDBY 状态"消息
  if (state == RecoveryState.STANDBY) {
    context.reply(MasterInStandby)
  } else if (idToWorker.contains(id)) {
    context.reply(RegisterWorkerFailed("Duplicate worker ID"))
  } else {
    val worker = new WorkerInfo(id, workerHost, workerPort, cores, memory,
      workerRef, workerWebUiUrl)
    //registerWorker 方法中注册 Worker，该方法中会把 Worker 放到列表中，
    //用于后续运行任务时使用
    if (registerWorker(worker)) {
      persistenceEngine.addWorker(worker)
      context.reply(RegisteredWorker(self, masterWebUiUrl))
      schedule()
    } else {
      val workerAddress = worker.endpoint.address
      context.reply(RegisterWorkerFailed("Attempted to re-register worker at
```

```
        same address: "+ workerAddress))
      }
    }
  }
```

（3）当Worker接收到注册成功后，会定时发送心跳信息Heartbeat给Master，以便Master了解Worker的实时状态。间隔时间可以在spark.worker.timeout中设置，注意的是，该设置值的1/4为心跳间隔。

```
private val HEARTBEAT_MILLIS = conf.getLong("spark.worker.timeout", 60) * 1000 / 4
```

当Worker获取到注册成功消息后，先记录日志并更新Master信息，然后启动定时调度进程发送心跳信息，该调度进程时间间隔为上面所定义的HEARTBEAT_MILLIS值。

```
case RegisteredWorker(masterRef, masterWebUiUrl) =>
  logInfo("Successfully registered with master " + masterRef.address.toSparkURL)
  registered = true
  changeMaster(masterRef, masterWebUiUrl)
  forwordMessageScheduler.scheduleAtFixedRate(new Runnable {
    override def run(): Unit = Utils.tryLogNonFatalError {
      self.send(SendHeartbeat)
    }
  }, 0, HEARTBEAT_MILLIS, TimeUnit.MILLISECONDS)
  //如果设置清理以前应用使用的文件夹，则进行该动作
  if (CLEANUP_ENABLED) {
    logInfo(s"Worker cleanup enabled; old application directories will be deleted in: $workDir")
    forwordMessageScheduler.scheduleAtFixedRate(new Runnable {
      override def run(): Unit = Utils.tryLogNonFatalError {self.send(WorkDirCleanup)}
    }, CLEANUP_INTERVAL_MILLIS, CLEANUP_INTERVAL_MILLIS, TimeUnit.MILLISECONDS)
  }
  //向Master汇报Worker中Executor最新状态
  val execs = executors.values.map { e =>
    new ExecutorDescription(e.appId, e.execId, e.cores, e.state)
  }
  masterRef.send(WorkerLatestState(workerId, execs.toList, drivers.keys.toSeq))
```

4.1.3　Spark运行时消息通信

用户提交应用程序时，应用程序的SparkContext会向Master发送应用注册消息，并由Master给该应用分配Executor，Executor启动后，Executor会向SparkContext发送注册成功消息；当SparkContext的RDD触发行动操作后，将创建RDD的DAG，通过DAGScheduler进行划分Stage并将Stage转化为TaskSet；接着由TaskScheduler向注册的Executor发送执行消息，Executor

接收到任务消息后启动并运行；最后当所有任务运行时，由 Driver 处理结果并回收资源。图 4-4 为 Spark 运行消息通信的交互过程。

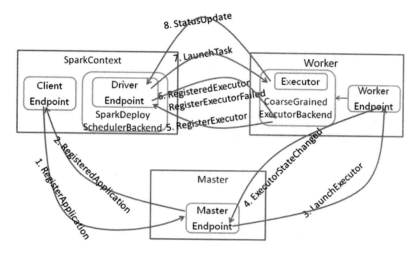

图 4-4　Spark 运行消息通信的交互过程

详细过程如下。

（1）执行应用程序需启动 SparkContext，在 SparkContext 启动过程中会先实例化 SchedulerBackend 对象，在独立运行（Standalone）模式中实际创建的是 SparkDeploySchedulerBackend 对象，在该对象的启动中会继承父类 DriverEndpoint 和创建 AppClient 的 ClientEndpoint 的两个终端点。

在 ClientEndpoint 的 tryRegisterAllMasters 方法中创建注册线程池 registerMasterThreadPool，在该线程池中启动注册线程并向 Master 发送 RegisterApplication 注册应用的消息，代码如下所示：

```
private def tryRegisterAllMasters(): Array[JFuture[_]] = {
  //由于 HA 等环境中有多个 Master，需要遍历所有 Master 发送消息
  for (masterAddress <- masterRpcAddresses) yield {
    //向线程池中启动注册线程，当该线程读到应用注册成功标识 registered=true 时退出注册线程
    registerMasterThreadPool.submit(new Runnable {
      override def run(): Unit = try {
        if (registered) {
          return
        }
        //获取 Master 终端点的引用，发送注册应用的消息
        val masterRef =rpcEnv.setupEndpointRef(Master.SYSTEM_NAME,
          masterAddress, Master.ENDPOINT_NAME)
```

```
      masterRef.send(RegisterApplication(appDescription, self))
    } catch {..}
  })
}
```

当 Master 接收到注册应用的消息时,在 registerApplication 方法中记录应用信息并把该应用加入到等待运行应用列表中,注册完毕后发送成功消息 RegisteredApplication 给 ClientEndpoint,同时调用 startExecutorsOnWorkers 方法运行应用。在执行前需要获取运行应用的 Worker,然后发送 LaunchExecutor 消息给 Worker,通知 Worker 启动 Executor。其中 Master.startExecutorsOnWorkers 方法代码如下:

```
private def startExecutorsOnWorkers(): Unit = {
  //使用 FIFO 调度算法运行应用,即先注册的应用先运行
  for (app <- waitingApps if app.coresLeft > 0) {
    val coresPerExecutor: Option[Int] = app.desc.coresPerExecutor

    //找出符合剩余内存大于等于启动 Executor 所需大小、核数大于等于 1 条件的 Worker
    val usableWorkers = workers.toArray.filter(_.state == WorkerState.ALIVE)
      .filter(worker => worker.memoryFree >= app.desc.memoryPerExecutorMB &&
        worker.coresFree >= coresPerExecutor.getOrElse(1))
      .sortBy(_.coresFree).reverse

    //确定运行在哪些 Worker 上和每个 Worker 分配用于运行的核数,分配算法有两种,一种是把应用
    //运行在尽可能多的 Worker 上,相反,另一种是运行在尽可能少的 Worker 上
    val assignedCores=scheduleExecutorsOnWorkers(app,usableWorkers,spreadOutApps)

    //通知分配的 Worker 启动 Worker
    for (pos <- 0 until usableWorkers.length if assignedCores(pos) > 0) {
      allocateWorkerResourceToExecutors(app, assignedCores(pos), coresPerExecutor,
        usableWorkers(pos))
    }
  }
}
```

(2) AppClient.ClientEndpoint 接收到 Master 发送 RegisteredApplication 消息,需要把注册标识 registered 置为 true,Master 注册线程获取状态变化后,完成注册 Application 进程,代码如下:

```
case RegisteredApplication(appId_, masterRef) =>
  appId.set(appId_)
  registered.set(true)
  master = Some(masterRef)
  listener.connected(appId.get)
```

(3) 在 Master 类的 startExecutorsOnWorkers 方法中分配资源运行应用程序时,调用

allocateWorkerResourceToExecutors 方法实现在 Worker 中启动 Executor。当 Worker 收到 Master 发送过来的 LaunchExecutor 消息，先实例化 ExecutorRunner 对象，在 ExecutorRunner 启动中会创建进程生成器 ProcessBuilder，然后由该生成器使用 command 创建 CoarseGrainedExecutorBackend 对象，该对象是 Executor 运行的容器，最后 Worker 发送 ExecutorStateChanged 消息给 Master，通知 Executor 容器已经创建完毕。

当 Worker 接收到启动 Executor 消息，执行代码如下：

```
case LaunchExecutor(masterUrl, appId, execId, appDesc, cores_, memory_) =>
  if (masterUrl != activeMasterUrl) {
    logWarning("Invalid Master (" + masterUrl + ") attempted to launch executor.")
  } else {
    try {
      //创建 Executor 执行目录
      val executorDir = new File(workDir, appId + "/" + execId)
      if (!executorDir.mkdirs()) {
        throw new IOException("Failed to create directory " + executorDir)
      }

      //通过 SPARK_EXECUTOR_DIRS 环境变量，在 Worker 中创建 Executor 执行目录，当程序执行
      //完毕后由 Worker 进行删除
      val appLocalDirs = appDirectories.getOrElse(appId,
        Utils.getOrCreateLocalRootDirs(conf).map { dir =>
          val appDir = Utils.createDirectory(dir, namePrefix = "executor")
          Utils.chmod700(appDir)
          appDir.getAbsolutePath()
        }.toSeq)
      appDirectories(appId) = appLocalDirs

      //在 ExecutorRunner 中创建 CoarseGrainedExecutorBackend 对象，创建的是使用应用
      //信息中的 command,而 command 在 SparkDeploySchedulerBackend 的 start 方法中构建
      val manager = new ExecutorRunner(appId,execId,appDesc.copy(command =
        Worker.maybeUpdateSSLSettings(appDesc.command,
        conf)),cores_,memory_,self,workerId,host,webUi.boundPort,publicAddress,
        sparkHome,executorDir,workerUri,conf,appLocalDirs,
        ExecutorState.RUNNING)
      executors(appId + "/" + execId) = manager
      manager.start()
      coresUsed += cores_
      memoryUsed += memory_

      //向 Master 发送消息，表示 Executor 状态已经更改为 ExecutorState.RUNNING
      sendToMaster(ExecutorStateChanged(appId, execId, manager.state, None,
```

```
        None))
    } catch {……}
}
```

在 ExecutorRunner 创建中调用了 fetchAndRunExecutor 方法进行实现，在该方法中 command 内容在 SparkDeploySchedulerBackend 中定义，指定构造 Executor 运行容器 CoarseGrainedExecutorBackend，其中创建过程代码如下：

```
private def fetchAndRunExecutor() {
  try {
    //通过应用程序的信息和环境配置创建构造器 builder
    val builder = CommandUtils.buildProcessBuilder(appDesc.command, new
      SecurityManager(conf),memory, sparkHome.getAbsolutePath,
      substituteVariables)
    val command = builder.command()
    val formattedCommand = command.asScala.mkString("\"", "\" \"", "\"")

    //在构造器 builder 中添加执行目录等信息
    builder.directory(executorDir)
    builder.environment.put("SPARK_EXECUTOR_DIRS",
      appLocalDirs.mkString(File.pathSeparator))
    builder.environment.put("SPARK_LAUNCH_WITH_SCALA", "0")

    //在构造器 builder 中添加监控页面输入日志地址信息
    val baseUrl = s"http://$publicAddress:$webUiPort/logPage/?appId=
      $appId&executorId=$execId&logType="
    builder.environment.put("SPARK_LOG_URL_STDERR", s"${baseUrl}stderr")
    builder.environment.put("SPARK_LOG_URL_STDOUT", s"${baseUrl}stdout")

    //启动构造器，创建 CoarseGrainedExecutorBackend 实例
    process = builder.start()
    val header = "Spark Executor Command: %s\n%s\n\n".format(formattedCommand,
      "=" * 40)
    //输出 CoarseGrainedExecutorBackend 实例运行信息
    val stdout = new File(executorDir, "stdout")
    stdoutAppender = FileAppender(process.getInputStream, stdout, conf)
    val stderr = new File(executorDir, "stderr")
    Files.write(header, stderr, StandardCharsets.UTF_8)
    stderrAppender = FileAppender(process.getErrorStream, stderr, conf)

    //等待 CoarseGrainedExecutorBackend 运行结果，当结束时向 Worker 发送退出状态信息
    val exitCode = process.waitFor()
    state = ExecutorState.EXITED
    val message = "Command exited with code " + exitCode
```

```
    worker.send(ExecutorStateChanged(appId, execId, state, Some(message),
      Some(exitCode)))
  } catch {……}
}
```

(4) Master 接收到 Worker 发送的 ExecutorStateChanged 消息，根据 ExecuteState。

```
case StatusUpdate(executorId, taskId, state, data) =>
  val execOption = idToApp.get(appId).flatMap(app => app.executors.get(execId))
  execOption match {
case Some(exec) => {
……
//向 Driver 发送 ExecutorUpdated 消息
exec.application.driver.send(ExecutorUpdated(execId, state, message, exitStatus))
……
}
```

（5）在步骤 3 的 CoarseGrainedExecutorBackend 启动方法 onStart 中，会发送注册 Executor 消息 RegisterExecutor 给 DriverEndpoint，DriverEndpoint 终端点，先判断该 Executor 是否已经注册，如果已经存在发送注册失败 RegisterExecutorFailed 消息，否则 Driver 终端点会记录该 Executor 信息，发送注册成功 RegisteredExecutor 消息，在 makeOffers()方法中分配运行任务资源，最后发送 LaunchTask 消息执行任务。

其中在 DriverEndpoint 终端点进行注册 Executor 的过程如下：

```
case RegisterExecutor(executorId, executorRef, hostPort, cores, logUrls) =>
  if (executorDataMap.contains(executorId)) {
    executorRef.send(RegisterExecutorFailed("Duplicate executor ID: " + executorId))
    context.reply(true)
  } else {
    ……
    //记录 Executor 编号以及该 Executor 需要使用的核数
    addressToExecutorId(executorRef.address) = executorId
    totalCoreCount.addAndGet(cores)
    totalRegisteredExecutors.addAndGet(1)
    val data = new ExecutorData(executorRef, executorRef.address, host, cores,
      cores, logUrls)
    //创建 Executor 编号和其具体信息的键值列表
    CoarseGrainedSchedulerBackend.this.synchronized {
      executorDataMap.put(executorId, data)
      if (currentExecutorIdCounter < executorId.toInt) {
        currentExecutorIdCounter = executorId.toInt
      }
      if (numPendingExecutors > 0) {
        numPendingExecutors -= 1      }
```

```
    }
    //回复Executor完成注册消息并在监听总线中加入添加Executor事件
    executorRef.send(RegisteredExecutor(executorAddress.host))
    listenerBus.post(SparkListenerExecutorAdded(System.currentTimeMillis(),
      executorId, data))

    //分配运行任务资源并发送LaunchTask消息执行任务
    makeOffers()
  }
```

（6）当CoarseGrainedExecutorBackend接收到Executor注册成功RegisteredExecutor消息时，在CoarseGrainedExecutorBackend容器中实例化Executor对象。启动完毕后，会定时向Driver发送心跳信息，等待接收从DriverEndpoint终端点发送执行任务的消息。CoarseGrainedExecutorBackend处理注册成功代码如下：

```
case RegisteredExecutor =>
  logInfo("Successfully registered with driver")
  //根据环境变量的参数启动Executor，在Spark中它是真正任务的执行者
  executor = new Executor(executorId, hostname, env, userClassPath, isLocal =
    false)
```

该Executor会定时向Driver发送心跳信息，等待Driver下发任务：

```
private val heartbeater = ThreadUtils.newDaemonSingleThreadScheduledExecutor
  ("driver-heartbeater")
private def startDriverHeartbeater(): Unit = {
  //设置间隔时间为10s
  val intervalMs = conf.getTimeAsMs("spark.executor.heartbeatInterval", "10s")

  //等待随机的时间间隔，这样心跳在同步中不会结束
  val initialDelay = intervalMs + (math.random * intervalMs).asInstanceOf[Int]
  val heartbeatTask = new Runnable() {
    override def run(): Unit = Utils.logUncaughtExceptions(reportHeartBeat())
  }
  //发送心跳信息给Driver
  heartbeater.scheduleAtFixedRate(heartbeatTask, initialDelay, intervalMs,
    TimeUnit.MILLISECONDS)
}
```

（7）CoarseGrainedExecutorBackend的Executor启动后，接收从DriverEndpoint终端点发送LaunchTask执行任务消息，任务执行是在Executor的launchTask方法实现的。在执行时会创建TaskRunner进程，由该进程进行任务的处理，处理完毕后发送StatusUpdate消息返回给CoarseGrainedExecutorBackend。

```scala
case LaunchTask(data) =>
  if (executor == null) {
    //当executor没有成功启动时，输出异常日志并关闭Executor
    logError("Received LaunchTask command but executor was null")
    System.exit(1)
  } else {
    val taskDesc = ser.deserialize[TaskDescription](data.value)
    logInfo("Got assigned task " + taskDesc.taskId)

    //启动TaskRunner进程执行任务
    executor.launchTask(this, taskId = taskDesc.taskId, attemptNumber =
      taskDesc.attemptNumber,taskDesc.name, taskDesc.serializedTask)
  }
```

调用 Executor 的 launchTask 方法，在该方法中创建 TaskRunner 进程，然后把该进程加入到执行池 threadPool 中，由 Executor 进行统一调度：

```scala
def launchTask(context: ExecutorBackend,taskId: Long,
    attemptNumber: Int,taskName: String,serializedTask: ByteBuffer): Unit = {
  val tr = new TaskRunner(context, taskId = taskId, attemptNumber = attemptNumber,
    taskName,
    serializedTask)
  runningTasks.put(taskId, tr)
  threadPool.execute(tr)
}
```

任务执行过程和获取执行结果参见 4.2.5 和 4.2.6 节内容。

（8）在 TaskRunner 执行任务完成时，会由向 DriverEndPoint 终端点发送状态变更 StatusUpdate 消息，当 DriverEndPoint 终端点接收到该消息时，调用 TaskSchedulerImpl 的 statusUpdate 方法，根据任务执行不同的结果进行处理，处理完毕后再给该 Executor 分配执行任务。其中，在 DriverEndPoint 终端点处理状态变更代码如下所示：

```scala
case StatusUpdate(executorId, taskId, state, data) =>
  //调用TaskSchedulerImpl的statusUpdate()方法，根据任务执行不同结果进行处理
  scheduler.statusUpdate(taskId, state, data.value)
  if (TaskState.isFinished(state)) {
    executorDataMap.get(executorId) match {
      //任务执行成功后，回收该Executor运行该任务的CPU，再根据实际情况分配任务
      case Some(executorInfo) =>
        executorInfo.freeCores += scheduler.CPUS_PER_TASK
        makeOffers(executorId)
      case None =>
    }
  }
```

4.2 作业执行原理

Spark 的作业和任务调度系统是其核心，它能够有效地进行调度根本原因是对任务划分 DAG 和容错，使得它对低层到顶层的各个模块之间的调用和处理显得游刃有余。下面介绍一些相关术语。

- **作业（Job）**：RDD 中由行动操作所生成的一个或多个调度阶段。
- **调度阶段（Stage）**：每个作业会因为 RDD 之间的依赖关系拆分成多组任务集合，称为调度阶段，也叫做任务集（TaskSet）。调度阶段的划分是由 DAGScheduler 来划分的，调度阶段有 Shuffle Map Stage 和 Result Stage 两种。
- **任务（Task）**：分发到 Executor 上的工作任务，是 Spark 实际执行应用的最小单元。
- **DAGScheduler**：DAGScheduler 是面向调度阶段的任务调度器，负责接收 Spark 应用提交的作业，根据 RDD 的依赖关系划分调度阶段，并提交调度阶段给 TaskScheduler。
- **TaskScheduler**：TaskScheduler 是面向任务的调度器，它接受 DAGScheduler 提交过来的调度阶段，然后以把任务分发到 Work 节点运行，由 Worker 节点的 Executor 来运行该任务。

4.2.1 概述

Spark 的作业调度主要是指基于 RDD 的一系列操作构成一个作业，然后在 Executor 中执行。这些操作算子主要分为转换操作和行动操作，对于转换操作的计算是 lazy 级别的，也就是延迟执行，只有出现了行动操作才触发了作业的提交。在 Spark 调度中最重要的是 DAGScheduler 和 TaskScheduler 两个调度器，其中，DAGScheduler 负责任务的逻辑调度，将作业拆分成不同阶段的具有依赖关系的任务集，而 TaskScheduler 则负责具体任务的调度执行。通过图 4-5 能从整体上对 Spark 的作业和任务调度系统做一下分析。

（1）Spark 应用程序进行各种转换操作，通过行动操作触发作业运行。提交之后根据 RDD 之间的依赖关系构建 DAG 图，DAG 图提交给 DAGScheduler 进行解析。

（2）DAGScheduler 是面向调度阶段的高层次的调度器，DAGScheduler 把 DAG 拆分成相互依赖的调度阶段，拆分调度阶段是以 RDD 的依赖是否为宽依赖，当遇到宽依赖就划分为新的调度阶段。每个调度阶段包含一个或多个任务，这些任务形成任务集，提交给底层调度器 TaskScheduler 进行调度运行。另外，DAGScheduler 记录哪些 RDD 被存入磁盘等物化动作，同时要寻求任务的最优化调度，例如数据本地性等；DAGScheduler 监控运行调度阶段过程，如果某个调度阶段运行失败，则需要重新提交该调度阶段。

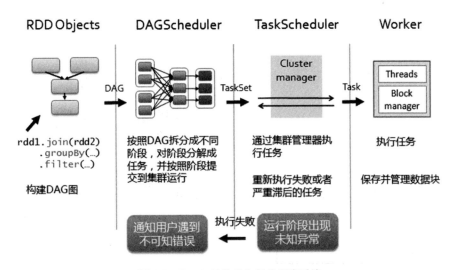

图 4-5 Spark 的作业和任务调度系统

（3）每 TaskScheduler 只为一个 SparkContext 实例服务，TaskScheduler 接收来自 DAGScheduler 发送过来的任务集，TaskScheduler 收到任务集后负责把任务集以任务的形式一个个分发到集群 Worker 节点的 Executor 中去运行。如果某个任务运行失败，TaskScheduler 要负责重试。另外，如果 TaskScheduler 发现某个任务一直未运行完，就可能启动同样的任务运行同一个任务，哪个任务先运行完就用哪个任务的结果。

（4）Worker 中的 Executor 收到 TaskScheduler 发送过来的任务后，以多线程的方式运行，每一个线程负责一个任务。任务运行结束后要返回给 TaskScheduler，不同类型的任务，返回的方式也不同。ShuffleMapTask 返回的是一个 MapStatus 对象，而不是结果本身；ResultTask 根据结果大小的不同，返回的方式又可以分为两类，具体参见 4.2.7 节和内容。

Spark 系统实现类图如图 4-6 所示，在该类图中展示了作业和任务调度方法之间的调用关系，需要注意的是，在类之间既有直接调用，也有通过 RPC 远程调用，在图中使用了虚箭号进行标记，消息调用可以参见 4.1 节的内容。

通过以上分析，对 Spark 运行调度流程和内部方法调用有了总体的了解，下面将对 Spark 运行各个环节按照运行顺序进行分析。

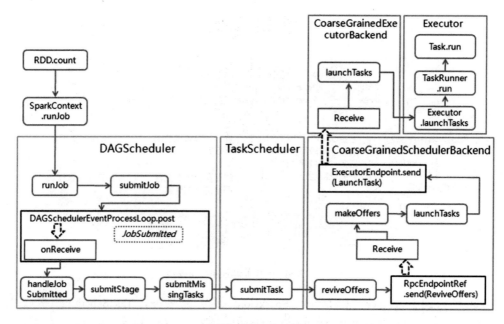

图 4-6　Spark 独立运行模式作业执行类调用关系图

4.2.2　提交作业

我们以经典的 WordCount 为例来分析提交作业的情况：

```
scala> val line = sc.textFile("hdfs://SparkMaster:9000/data/README.md")
scala> val wordcount = line.flatMap(_.split(" ")).map(x => (x,1)).
    reduceByKey(_+_).count
```

这个作业的真正提交是从"count"这个行动操作出现开始的，在 RDD 的源码的 count 方法触发了 SparkContext 的 runJob 方法来提交作业，而这个提交是在其内部隐性调用 runJob 方法进行的，对于用户来说不用显性地去提交作业。

```
// 返回 RDD 中元素的个数
def count(): Long = sc.runJob(this, Utils.getIteratorSize _).sum
```

对于 RDD 来说，它们会根据彼此之间的依赖关系形成一个有向无环图（DAG），然后把这个 DAG 图交给 DAGScheduler 来处理。从源代码来看，SparkContext 的 runJob 方法经过几次调用后，进入 DAGScheduler 的 runJob 方法，其中 SparkContext 中调用 DAGScheduler 类 runJob 方法代码如下：

```
def runJob[T, U: ClassTag]( rdd: RDD[T],func: (TaskContext, Iterator[T]) => U,
    partitions: Seq[Int],resultHandler: (Int, U) => Unit): Unit = {
    …………
```

```
    val callSite = getCallSite
    val cleanedFunc = clean(func)
    ..........
    //调用 DAGScheduler 的 runJob 进行处理
    dagScheduler.runJob(rdd, cleanedFunc, partitions, callSite, resultHandler,
        localProperties.get)
    progressBar.foreach(_.finishAll())
    rdd.doCheckpoint()
}
```

在 DAGScheduler 类内部会进行一系列的方法调用,首先是在 runJob 方法里,调用 submitJob 方法来继续提交作业,这里会发生阻塞,直到返回作业完成或失败的结果;然后在 submitJob 方法里,创建了一个 JobWaiter 对象,并借助内部消息处理进行把这个对象发送给 DAGScheduler 的内嵌类 DAGSchedulerEventProcessLoop 进行处理;最后在 DAGSchedulerEventProcessLoop 消息接收方法 OnReceive 中,接收到 JobSubmitted 样例类完成模式匹配后,继续调用 DAGScheduler 的 handleJobSubmitted 方法来提交作业,在该方法中将进行划分阶段。

```
def submitJob[T, U](rdd: RDD[T], func: (TaskContext, Iterator[T]) => U,
    partitions: Seq[Int], callSite: CallSite,resultHandler: (Int, U) =>
    Unit,properties: Properties): JobWaiter[U] = {
    //判断任务处理的分区是否存在,如果不存在,则抛出异常
    val maxPartitions = rdd.partitions.length
    partitions.find(p => p >= maxPartitions || p < 0).foreach { p =>
        throw new IllegalArgumentException("Attempting to access a non-existent
    partition: " + p + ". " + "Total number of partitions: " + maxPartitions)
    }

    //如果作业只包含 0 个任务,则创建 0 个任务的 JobWaiter,并立即返回
    val jobId = nextJobId.getAndIncrement()
    if (partitions.size == 0) {
        return new JobWaiter[U](this, jobId, 0, resultHandler)
    }

    //创建 JobWaiter 对象,等待作业运行完毕,使用内部类提交作业
    val func2 = func.asInstanceOf[(TaskContext, Iterator[_]) => _]
    val waiter = new JobWaiter(this, jobId, partitions.size, resultHandler)
    eventProcessLoop.post(JobSubmitted( jobId, rdd, func2, partitions.toArray,
        callSite, waiter,SerializationUtils.clone(properties)))
    waiter
}
```

在 Spark 应用程序中,会拆分多个作业,然后对于多个作业之间的调度,Spark 目前提供了两种调度策略:一种是 FIFO 模式,这也是目前默认的模式;另一种是 FAIR 模式,详细介绍参

见 4.3.2 节的内容。

4.2.3 划分调度阶段

Spark 调度阶段的划分是由 DAGScheduler 实现的，DAGScheduler 会从最后一个 RDD 出发使用广度优先遍历整个依赖树，从而划分调度阶段，调度阶段划分依据是以操作是否为宽依赖（ShuffleDependency）进行的，即当某个 RDD 的操作是 Shuffle 时，以该 Shuffle 操作为界限划分成前后两个调度阶段。

代码实现是在 DAGScheduler 的 handleJobSubmitted 方法中根据最后一个 RDD 生成 ResultStage 开始的，具体方法从 finalRDD 使用 getParentStages 找出其依赖的祖先 RDD 是否存在 Shuffle 操作，如果没有存在 Shuffle 操作，则本次作业仅有一个 ResultStage，该 ResultStage 不存在父调度阶段；如果存在 Shuffle 操作，则本次作业存在一个 ResultStage 和至少一个 ShuffleMapStage，该 ResultStage 存在父调度阶段。其中，handleJobSubmitted 方法代码如下：

```
private[scheduler] def handleJobSubmitted(jobId: Int, finalRDD: RDD[_], func:
    (TaskContext, Iterator[_]) => _,partitions: Array[Int], callSite: CallSite,
    listener: JobListener, properties: Properties) {
    //根据最后一个 RDD 回溯，获取最后一个调度阶段 finalStage
    var finalStage: ResultStage = null
    try {
        finalStage = newResultStage(finalRDD, func, partitions, jobId, callSite)
    } catch {……}

    //根据最后一个调度阶段 finalStage 生成作业
    val job = new ActiveJob(jobId, finalStage, callSite, listener, properties)
    clearCacheLocs()

    val jobSubmissionTime = clock.getTimeMillis()
    jobIdToActiveJob(jobId) = job
    activeJobs += job
    finalStage.setActiveJob(job)
    val stageIds = jobIdToStageIds(jobId).toArray
    val stageInfos=stageIds.flatMap(id=>stageIdToStage.get(id).map(_.latestInfo))
    listenerBus.post(SparkListenerJobStart(job.jobId, jobSubmissionTime, stageInfos,
        properties))

    //提交执行
    submitStage(finalStage)
    submitWaitingStages()
}
```

在上面的代码中,把最后一个 RDD 传入到 getParentStagesAndId 方法中,在该方法中调用 getParentStages,生成最后一个调度阶段 finalStage,其中 DAGScheduler.getParentStages 代码如下:

```
private def getParentStages(rdd: RDD[_], firstJobId: Int): List[Stage] = {
  val parents = new HashSet[Stage]
  val visited = new HashSet[RDD[_]]

  //存放等待访问的堆栈,存放的是非 ShuffleDependency 的 RDD
  val waitingForVisit = new Stack[RDD[_]]

  //定义遍历处理方法,先对访问过的 RDD 标记,然后根据当前 RDD 所依赖 RDD 操作类型进行不同处理
  def visit(r: RDD[_]) {
    if (!visited(r)) {
      visited += r
      for (dep <- r.dependencies) {
        dep match {
         //所依赖 RDD 操作类型是 ShuffleDependency,需要划分 ShuttleMap 调度阶段,
         //以调度 getShuffleMapStage 方法为入口,向前遍历划分调度阶段
          case shufDep: ShuffleDependency[_, _, _] =>
            parents += getShuffleMapStage(shufDep, firstJobId)

         //所依赖 RDD 操作类型是非 ShuffleDependency,把该 RDD 压入等待访问的堆栈
          case _ =>
            waitingForVisit.push(dep.rdd)
        }
      }
    }
  }

  //以最后一个 RDD 开始向前遍历整个依赖树,如果该 RDD 依赖树存在 ShuffleDependency 的 RDD,
  //则父调度阶段存在,反之,则不存在
  waitingForVisit.push(rdd)
  while (waitingForVisit.nonEmpty) {
    visit(waitingForVisit.pop())
  }
  parents.toList
}
```

当 finalRDD 存在父调度阶段,需要从发生 Shuffle 操作的 RDD 往前遍历,找出所有的 ShuffleMapStage。这是调度阶段划分的最关键的部分,该算法和 getParentStages 类似,由 getAncestorShuffleDependencies 方法中实现。在该方法中找出所有的操作类型是宽依赖的 RDD,然后通过 registerShuffleDependencies 和 newOrUsedShuffleStage 两个方法划分所有

ShuffleMapStage。

```
    private def getAncestorShuffleDependencies(rdd: RDD[_]): Stack
      [ShuffleDependency[_, _, _]] = {
    val parents = new Stack[ShuffleDependency[_, _, _]]
    val visited = new HashSet[RDD[_]]

    //存放等待访问的堆栈，存放的是非ShuffleDependency的RDD
    val waitingForVisit = new Stack[RDD[_]]
    def visit(r: RDD[_]) {
      if (!visited(r)) {
        visited += r
        for (dep <- r.dependencies) {
          dep match {
            //所依赖RDD操作类型是ShuffleDependency，作为划分ShuttleMap调度阶段界限
            case shufDep: ShuffleDependency[_, _, _] =>
              if (!shuffleToMapStage.contains(shufDep.shuffleId)) {
                parents.push(shufDep)
              }
            case _ =>
          }
          waitingForVisit.push(dep.rdd)
        }
      }
    }

    //向前遍历依赖树，获取所有的操作类型是ShuffleDependency的RDD，作为划分阶段依据
    waitingForVisit.push(rdd)
    while (waitingForVisit.nonEmpty) {
      visit(waitingForVisit.pop())
    }
    parents
  }
```

当所有调度阶段划分完毕时，这些调度阶段建立起依赖关系。该依赖关系是通过调度阶段其中属性parents: List[Stage]来定义的，通过该属性可以获取当前阶段所有祖先阶段，可以根据这些信息按顺序提交调度阶段进行运行。

调度阶段划分是Spark作业执行的重要部分，我们将以图4-7为例讲解阶段的划分。在图中有7个RDD，分别是rddA～rddG，它们之间有5个操作，其划分调度阶段详细步骤如下：

（1）在SparkContext中提交运行时，会调用DAGScheduler的handleJobSubmitted进行处理，在该方法中会先找到最后一个RDD（即rddG），并调用getParentStages方法。

（2）在getParentStages方法判断rddG的依赖RDD树中是否存在Shuffle操作，在该例子中

发现 join 操作为 Shuffle 操作，则获取进行该操作的 RDD 为 rddB 和 rddF。

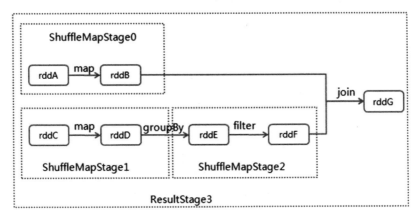

图 4-7 Spark 调度阶段划分

（3）使用 getAncestorShuffleDependencies 方法从 rddB 向前遍历，发现该依赖分支上没有其他的宽依赖，调用 newOrUsedShuffleStage 方法生成调度阶段 ShuffleMapStage0。

（4）使用 getAncestorShuffleDependencies 方法从 rddF 向前遍历，寻找该依赖分支存在宽依赖操作 groupBy，以此为分界划分 rddD 和 rddE 为 ShuffleMapStage1，rddE 和 rddF 为 ShuffleMapStage2；

（5）最后生成 rddG 的 ResultStage3。在该划分调度阶段中，共划分 4 个调度阶段，分别为 ShuffleMapStage0～3。

4.2.4 提交调度阶段

在 DAGScheduler 的 handleJobSubmitted 方法中，生成 finalStage 的同时建立起所有调度阶段的依赖关系，然后通过 finalStage 生成一个作业实例，在该作业实例中按照顺序提交调度阶段进行执行，在执行过程中通过监听总线获取作业、阶段执行情况。

在作业提交调度阶段开始时，在 submitStage 方法中调用 getMissingParentStages 方法获取 finalStage 父调度阶段，如果不存在父调度阶段，则使用 submitMissingTasks 方法提交执行；如果存在父调度阶段，则把该调度阶段存放到 waitingStages 列表中，同时递归调用 submitStage。通过该算法把存在父调度阶段的等待调度阶段放入列表 waitingStages 中，不存在父调度阶段的调度阶段作为作业运行的入口，代码如下：

```
private def submitStage(stage: Stage) {
    val jobId = activeJobForStage(stage)
```

```
    if (jobId.isDefined) {
      logDebug("submitStage(" + stage + ")")
      if (!waitingStages(stage) && !runningStages(stage) && !failedStages(stage)) {
        //在该方法中，获取该调度阶段的父调度阶段，获取的方法是通过 RDD 的依赖关系向前遍历看
        //是否存在 Shuffle 操作，这里并没有使用调度阶段的依赖关系进行获取
        val missing = getMissingParentStages(stage).sortBy(_.id)
        if (missing.isEmpty) {
          //如果不存在父调度阶段，直接把该调度阶段提交执行
          logInfo("Submitting " + stage + " (" + stage.rdd + "), which has no missing
            parents")
          submitMissingTasks(stage, jobId.get)
        } else {
          //如果存在父调度阶段，把该调度阶段加入到等待运行调度阶段列表中，
          //同时递归调用 submitStage 方法，直至找到开始的调度阶段，即该调度阶段没有父调度阶段
          for (parent <- missing) {
            submitStage(parent)
          }
          waitingStages += stage
        }
      }
    } else {
      abortStage(stage, "No active job for stage " + stage.id, None)
    }
  }
```

当入口的调度阶段运行完成后相继提交后续调度阶段，在调度前先判断该调度阶段所依赖的父调度阶段的结果是否可用（即运行是否成功）。如果结果都可用，则提交该调度阶段；如果结果存在不可用的情况，则尝试提交结果不可用的父调度阶段。对于调度阶段是否可用的判断是在 ShuffleMapTask 完成时进行，DAGScheduler 会检查调度阶段的所有任务是否都完成了。如果存在执行失败的任务，则重新提交该调度阶段；如果所有任务完成，则扫描等待运行调度阶段列表，检查它们的父调度阶段是否存在未完成，如果不存在则表明该调度阶段准备就绪，生成实例并提交运行。代码参见 DAGScheduler 的 handleTaskCompletion 方法。该方法是在 Executor.run 任务执行完成时发送消息，通知 DAGScheduler 等调度器更新状态，具体实现如下：

```
  case smt: ShuffleMapTask =>
    val shuffleStage = stage.asInstanceOf[ShuffleMapStage]
    val status = event.result.asInstanceOf[MapStatus]
    val execId = status.location.executorId
    ……

    //如果当前调度阶段处在运行调度阶段列表中，并且没有任务处于挂起状态（均已完成），则标记
    //该调度阶段已经完成并注册输出结果的位置
```

```
if (runningStages.contains(shuffleStage)&&shuffleStage.pendingTasks.isEmpty){
  markStageAsFinished(shuffleStage)
  mapOutputTracker.registerMapOutputs(shuffleStage.shuffleDep.shuffleId,
    shuffleStage.outputLocs.map(list => if (list.isEmpty) null else
      list.head),changeEpoch = true)
  clearCacheLocs()

  if (shuffleStage.outputLocs.contains(Nil)) {
    //当调度阶段中存在部分任务执行失败,则重新提交运行
    submitStage(shuffleStage)
  } else {
    //当该调度阶段没有等待运行的任务,则设置该调度阶状态为完成
    if (shuffleStage.mapStageJobs.nonEmpty) {
      val stats = mapOutputTracker.getStatistics(shuffleStage.shuffleDep)
      for (job <- shuffleStage.mapStageJobs) {
        markMapStageJobAsFinished(job, stats)
      }
    }
  }
}
```

在前一节例子中对作业划分调度阶段,在本节中将按顺序提交调度阶段运行,具体步骤如图 4-8 所示。

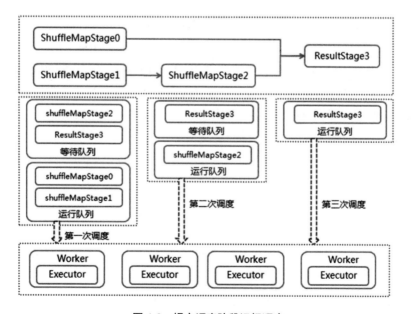

图 4-8 提交调度阶段运行顺序

（1）在 handleJobSubmitted 方法中获取了该例子最后一个调度阶段 ResultStage3，通过 submitStage 方法提交运行该调度阶段。

（2）在 submitStage 方法中，先创建作业实例，然后判断该调度阶段是否存在父调度阶段，由于 ResultStage3 有两个父调度阶段 ShuffleMapStage0 和 ShuffleMapStage2，所以并不能立即提交调度阶段运行，把 ResultStage3 加入到等待执行调度阶段列表 waitingStages 中。

（3）递归调用 submitStage 方法可以知道 ShuffleMapStage0 不存在父调度阶段，而 ShuffleMapStage2 存在父调度阶段 ShuffleMapStage1，这样 ShuffleMapStage2 加入到等待执行调度阶段列表 waitingStages 中，而 ShuffleMapStage0 和 ShtffleMapStage1 两个调度阶段作为第一次调度使用 submitMissingTasks 方法提交运行。

（4）Executor 任务执行完成时发送消息，通知 DAGScheduler 等调度器更新状态时，检查调度阶段运行情况，如果存在执行失败的任务，则重新提交该调度阶段；如果所有任务完成，则继续提交调度阶段运行。由于 ResultStage3 的父调度阶段没有全部完成，在第二次调度只提交 ShuffleMapStage2 运行。

（5）当 ShuffleMapStage2 运行完毕后，此时 ResultStage3 的父调度阶段全部完成，提交该调度阶段运行。

4.2.5 提交任务

当调度阶段提交运行后，在 DAGScheduler 的 submitMissingTasks 方法中，会根据调度阶段 Partition 个数拆分对应个数任务，这些任务组成一个任务集提交到 TaskScheduler 进行处理。对于 ResultStage（作业中最后的调度阶段）生成 ResultTask，对于 ShuffleMapStage 生成 ShuffleMapTask。对于每一个任务集包含了对应调度阶段的所有任务，这些任务处理逻辑完全一样，不同的是对应处理的数据，而这些数据是其对应的数据分片（Partition）。DAGScheduler 的 submitMissingTasks 方法具体代码如下：

```
private def submitMissingTasks(stage: Stage, jobId: Int) {
  ……
  val tasks: Seq[Task[_]] = try {
    stage match {
      //对于ShuffleMapStage生成ShuffleMapTask任务
      case stage: ShuffleMapStage =>
        partitionsToCompute.map { id =>
          val locs = taskIdToLocations(id)
          val part = stage.rdd.partitions(id)
          new ShuffleMapTask(stage.id, stage.latestInfo.attemptId, taskBinary,
           part, locs, stage.internalAccumulators)
```

```scala
        }
        //对于ResultStage生成ResultTask任务
        case stage: ResultStage =>
          val job = stage.resultOfJob.get
          partitionsToCompute.map { id =>
            val p: Int = job.partitions(id)
            val part = stage.rdd.partitions(p)
            val locs = taskIdToLocations(id)
            new ResultTask(stage.id, stage.latestInfo.attemptId, taskBinary, part,
              locs, id, stage.internalAccumulators)
          }
      }
    } catch {……}

    if (tasks.size > 0) {
      //把这些任务以任务集的方式提交到taskScheduler
      stage.pendingPartitions ++= tasks.map(_.partitionId)
      taskScheduler.submitTasks(new TaskSet(tasks.toArray, stage.id,
          stage.latestInfo.attemptId, jobId, properties))
      stage.latestInfo.submissionTime = Some(clock.getTimeMillis())
    } else {
      //如果调度阶段中不存在任务标记，则表示该调度阶段已经完成
      markStageAsFinished(stage, None)
      ……
    }
  }
```

当 TaskScheduler 收到发送过来的任务集时，在 submitTasks 方法中（在 TaskSchedulerImpl 类中进行实现）构建一个 TaskSetManager 的实例，用于管理这个任务集的生命周期，而该 TaskSetManager 会放入系统的调度池中，根据系统设置的调度算法进行调度。TaskSchedulerImpl.submitTasks 方法代码如下：

```scala
    override def submitTasks(taskSet: TaskSet) {
      val tasks = taskSet.tasks
      this.synchronized {
        //创建任务集的管理器，用于管理这个任务集的生命周期
        val manager = createTaskSetManager(taskSet, maxTaskFailures)
        val stage = taskSet.stageId
        val stageTaskSets = taskSetsByStageIdAndAttempt.getOrElseUpdate(stage, new
          HashMap[Int, TaskSetManager])
        stageTaskSets(taskSet.stageAttemptId) = manager
      }
      val conflictingTaskSet = stageTaskSets.exists { case (_, ts) =>
```

```
        ts.taskSet != taskSet && !ts.isZombie
    }

    //将该任务集的管理器加入到系统调度池中,由系统统一调配,该调度器属于应用级别,
    //支持FIFO和FAIR(公平调度)两种
    schedulableBuilder.addTaskSetManager(manager, manager.taskSet.properties)
    ……
}

    //调用调度器后台进程SparkDeploySchedulerBackend的reviveOffers方法分配资源并运行
    backend.reviveOffers()
}
```

参见 Spark 作业执行类图,SparkDeploySchedulerBackend 的 reviveOffers 方法是继承于父类 CoarseGrainedSchedulerBackend,该方法会向 DriverEndPoint 终端点发送消息,调用其 makeOffers 方法。在该方法中先会获取集群中可用的 Executor,然后发送到 TaskSchedulerImpl 中进行对任务集的任务分配运行资源,最后提交到 launchTasks 方法中。其中 CoarseGrainedSchedulerBackend.makeOffers 代码如下:

```
private def makeOffers() {
    //获取集群中可用的Executor列表
    val activeExecutors = executorDataMap.filterKeys(!executorsPendingToRemove.
        contains(_))
    val workOffers = activeExecutors.map { case (id, executorData) =>
        new WorkerOffer(id, executorData.executorHost, executorData.freeCores)
    }.toSeq
    //对任务集的任务分配运行资源,并把这些任务提交运行
    launchTasks(scheduler.resourceOffers(workOffers))
}
```

在 TaskSchedulerImpl 的 resourceOffers 方法中进行非常重要的步骤——资源分配,在分配的过程中会根据调度策略对 TaskSetManager 进行排序,然后依次对这些 TaskSetManager 按照就近原则分配资源,按照顺序为 PROCESS_LOCAL、NODE_LOCAL、NO_PREF、RACK_LOCAL 和 ANY,具体实现代码如下:

```
def resourceOffers(offers: Seq[WorkerOffer]): Seq[Seq[TaskDescription]] =
    synchronized {
    //对传入的可用Executor列表进行处理,记录其信息,如有新的Executor加入,则进行标记
    var newExecAvail = false
    for (o <- offers) {
        executorIdToHost(o.executorId) = o.host
        executorIdToTaskCount.getOrElseUpdate(o.executorId, 0)
        if (!executorsByHost.contains(o.host)) {
            executorsByHost(o.host) = new HashSet[String]()
```

```
      executorAdded(o.executorId, o.host)
      newExecAvail = true
    }
    for (rack <- getRackForHost(o.host)) {
      hostsByRack.getOrElseUpdate(rack, new HashSet[String]()) += o.host
    }
  }

  //为任务随机分配Executor，避免将任务集中分配到Worker上
  val shuffledOffers = Random.shuffle(offers)

  //用于存储分配好资源的任务
  val tasks = shuffledOffers.map(o => new ArrayBuffer[TaskDescription](o.cores))
  val availableCpus = shuffledOffers.map(o => o.cores).toArray

  //获取按照调度策略排序好的TaskSetManager
  val sortedTaskSets = rootPool.getSortedTaskSetQueue

  //如果有新加入的Executor，需要重新计算数据本地性
  for (taskSet <- sortedTaskSets) {
    if (newExecAvail) {
      taskSet.executorAdded()
    }
  }

  //为排好序的 TaskSetManager 列表进行分配资源，分配的原则是就近原则，按照顺序为
    PROCESS_LOCAL, NODE_LOCAL, NO_PREF, RACK_LOCAL, ANY
  var launchedTask = false
  for (taskSet <- sortedTaskSets; maxLocality <- taskSet.myLocalityLevels) {
    do {
      launchedTask = resourceOfferSingleTaskSet(taskSet, maxLocality,
        shuffledOffers, availableCpus, tasks)
    } while (launchedTask)
  }

  if (tasks.size > 0) {
    hasLaunchedTask = true
  }
  return tasks
}
```

分配好资源的任务提交到 CoarseGrainedSchedulerBackend 的 launchTasks 方法中，在该方法中会把任务一个个发送到 Worker 节点上的 CoarseGrainedExecutorBackend，然后通过其内部的

Executor 来执行任务，具体代码如下：

```
private def launchTasks(tasks: Seq[Seq[TaskDescription]]) {
  for (task <- tasks.flatten) {
    //序列化每一个 task
    val serializedTask = ser.serialize(task)
    if (serializedTask.limit >= maxRpcMessageSize) {
      scheduler.taskIdToTaskSetManager.get(task.taskId).foreach { taskSetMgr =>
        try {
          taskSetMgr.abort(msg)
        } catch {
          case e: Exception => logError("Exception in error callback", e)
        }
      }
    }else {
      val executorData = executorDataMap(task.executorId)
      executorData.freeCores -= scheduler.CPUS_PER_TASK
      //向 Worker 节点的 CoarseGrainedExecutorBackend 发送消息执行 Task
      executorData.executorEndpoint.send(LaunchTask(new
        SerializableBuffer(serializedTask)))
    }
  }
}
```

在前一节例子中按顺序提交调度阶段，在本节中对这些调度阶段拆分成任务集，然后对这些任务分配资源，并提交 Executor 运行，具体步骤如图 4-9 所示。

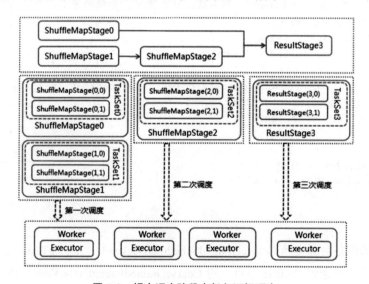

图 4-9　提交调度阶段中任务运行顺序

（1）在提交调度阶段中，第一次调度的是 ShuffleMapStage0 和 ShuffleMapStage1，这两个调度阶段在 DAGScheduler 的 submitMissingTasks 方法中，会根据 Partition 个数拆分任务。由于这两个调度阶段分别有两个分片，ShuffleMapStage0 分别拆分成 ShuffleMapStage(0,0) 和 ShuffleMapStage(0,1) 两个任务，这两个任务组成任务集 TaskSet0，而 ShuffleMapStage1 分别拆分成 ShuffleMapStage(1,0) 和 ShuffleMapStage(1,1) 两个任务，这两个任务组成任务集 TaskSet1。

（2）TaskScheduler 收到发送过来的任务集 TaskSet0 和 TaskSet1，在 submitTasks 方法中分别构建 TaskSetManager 的实例 TaskSetManager0 和 TaskSetManager1 进行管理，这两个 TaskSetManager 放到系统的调度池中，根据系统设置的调度算法进行调度。

（3）在 TaskSchedulerImpl 的 resourceOffers 方法中按照就近原则进行资源分配，到该步骤每个任务均分配运行代码、数据分片和处理资源等，使用 CoarseGrainedSchedulerBackend 的 launchTasks 方法把任务发送到 Worker 节点上的 CoarseGrainedExecutorBackend 调用其 Executor 来执行任务。

（4）当 ShuffleMapStage0 和 ShuffleMapStage1 执行完毕后，随后的 ShuffleMapStage2 和 ResultStage3 会按照步骤 1～3 运行，不同的是，ResultStage3 生成的任务类型是 ResultTask。

4.2.6　执行任务

当 CoarseGrainedExecutorBackend 接收到 LaunchTask 消息时，会调用 Executor 的 launchTask 方法进行处理。在 Executor 的 launchTask 方法中，初始化一个 TaskRunner 来封装任务，它用于管理任务运行时的细节，再把 TaskRunner 对象放入到 ThreadPool（线程池）中去执行。

在 TaskRunner 的 run 方法里，首先会对发送过来的 Task 本身以及它所依赖的 Jar 等文件的反序列，然后对反序列化的任务调用 Task 的 runTask 方法。由于 Task 本身是一个抽象类，具体的 runTask 方法是由它的两个子类 ShuffleMapTask 和 RedultTask 来实现的。

具体任务执行在 TaskRunner 的 run 方法前半部分实现，代码如下：

```
override def run(): Unit = {
  //生成内存管理taskMemoryManager实例，用于任务运行期间内存管理
  val taskMemoryManager = new TaskMemoryManager(env.memoryManager, taskId)
  val deserializeStartTime = System.currentTimeMillis()
  Thread.currentThread.setContextClassLoader(replClassLoader)
  val ser = env.closureSerializer.newInstance()

  //向Driver终端点发送任务运行开始消息
  execBackend.statusUpdate(taskId, TaskState.RUNNING, EMPTY_BYTE_BUFFER)
  var taskStart: Long = 0
  startGCTime = computeTotalGcTime()
```

```
try {
  //对任务运行时所需要的文件、Jar 包、代码等反序列化
  val (taskFiles, taskJars, taskBytes) =
    Task.deserializeWithDependencies(serializedTask)
  updateDependencies(taskFiles, taskJars)
  task = ser.deserialize[Task[Any]](taskBytes,
    Thread.currentThread.getContextClassLoader)
  task.setTaskMemoryManager(taskMemoryManager)

  //任务在反序列化之前被杀死,则抛出异常并退出
  if (killed) {
    throw new TaskKilledException
  }
  env.mapOutputTracker.updateEpoch(task.epoch)

  //调用 Task 的 runTask 方法,由于 Task 本身是一个抽象类,具体的 runTask 方法是由它的
  //两个子类 ShuffleMapTask 和 RedultTask 来实现的
  taskStart = System.currentTimeMillis()
  var threwException = true
  val value = try {
    val res = task.run(
      taskAttemptId = taskId,
      attemptNumber = attemptNumber,
      metricsSystem = env.metricsSystem)
    res
  }finally {……}
   ……
 }
}
```

对于 ShuffleMapTask 而言,它的计算结果会写到 BlockManager 之中,最终返回给 DAGScheduler 的是一个 MapStatus 对象。该对象中管理了 ShuffleMapTask 的运算结果存储到 BlockManager 里的相关存储信息,而不是计算结果本身,这些存储信息将会成为下一阶段的任务需要获得的输入数据时的依据。其中 ShuffleMapTask.runTask 代码如下:

```
override def runTask(context: TaskContext): MapStatus = {
  val deserializeStartTime = System.currentTimeMillis()
  //反序列化获取 RDD 和 RDD 的依赖
  val ser = SparkEnv.get.closureSerializer.newInstance()
  val (rdd, dep) = ser.deserialize[(RDD[_], ShuffleDependency[_, _, _])](
    ByteBuffer.wrap(taskBinary.value),Thread.currentThread.
      getContextClassLoader)
  _executorDeserializeTime = System.currentTimeMillis() - deserializeStartTime
```

```
  metrics = Some(context.taskMetrics)
  var writer: ShuffleWriter[Any, Any] = null
  try {
    val manager = SparkEnv.get.shuffleManager
    writer=manager.getWriter[Any, Any](dep.shuffleHandle,partitionId,context)
    //首先调用 rdd.iterator,如果该 RDD 已经 Cache 或 Checkpoint,那么直接读取结果,
    //否则计算,计算结果会保存在本地系统的 BlockManager 中
    writer.write(rdd.iterator(partition, context).asInstanceOf[Iterator[_ <:
      Product2[Any, Any]]])
    //关闭 writer,返回计算结果,返回包含了数据的 location 和 size 等元数据信息的
    //MapStatus 信息
    writer.stop(success = true).get
  } catch {……}
}
```

对于 ResultTask 的 runTask 方法而言，它最终返回的是 func 函数的计算结果。

```
override def runTask(context: TaskContext): U = {
  //反序列化广播变量得到 RDD
  val deserializeStartTime = System.currentTimeMillis()
  val ser = SparkEnv.get.closureSerializer.newInstance()
  val (rdd, func) = ser.deserialize[(RDD[T], (TaskContext, Iterator[T]) => U)](
    ByteBuffer.wrap(taskBinary.value), Thread.currentThread.getContextClassLoader)
  _executorDeserializeTime = System.currentTimeMillis() - deserializeStartTime

  metrics = Some(context.taskMetrics)
  //ResultTask 的 runTask 方法返回的是计算结果
  func(context, rdd.iterator(partition, context))
}
```

4.2.7 获取执行结果

对于 Executor 的计算结果，会根据结果的大小有不同的策略。

（1）生成结果大小在(∞, 1GB)：结果直接丢弃，该配置项可以通过 spark.driver.maxResultSize 进行设置。

（2）生成结果大小在[1GB，128MB-200KB]：如果生成的结果大于等于（128MB-200KB）时，会把该结果以 taskId 为编号存入到 BlockManager 中，然后把该编号通过 Netty 发送给 Driver 终端点，该阈值是 Netty 框架传输的最大值 spark.akka.frameSize(默认为 128MB)和 Netty 的预留空间 reservedSizeBytes（200KB）差值。

（3）生成结果大小在(128MB-200KB，0)：通过 Netty 直接发送到 Driver 终端点。

具体任务执行在 TaskRunner 的 run 方法后半部分实现，代码如下：

```
override def run(): Unit = {
  try {
    //执行任务…….
    val taskFinish = System.currentTimeMillis()
    if (task.killed) {
      throw new TaskKilledException
    }

    //对生成的结果序列化，并把结果放到 DirectTaskResult 中
    val valueBytes = resultSer.serialize(value)
    val directResult = new DirectTaskResult(valueBytes, accumUpdates,
      task.metrics.orNull)
    val serializedDirectResult = ser.serialize(directResult)
    val resultSize = serializedDirectResult.limit

    //对生成的结果序列化，并把结果放到 DirectTaskResult 中
    val resultSer = env.serializer.newInstance()
    val beforeSerialization = System.currentTimeMillis()
    val valueBytes = resultSer.serialize(value)
    val afterSerialization = System.currentTimeMillis()
    ……
    val accumUpdates = task.collectAccumulatorUpdates()
    val directResult = new DirectTaskResult(valueBytes, accumUpdates)
    val serializedDirectResult = ser.serialize(directResult)
    val resultSize = serializedDirectResult.limit

    // directSend = sending directly back to the driver
    val serializedResult: ByteBuffer = {
      if (maxResultSize > 0 && resultSize > maxResultSize) {
        //生成结果序列化结果大于最大值（默认为1GB）直接丢弃
        ser.serialize(new   IndirectTaskResult[Any](TaskResultBlockId(taskId),
          resultSize))
      } else if (resultSize >= maxDirectResultSize) {
        //生成结果序列化结果在[1GB, 128MB-200KB]之间，存放到 BlockManager 中，
        //然后把该编号通过 Netty 发送给 Driver 终端点
        val blockId = TaskResultBlockId(taskId)
          env.blockManager.putBytes(blockId,newChunkedByteBuffer(
          serializedDirectResult.duplicate()),StorageLevel.MEMORY_AND_DISK_SER)
          ser.serialize(new IndirectTaskResult[Any](blockId, resultSize))
      } else {
        //通过 Netty 直接发送到 Driver 终端点
        serializedDirectResult
```

```
      }
    }
    //向 Driver 终端点发送任务运行完毕消息
    execBackend.statusUpdate(taskId, TaskState.FINISHED, serializedResult)
  }
}
```

任务执行完毕后，TaskRunner 将任务的执行结果发送给 DriverEndPoint 终端点。该终端点会转给 TaskSchedulerImpl 的 statusUpdate 方法进行处理，在该方法中对于不同的任务状态有不同的处理。

（1）如果类型是 TaskState.FINISHED，那么调用 TaskResultGetter 的 enqueueSuccessfulTask 方法进行处理。enqueueSuccessfulTask 方法的逻辑也比较简单，就是如果是 IndirectTaskResult，那么需要通过 blockid 来获取结果：sparkEnv.blockManager.getRemoteBytes(blockId)；如果是 DirectTaskResult，那么结果就无需远程获取了。

（2）如果类型是 TaskState.FAILED 或者 TaskState.KILLED 或者 TaskState.LOST，调用 TaskResultGetter 的 enqueueFailedTask 进行处理。对于 TaskState.LOST，还需要将其所在的 Executor 标记为 failed，并且根据更新后的 Executor 重新调度。

TaskSchedulerImpl 的 handleSuccessfulTask 方法中连续调用（如图 4-10 所示），最终调用 DAGScheduler 的 handleTaskCompletion 方法。

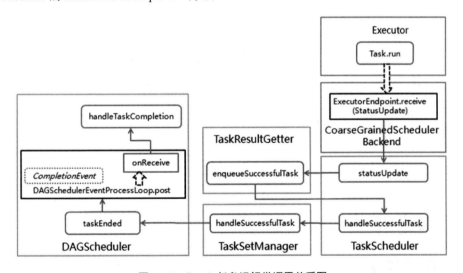

图 4-10　Spark 任务运行类调用关系图

如果任务是 ShuffleMapTask，那么它需要将结果通过某种机制告诉下游的调度阶段，以便其可以作为后续调度阶段的输入，这个机制是怎么实现的？实际上，对于 ShuffleMapTask 来说，其结果实际上是 MapStatus；其序列化后存入了 DirectTaskResult 或者 IndirectTaskResult 中。而 DAGScheduler 的 handleTaskCompletion 方法获取这个结果，并把这个 MapStatus 注册到 MapOutputTrackerMaster 中，从而完成 ShuffleMapTask 的处理。

```
case smt: ShuffleMapTask =>
  val shuffleStage = stage.asInstanceOf[ShuffleMapStage]
  updateAccumulators(event)
  ……
  mapOutputTracker.registerMapOutputs(
    shuffleStage.shuffleDep.shuffleId,
    shuffleStage.outputLocs.map(list   =>   if   (list.isEmpty)   null   else
      list.head),changeEpoch = true)
```

如果任务是 ResultTask，判断该作业是否完成，如果完成，则标记该作业已经完成，清除作业依赖的资源并发送消息给系统监听总线告知作业执行完毕。

```
case rt: ResultTask[_, _] =>
val resultStage = stage.asInstanceOf[ResultStage]
resultStage.resultOfJob match {
  case Some(job) =>
    if (!job.finished(rt.outputId)) {
      updateAccumulators(event)
      job.finished(rt.outputId) = true
      job.numFinished += 1
      //如果作业执行完毕，标记该作业已经完成，清除作业依赖的资源并发送消息给系统消息总线
      //告知作业执行完毕
      if (job.numFinished == job.numPartitions) {
        markStageAsFinished(resultStage)
        cleanupStateForJobAndIndependentStages(job)
        listenerBus.post(SparkListenerJobEnd(job.jobId, clock.getTimeMillis(),
          JobSucceeded))
      }
    }
  }
}
```

4.3 调度算法

4.3.1 应用程序之间

前面介绍了在 Spark 集群中，ClusterManager 提供了资源的分配和管理，而在独立运行模式中 Master 提供了资源管理调度功能。在调度过程中，Master 先启动等待列表中应用程序的 Driver，

这些 Driver 尽可能分散在集群的 Worker 节点上，然后根据集群的内存和 CPU 使用情况，对等待运行的应用程序进行资源分配，在分配算法上根据先来先分配，先分配的应用程序会尽可能多地获取满足条件的资源，后分配的应用程序只能在剩余资源中再次筛选。如果没有合适资源的应用程序只能等待，直到其他应用程序释放。该策略可以认为是有条件的 FIFO 策略。该调度在 Master.schedule 方法实现，代码如下：

```
private def schedule(): Unit = {
  if (state != RecoveryState.ALIVE) { return }

  //对 Worker 节点进行随机排序，能够使 Driver 更加均衡分布在集群中
  val shuffledAliveWorkers = Random.shuffle(workers.toSeq.filter(_.state ==
    WorkerState.ALIVE))
  val numWorkersAlive = shuffledAliveWorkers.size
  var curPos = 0

  //按照顺序在集群中启动 Driver, Driver 尽可能在不同的 Worker 节点上运行
  for (driver <- waitingDrivers.toList) {
    var launched = false
    var numWorkersVisited = 0
    while (numWorkersVisited < numWorkersAlive && !launched) {
      val worker = shuffledAliveWorkers(curPos)
      numWorkersVisited += 1
      if (worker.memoryFree >= driver.desc.mem && worker.coresFree >=
        driver.desc.cores) {
        launchDriver(worker, driver)
        waitingDrivers -= driver
        launched = true
      }
      curPos = (curPos + 1) % numWorkersAlive
    }
  }
  //对等待的应用程序按照顺序分配运行资源
  startExecutorsOnWorkers()
}
```

默认情况下，在独立运行模式中每个应用程序所能分配到的 CPU 核数可以由 spark.deploy.defaultCores 进行设置，但是该配置项默认情况为 Int.Max，也就是不限制，从而应用程序会尽可能获取 CPU 资源。为了限制每个应用程序使用 CPU 资源，用户一方面可以设置 spark.cores.max 配置项，约束每个应用程序所能申请最大的 CPU 核数；另一方面可以设置 spark.executor.cores 配置项，用于设置在每个 Executor 上启动的 CPU 核数。

另外在分配应用程序资源的时候，会根据 Worker 的分配策略进行。分配算法有两种：一种

是把应用程序运行在尽可能多的 Worker 上,这种分配算法不仅能够充分使用集群资源,而且有利于数据处理的本地性;相反,另一种是应用程序运行在尽可能少的 Worker 上,该情况适合 CPU 密集型而内存使用较少的场景。该策略由 spark.deploy.spreadOut 配置项进行设置,默认情况下为 true,也就是第一种尽可能分散运行。代码由 Master. scheduleExecutorsOnWorkers 方法实现,代码如下:

```
//该方法返回集群中 Worker 节点所能提供 CPU 核数数组
private def scheduleExecutorsOnWorkers(app: ApplicationInfo,
    usableWorkers: Array[WorkerInfo],spreadOutApps: Boolean): Array[Int] = {
  //应用程序中每个 Executor 所需要 CPU 核数
  val coresPerExecutor = app.desc.coresPerExecutor

  //每次分配时分配给 Executor 所需的最少 CPU 核数,如果为应用程序设置了每个 Executor
  //所需要 CPU 核数,则为该值,否则默认为 1
  val minCoresPerExecutor = coresPerExecutor.getOrElse(1)

  //如果没有设置,则表示该应用程序在 Worker 节点上只启动 1 个 Executor,并尽可能分配资源
  val oneExecutorPerWorker = coresPerExecutor.isEmpty
  val memoryPerExecutor = app.desc.memoryPerExecutorMB

  val numUsable=usableWorkers.length              //集群中可用 Worker 节点数
  val assignedCores=new Array[Int](numUsable)     //Worker 节点所能提供 CPU 核数数组
  val assignedExecutors=new Array[Int](numUsable)//Worker 分配 Executor 个数数组

  //需要分配 CPU 核数,为应用程序所需 CPU 核数和可用 CPU 核数最小值
  var coresToAssign=math.min(app.coresLeft,usableWorkers.map(_.coresFree).sum)

  //返回指定的 Worker 节点是否能够启动 Executor,满足的条件为:
  //  1)应用程序需要分配 CPU 核数>=每个 Executor 所需的最少 CPU 核数。
  //  2)是否有足够的 CPU 核数,判断条件为 Worker 节点可用 CPU 核数-该 Worker 节点已分配的
  //     CPU 核数>=每个 Executor 所需的最少 CPU 核数。
  //如果在该 Worker 节点上允许启动新的 Executor,需要追加下面两个条件:
  //  1)判断内存是否足够启动 Executor,其方法是:当前 Worker 节点可用内存-该 Worker 节点
  //     已经分配的内存>=每个 Executor 分配的内存大小,其中已经分配的内存为
  //     已分配的 Executor 数乘以每个 Executor 所分配的内存数。
  //  2)已经分配给该应用程序的 Executor 数量+已经运行该应用程序的 Executor 数量<
  //     该应用程序 Executor 设置的最大数量。
  def canLaunchExecutor(pos: Int): Boolean = {
    val keepScheduling = coresToAssign >= minCoresPerExecutor
    val enoughCores = usableWorkers(pos).coresFree - assignedCores(pos) >=
      minCoresPerExecutor
```

```scala
//启动新的 Executor 条件为：该 Worker 节点允许启动多个 Executor 或者在该 Worker 节点
//没有为该应用程序分配 Executor。其中允许启动多个 Executor 判断条件为是否设置
//每个 Executor 所需 CPU 核数，如果没有设置则表示允许，否则在该 Worker 节点只允许
//启动 1 个 Executor
val launchingNewExecutor = !oneExecutorPerWorker || assignedExecutors(pos) == 0
//如果在该 Worker 节点上允许启动多个 Executor，那么该 Worker 节点满足启动条件就可以
//启动新的 Executor，否则该 Worker 节点只能启动一个 Executor 并尽可能多分配 CPU 核数。
if (launchingNewExecutor) {
  val assignedMemory = assignedExecutors(pos) * memoryPerExecutor
  val enoughMemory = usableWorkers(pos).memoryFree - assignedMemory >=
    memoryPerExecutor
  val underLimit=assignedExecutors.sum+app.executors.size<app.executorLimit
  keepScheduling && enoughCores && enoughMemory && underLimit
} else {
  keepScheduling && enoughCores
}
}

//在可用的 Worker 节点中启动 Executor，在 Worker 节点每次分配资源时，分配给 Executor
//所需的最少 CPU 核数，该过程是通过多次轮询进行，直到没有 Worker 节点能够没有满足启动
//Executor 条件或者已经达到应用程序限制。在分配过程中 Worker 节点可能多次分配，
//如果该 Worker 节点可以启动多个 Executor,则每次分配的时候启动新的 Executor 并赋予资源；
//如果 Worker 节点只能启动一个 Executor，则每次分配的时候把资源追加到该 Executor
var freeWorkers = (0 until numUsable).filter(canLaunchExecutor)
while (freeWorkers.nonEmpty) {
  freeWorkers.foreach { pos =>
    var keepScheduling = true

    //满足 keepScheduling 标志为真（第一次分配或者集中运行）和该 Worker 节点满足
    //启动 Executor 条件时，进行资源分配
    while (keepScheduling && canLaunchExecutor(pos)) {
      //每次分配 CPU 核数为 Executor 所需的最少 CPU 核数
      coresToAssign -= minCoresPerExecutor
      assignedCores(pos) += minCoresPerExecutor

      //如果未设置每个 Executor 启动 CPU 核数，则该 Worker 只能为该应用程序
      //启动 1 个 Executor，否则在每次分配中启动 1 个新的 Executor
      if (oneExecutorPerWorker) {
        assignedExecutors(pos) = 1
      } else {
        assignedExecutors(pos) += 1
      }

      //如果是分散运行，则在某一 Worker 节点上做完资源分配立即移到下一个 Worker 节点，
```

```
            //如果是集中运行,则持续在某一Worker节点上做资源分配,直到使用完该Worker节点
            //所有资源。由于传入的Worker列表是按照可用CPU核数倒序排列,在集中运行时,
            //会尽可能少的使用Worker节点。
            if (spreadOutApps) {
              keepScheduling = false
            }
          }
        }
        //继续从上次分配完的可用Worker节点列表获取满足启动Executor的Worker节点列表
        freeWorkers = freeWorkers.filter(canLaunchExecutor)
      }
      assignedCores
    }
```

该算法在 Spark 1.4.2 版本中进行了优化:

(1) 在 Spark 1.4.2 及以前版本中,在 Worker 节点中只能为某应用程序启动 1 个 Executor,在后续版本中,在 Worker 节点中可以为某应用程序启动多个 Executor。

(2) 在 Spark 1.4.2 及以前版本中,轮询分配资源时,Worker 节点每次分配 1 个 CPU 核数,这样有可能会造成某 Worker 节点最终分配 CPU 核数小于每个 Executor 所需 CPU 核数,那么该 Worker 节点将不启动该 Executor。例如,在某一集群中有 4 个 Worker 节点,每个节点拥有 16 个 CPU 核数,其中设置了 spark.cores.max = 48 和 spark.executor.cores = 16,按照每次分配 1 个 CPU 核数,则每个 Worker 节点的 Executor 将分配到 12 个 CPU 核数,由于 12<16,没有满足 Executor 条件,则没有 Executor 能够启动,参见 SPARK-8881 问题说明。而在后续版本中,每次分配 CPU 核数为 Executor 指定的 CPU 核数,如果没有指定默认情况为 1,这样在前面的例子中,按照该分配方式将在 3 个 Worker 节点中的 Executor 分配 16 个 CPU 核数,这样就能够正常启动 Executor。

4.3.2 作业及调度阶段之间

Spark 应用程序提交执行时,会根据 RDD 依赖关系形成有向无环图(DAG),然后交给 DAGScheduler 进行划分作业和调度阶段。这些作业之间可以没有任何依赖关系,对于多个作业之间的调度,Spark 目前提供了两种调度策略:一种是 FIFO 模式,这也是目前默认的模式;另一种是 FAIR 模式,该模式的调度可以通过两个参数的配置来决定 Job 执行的优先模式,两个参数分别是 minShare(最小任务数)和 weight(任务的权重)。该调度策略执行过程如下:

1. 创建调度池

在 TaskSchedulerImpl.initialize 方法中先创建根调度池 rootPool 对象,然后根据系统配置的

调度模式创建调度创建器，针对两种调度策略具体实例化 FIFOSchedulableBuilder 或 FairSchedulableBuilder，最终使用调度创建器的 buildPools 方法在根调度池 rootPool 下创建调度池。

```
def initialize(backend: SchedulerBackend) {
  this.backend = backend
  rootPool = new Pool("", schedulingMode, 0, 0)
  schedulableBuilder = {
  //根据调度模式配置调度池
    schedulingMode match {
      //使用 FIFO 调度方式
      case SchedulingMode.FIFO =>
        new FIFOSchedulableBuilder(rootPool)
      //使用 FAIR 调度方式
      case SchedulingMode.FAIR =>
        new FairSchedulableBuilder(rootPool, conf)
    }
  }
  schedulableBuilder.buildPools()
}
```

2. 调度池加入调度内容

在 TaskSchedulerImpl.submitTasks 方法中，先把调度阶段拆分为任务集，然后把这些任务集交给管理器 TaskSetManager 进行管理，最后把该任务集的管理器加入到调度池中，等待分配执行。在 FIFO 中，由于创建器的 buildPools 方法为空，所以在根调度池 rootPool 中并没有下级调度池，而是直接包含了一组 TaskSetManager；而在 FAIR 中，根调度池 rootPool 中包含了下级调度池 Pool，在这些下级调度池 Pool 包含一组 TaskSetManager。

```
override def submitTasks(taskSet: TaskSet) {
  val tasks = taskSet.tasks
  this.synchronized {
    //创建任务集的管理器，用于管理这个任务集的生命周期
    val manager = createTaskSetManager(taskSet, maxTaskFailures)
    val stage = taskSet.stageId
    val stageTaskSets = taskSetsByStageIdAndAttempt.getOrElseUpdate(stage, new
      HashMap[Int, TaskSetManager])
    stageTaskSets(taskSet.stageAttemptId) = manager
  }

    val conflictingTaskSet = stageTaskSets.exists { case (_, ts) =>
      ts.taskSet != taskSet && !ts.isZombie
    }
```

```
    if (conflictingTaskSet) {
      throw new IllegalStateException(s"more than one active taskSet for stage
      $stage:"+s" ${stageTaskSets.toSeq.map{_._2.taskSet.id}.mkString(",")}")
    }

    //将该任务集的管理器加入到系统调度池中，由系统统一调配，该调度器属于应用级别，
    //支持 FIFO 和 FAIR 两种
    schedulableBuilder.addTaskSetManager(manager, manager.taskSet.properties)
    ......
  }
  ......
}
```

3. 提供已排序的任务集管理器

在 TaskSchedulerImpl.resourceOffers 方法中进行资源分配时，会从根调度池 rootPool 获取已经排序的任务管理器，该排序算法由两种调度策略 FIFOSchedulingAlgorithm 和 FairSchedulingAlgorithm 的 comparator 方法提供。

```
def resourceOffers(offers: Seq[WorkerOffer]): Seq[Seq[TaskDescription]] =
  synchronized {
    ......
    //获取按照调度策略排序好的 TaskSetManager
    val sortedTaskSets = rootPool.getSortedTaskSetQueue
    ......
  }
```

在 FIFO 调度策略中，由于根调度池 rootPool 直接包含了多个作业的任务管理器，在比较时，首先需要比较作业优先级（根据作业编号判断，作业编号越小优先级越高），如果是同一个作业，会再比较调度阶段优先级（根据调度阶段编号判断，调度阶段编号越小优先级越高）。其中 FIFOSchedulingAlgorithm.comparator 代码如下：

```
override def comparator(s1: Schedulable, s2: Schedulable): Boolean = {
  //获取作业优先级，实际上是作业编号
  val priority1 = s1.priority
  val priority2 = s2.priority
  var res = math.signum(priority1 - priority2)

  //如果是同一个作业，再比较调度阶段优先级
  if (res == 0) {
    val stageId1 = s1.stageId
    val stageId2 = s2.stageId
    res = math.signum(stageId1 - stageId2)
  }
```

```
    if (res < 0) {
      true
    } else {
      false
    }
}
```

在FAIR调度策略中包含了两层调度,第一层中根调度池rootPool中包含了下级调度池Pool,第二层为下级调度池 Pool 包含多个 TaskSetManager。具体配置参见$SPARK_HOME/conf/fairscheduler.xml 文件,在该文件中包含多个下级调度池 Pool 配置项,其中 minShare(最小任务数)和 weight(任务的权重)两个参数是用来设置第一级调度算法,而 schedulingMode 参数是用来设置第二层调度算法。配置文件内容如下:

```
<?xml version="1.0"?>
<allocations>
  <pool name="ProductA">
    <schedulingMode>FIFO</schedulingMode>
    <weight>2</weight>
    <minShare>3</minShare>
  </pool>
  <pool name="ProductB">
    <schedulingMode>FAIR</schedulingMode>
    <weight>1</weight>
    <minShare>2</minShare>
  </pool>
</allocations>
```

在 FAIR 算法中,先获取两个调度的饥饿程度,饥饿程度为正在运行的任务是否小于最小任务,如果是,则表示该调度处于饥饿程度。获取饥饿程度后进行如下比较:

- 如果某个调度处于饥饿状态另外一个非饥饿状态,则先满足处于饥饿状态的调度;
- 如果两个调度都处于饥饿状态,则比较资源比,先满足资源比小的调度;
- 如果两个调度都处于非饥饿状态,则比较权重比,先满足权重比小的调度;
- 以上情况均相同的情况,根据调度的名称排序。

其中 FAIR 算法代码实现为 FairSchedulingAlgorithm.comparator,内容如下:

```
//比较两个调度优先级方法,返回 true 表示前者优先级高,false 表示后者优先级高
override def comparator(s1: Schedulable, s2: Schedulable): Boolean = {
  //最小任务数
  val minShare1 = s1.minShare
  val minShare2 = s2.minShare
  //正在运行的任务数
  val runningTasks1 = s1.runningTasks
```

```
val runningTasks2 = s2.runningTasks
//饥饿程序,判断标准为正在运行的任务数是否小于最小任务数
val s1Needy = runningTasks1 < minShare1
val s2Needy = runningTasks2 < minShare2
//资源比,正在运行的任务数/最小任务数
val minShareRatio1=runningTasks1.toDouble / math.max(minShare1, 1.0).toDouble
val minShareRatio2=runningTasks2.toDouble / math.max(minShare2, 1.0).toDouble
//权重比,正在运行的任务数/任务的权重
val taskToWeightRatio1 = runningTasks1.toDouble / s1.weight.toDouble
val taskToWeightRatio2 = runningTasks2.toDouble / s2.weight.toDouble
var compare: Int = 0

//判断执行
if (s1Needy && !s2Needy) {
  return true
} else if (!s1Needy && s2Needy) {
  return false
} else if (s1Needy && s2Needy) {
  compare = minShareRatio1.compareTo(minShareRatio2)
} else {
  compare = taskToWeightRatio1.compareTo(taskToWeightRatio2)
}

if (compare < 0) {
  true
} else if (compare > 0) {
  false
} else {
  s1.name < s2.name
}
}
```

4.3.3 任务之间

在介绍任务间调度算法之前,我们先详细了解数据本地性和延迟执行两个概念。

1. 数据本地性

数据的计算尽可能在数据所在的节点上进行,这样可以减少数据在网络上传输,毕竟移动计算比移动数据代价来得小些。进一步看,数据如果在运行节点的内存中,就能够进一步减少磁盘 I/O 的传输。在 Spark 中数据本地性优先级从高到低为 PROCESS_LOCAL>NODE_LOCAL>NO_PREF>RACK_LOCAL>ANY,即最好是任务运行的节点内存中存在数据、次好是

同一个 Node（即同一机器）上，再次是同机架，最后是任意位置。其中任务数据本地性通过以下情况来确定：

- 如果任务处于作业开始的调度阶段内，这些任务对应的 RDD 分区都有首选运行位置，该位置也是任务运行首选位置，数据本地性为 NODE_LOCAL。
- 如果任务处于非作业开头的调度阶段，可以根据父调度阶段运行的位置得到任务的首选位置，这种情况下，如果 Executor 处于活动状态，则数据本地性为 PROCESS_LOCAL；如果 Executor 不处于活动状态，但存在父调度阶段运行结果，则数据本地性为 NODE_LOCAL。
- 如果没有首选位置，则数据本地性为 NO_PREF。

2．延迟执行

在任务分配运行节点时，先判断任务最佳运行节点是否空闲，如果该节点没有足够的资源运行该任务，在这种情况下任务会等待一定时间；如果在等待时间内该节点释放出足够的资源，则任务在该节点运行，如果还是不足会找出次佳的节点进行运行。通过这样的方式进行能地让任务运行在更高级别数据本地性的节点，从而减少磁盘 I/O 和网络传输。一般来说只对 PROCESS_LOCAL 和 NODE_LOCAL 两个数据本地性级别进行等待，系统默认延迟时间为 3s。

Spark 任务分配的原则就是让任务运行在数据本地性优先级高的节点上，甚至可以为此等待一定的时间。该任务分配过程是由 TaskSchedulerImpl.resourceOffers 方法实现，在该方法中先对应用程序获取的资源（如 Worker 节点）进行混洗，以使任务能够更加均衡地分散在集群中运行，然后对任务集对应 TaskSetManager 根据设置的调度算法进行排序（在前一小节介绍），最后对 TaskSetManager 中的任务按照数据本地性分配任务运行节点，在分配时先根据任务集的本地性从优先级高到低进行分配任务，在分配的过程中动态地判断集群中节点运行的情况，通过延迟执行等待数据本地性更高的节点运行。

在任务分配中 TaskSetManager 是核心对象，先在其初始化的时使用 addPendingTask 方法，根据任务自身的首选位置得到 pendingTasksForExecutor、pendingTasksForHost、pendingTasksWithNoPrefs 和 pendingTasksForRack 4 个列表；然后根据这 4 个列表在 computeValidLocalityLevels 方法中得到该任务集的数据本地性列表，按照获取的数据本地性从高到低匹配可用的 Worker 节点，在匹配前使用 getAllowedLocalityLevel 方法得到数据集允许的数据本地性，比较该数据本地性和指定数据本地性优先级，取优先级高的数据本地性；最后在指定的 Worker 节点中判断比较获得的数据优先级是否存在需要运行的任务，如果存在则返回该任务和数据本地性进行相关信息更新处理。其中 resourceOffers 方法代码如下：

```
def resourceOffers(offers: Seq[WorkerOffer]): Seq[Seq[TaskDescription]] =
  synchronized {
  ……
  //为了负载均衡，打乱offers顺序，Random.shuffle用于将一个集合中的元素打乱
  val shuffledOffers = Random.shuffle(offers)

  //创建用于存储每个Worker所对应运行任务列表的Map
  val tasks = shuffledOffers.map(o => new ArrayBuffer[TaskDescription](o.cores))
  val availableCpus = shuffledOffers.map(o => o.cores).toArray

  //获取按照调度策略排序好的TaskSetManager
  val sortedTaskSets = rootPool.getSortedTaskSetQueue

  //如果有新加入的Executor，需要重新计算数据本地性
  for (taskSet <- sortedTaskSets) {
    logDebug("parentName: %s, name: %s, runningTasks: %s".format(taskSet.parent.
      name, taskSet.name, taskSet. runningTasks))
    if (newExecAvail) {
      taskSet.executorAdded()
    }
  }

  //为排好序的TaskSetManager列表进行分配资源，分配的原则是就近原则，
  //按照顺序为PROCESS_LOCAL, NODE_LOCAL, NO_PREF, RACK_LOCAL, ANY
  var launchedTask = false
  //通过任务集管理器的myLocalityLevels变量获取任务集中任务自身数据本地性列表
  for (taskSet <- sortedTaskSets; maxLocality <- taskSet.myLocalityLevels) {
    do {
      launchedTask  =  resourceOfferSingleTaskSet(taskSet,  maxLocality,
        shuffledOffers, availableCpus, tasks)
    } while (launchedTask)
  }
  if (tasks.size > 0) {
    hasLaunchedTask = true
  }
  return tasks
}
```

对单个任务集的任务调度由 TaskSchedulerImpl.resourceOfferSingleTaskSet 方法实现，在该方法中会遍历所有 Worker，先判断 Worker 中的 CPU 核数是否满足任务运行的核数，如果满足则调用 resourceOffer 方法对该 Worker 的 Executor 分配运行任务，分配完毕后更新任务对应任务集管理器列表、任务对应 Executor 列表和 Executor 对应机器列表，并减去该任务使用的 CPU 核数等。其代码如下：

```scala
private def resourceOfferSingleTaskSet(taskSet:TaskSetManager,
  maxLocality : TaskLocality,shuffledOffers : Seq[WorkerOffer], availableCpus :
  Array[Int],tasks : Seq[ArrayBuffer [TaskDescription]]) : Boolean = {
  //遍历所有Worker，为每个Worker分配运行的任务
  var launchedTask = false
  for (i <- 0 until shuffledOffers.size) {
    val execId = shuffledOffers(i).executorId
    val host = shuffledOffers(i).host

    //当Worker的CPU核数满足任务运行核数
    if (availableCpus(i) >= CPUS_PER_TASK) {
      try {
        //对指定Executor分配运行的任务，分配后更新相关列表和递减可用CPU
        for (task <- taskSet.resourceOffer(execId, host, maxLocality)) {
          tasks(i) += task
          val tid = task.taskId
          taskIdToTaskSetManager(tid) = taskSet
          taskIdToExecutorId(tid) = execId
          executorIdToTaskCount(execId) += 1
          executorsByHost(host) += execId
          availableCpus(i) -= CPUS_PER_TASK
          assert(availableCpus(i) >= 0)
          launchedTask = true
        }
      } catch {case e: TaskNotSerializableException =>return launchedTask}
    }
  }
  return launchedTask
}
```

对指定Worker的Executor分配运行的任务调用TaskSetManager.resourceOffer方法实现。首先，在该方法中调用getAllowedLocalityLevel方法获取当前任务集允许执行的数据本地性，如果获取运行的数据本地性比指定的数据本地性优先级来得低，则使用指定的数据本地性。在比较数据本地性时比较的是数据本地性的索引值，而数据本地性优先级和数据本地性索引值相反，优先级越高的数据本地性对应的索引值越小，例如：PROCESS_LOCAL对应的索引值为0，而ANY对应的索引值为4，在比较数据本地性时，ANY>PROCESS_LOCAL。接着，对于指定的节点和获取的数据本地性，在dequeueTask方法中验证是否存在可以分配合适的任务，验证是通过对前面4个列表进行判断（这4个列表会随着时间进行更新），如果存在，则返回该任务和数据本地性等信息。最后，更新该任务集的相关记录信息，序列化任务并把该任务加入到运行列表中等。其代码如下：

```scala
def resourceOffer(execId: String,host: String,maxLocality: TaskLocality.
  TaskLocality) : Option[TaskDescription] = {
  if (!isZombie) {
    val curTime = clock.getTimeMillis()
    var allowedLocality = maxLocality

    //如果资源有Locality特征
    if (maxLocality != TaskLocality.NO_PREF) {
      //获取当前任务集允许执行的Locality, getAllowedLocalityLevel随时间变化而变化
      allowedLocality = getAllowedLocalityLevel(curTime)

      //如果允许的Locality级别低于maxLocality,则使用maxLocality覆盖允许的Locality
      if (allowedLocality > maxLocality) {
        allowedLocality = maxLocality
      }
    }

    dequeueTask(execId, host, allowedLocality) match {
      case Some((index, taskLocality, speculative)) => {
      ……

        //更新相关信息,并对任务进行序化
        val serializedTask: ByteBuffer = try {
          Task.serializeWithDependencies(task, sched.sc.addedFiles,
            sched.sc.addedJars, ser)
        } catch {……}

        //把该任务加入到运行任务列表中
        addRunningTask(taskId)
        val taskName = s"task ${info.id} in stage ${taskSet.id}"
          sched.dagScheduler.taskStarted(task, info)

        //返回Worker中Executor运行任务相关信息
        return Some(new TaskDescription(taskId = taskId, attemptNumber =
          attemptNum, execId, taskName, index, serializedTask))
      }
      case _ =>
    }
  }
  None
}
```

我们再看看 TaskSetManager.getAllowedLocalityLevel 方法获取当前任务集允许执行的数据本地性实现,在该方法从上次获取到的数据本地性开始,根据优先级从高到低判断是否存在任

务需要运行，如果对于其中一级数据本地性没有存在需要运行的任务，则不进行延迟等待，而是进行下一级数据本地性处理。如果存在需要运行的任务，但延迟时间超过了该数据本地性设置的延迟时间，那么也进行下一级数据本地性处理。如果不满足前面两种情况，则返回数据本地性。而判断数据本地性是否存在需要运行任务的方法是：在正在运行任务 copiesRunning（任务加入到运行队列时会记录到该列表中）和成功运行任务 successful（任务成功时在 handleSuccessfulTask 会把记录写到该列表中）两个列表中检查是否包含指定的任务，如果这两个列表中不包含，则表示该任务需要处理，反之则该任务不需要处理，并需要在前面的 4 个 Pending 列表中移除对应的任务信息。

```
private def getAllowedLocalityLevel(curTime: Long): TaskLocality.TaskLocality = {
  //在正在运行任务 copiesRunning 和成功运行任务 successful 两个中检查是否包含指定的任务，
  //如果不包含，则表示这些任务需要运行；如果包含，则需要把该任务从前面 Pending4 个列表移除
  def tasksNeedToBeScheduledFrom(pendingTaskIds: ArrayBuffer[Int]): Boolean ={}
  def moreTasksToRunIn(pendingTasks:HashMap[String, ArrayBuffer[Int]]):Boolean={}

  while (currentLocalityIndex < myLocalityLevels.length - 1) {
      //获取指定的数据本地性是否包含需要运行的任务
    val moreTasks = myLocalityLevels(currentLocalityIndex) match {
     case TaskLocality.PROCESS_LOCAL=>moreTasksToRunIn(pendingTasksForExecutor)
     case TaskLocality.NODE_LOCAL=>moreTasksToRunIn(pendingTasksForHost)
     case TaskLocality.NO_PREF=>pendingTasksWithNoPrefs.nonEmpty
     case TaskLocality.RACK_LOCAL=>moreTasksToRunIn(pendingTasksForRack)
    }
    if (!moreTasks) {
       //如果没有包含需要运行的任务，则进入下一级数据本地性处理
       lastLaunchTime = curTime
       currentLocalityIndex += 1
    } else if (curTime - lastLaunchTime >= localityWaits(currentLocalityIndex)) {
       //如果存在需要运行的任务但延迟时间超过了该数据本地性设置的延迟时间，也进行下一级数据
       //本地性处理
       lastLaunchTime += localityWaits(currentLocalityIndex)
       currentLocalityIndex += 1
    } else {
       //返回满足条件的数据本地性
       return myLocalityLevels(currentLocalityIndex)
    }
  }
  myLocalityLevels(currentLocalityIndex)
}
```

4.4 容错及 HA

所谓容错是指一个系统的部分出现错误的情况还能够持续地提供服务,不会因为一些细微的错误导致系统性能严重下降或者出现系统瘫痪。在一个集群出现机器故障、网络问题等是常态,尤其集群达到较大规模后,很可能较频繁出现机器故障不能进行提供服务,因此对于分布式集群需要进行容错设计。Spark 在设计之初考虑到这种情况,所以它能够实现高容错,以下将从 Executor、Worker 和 Master 的异常处理来介绍。

4.4.1 Executor 异常

Spark 支持多种运行模式,这些运行模式中的集群管理器会为任务分配运行资源,在运行资源中启动 Executor,由 Executor 是负责执行任务的运行,最终把任务运行状态发送给 Driver。下面将以独立运行模式分析 Executor 出现异常的情况,其运行结构如图 4-11 所示,其中虚线为正常运行中进行消息通信线路,实线为异常处理步骤。

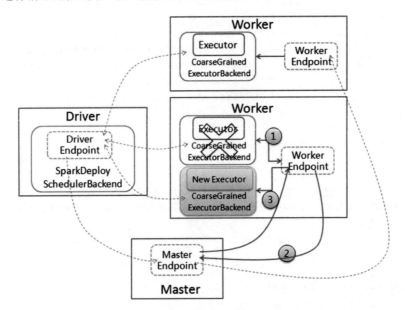

图 4-11 Executor 异常容错过程图

(1)首先看 Executor 的启动过程:在集群中由 Master 给应用程序分配运行资源后,然后在 Worker 中启动 ExecutorRunner,而 ExecutorRunner 根据当前的运行模式启动 CoarseGrainedExecutorBackend 进程,当该进程会向 Driver 发送注册 Executor 信息,如果注册成功,则 CoarseGrainedExecutorBackend 在其内部启动 Executor。Executor 由 ExecutorRunner 进行管理,当 Executor 出现异常时(如所运行容器 CoarseGrainedExecutorBackend 进程异常退出等),由 ExecutorRunner 捕获该异常并发送 ExecutorStateChanged 消息给 Worker。

(2) Worker 接收到 ExecutorStateChanged 消息时,在 Worker 的 handleExecutorStateChanged 方法中,根据 Executor 状态进行信息更新,同时把 Executor 状态信息转发给 Master。

(3) Master 接收到 Executor 状态变化消息后,如果发现 Executor 出现异常退出,则调用 Master.schedule 方法,尝试获取可用的 Worker 节点并启动 Executor,而这个 Worker 很可能不是失败之前运行 Executor 的 Worker 节点。该尝试系统会进行 10 次,如果超过 10 次,则标记该应用运行失败并移除集群中移除该应用。这种限定失败次数是为了避免提交的应用程序存在 Bug 而反复提交,进而挤占集群宝贵的资源。

4.4.2　Worker 异常

Spark 独立运行模式采用的是 Master/Slave 的结构,其中 Slave 是有 Worker 来担任的,在运行的时候会发送心跳给 Master,让 Master 知道 Worker 的实时状态,另一方面 Master 也会检测注册的 Worker 是否超时,因为在集群运行过程中,可能由于机器宕机或者进程被杀死等原因造成 Worker 进程异常退出。下面将分析 Spark 集群如何处理这种情况,其处理流程如图 4-12 所示。

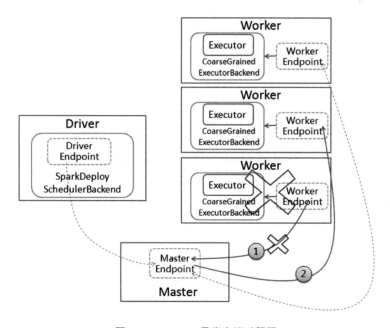

图 4-12　Worker 异常容错过程图

（1）这里需要了解 Master 是如何感知到 Worker 超时？在 Master 接收 Worker 心跳的同时，在其启动方法 onStart 中启动检测 Worker 超时的线程，其代码如下：

```
checkForWorkerTimeOutTask = forwardMessageThread.scheduleAtFixedRate(new Runnable {
  override def run(): Unit = Utils.tryLogNonFatalError {
    //非自身发送消息 CheckForWorkerTimeOut，调用 timeOutDeadWorkers 方法进行检测
    self.send(CheckForWorkerTimeOut)
  }
}, 0, WORKER_TIMEOUT_MS, TimeUnit.MILLISECONDS)
```

（2）当 Worker 出现超时时，Master 调用 timeOutDeadWorkers 方法进行处理，在处理时根据 Worker 运行的是 Executor 和 Driver 分别进行处理。

- 如果是 Executor，Master 先把该 Worker 上运行的 Executor 发送消息 ExecutorUpdated 给对应的 Driver，告知 Executor 已经丢失，同时把这些 Executor 从其应用程序运行列表中删除。另外，相关 Executor 的异常也需要按照前一小节进行处理。
- 如果是 Driver，则判断是否设置重新启动。如果需要，则调用 Master.schedule 方法进行调度，分配合适节点重启 Driver；如果不需要重启，则删除该应用程序。

4.4.3 Master 异常

Master 作为 Spark 独立运行模式中的核心，如果 Master 出现异常，则整个集群的运行情况和资源将无法进行管理，整个集群将处于"群龙无首"的状况。很幸运的是，Spark 在设计时考虑了这种情况，在集群运行的时候，Master 将启动一个或多个 Standby Master，当 Master 出现异常的时候，Standby Master 将根据一定规则确定其中一个接管 Master。在独立运行模式中，Spark 支持如下几种策略，可以在配置文件 spark-env.sh 配置项 spark.deploy.recoveryMode 进行设置，默认为 NONE。

- ZOOKEEPER：集群的元数据持久化到 ZooKeeper 中，当 Master 出现异常时，ZooKeeper 会通过选举机制选举出新的 Master，新的 Master 接管时需要从 ZooKeeper 获取持久化信息并根据这些信息恢复集群状态。具体结构如图 4-13 所示。
- FILESYSTEM：集群的元数据持久化到本地文件系统中，当 Master 出现异常时，只要在该机器上重新启动 Master，启动后新的 Master 获取持久化信息并根据这些信息恢复集群状态。
- CUSTOM：自定义恢复方式，对 StandaloneRecoveryModeFactory 抽象类进行实现并把该类配置到系统中，当 Master 出现异常时，会根据用户自定义的方式进行恢复集群状态态。

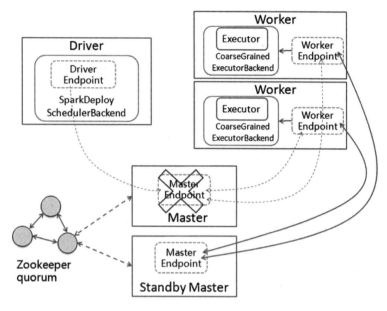

图 4-13 Master 异常容错过程图

- NONE：不持久化集群的元数据，当 Master 出现异常时，新启动的 Master 不进行恢复集群状态，而是直接接管集群。

4.5 监控管理

Spark 提供了 UI 监控、Spark Metrics 和 REST 3 种方式监控应用程序运行状态。其中，UI 监控以网页方式提供用户监控调度阶段、存储、运行环境和 Executor 参数等信息，Spark Metrics 通过定制的方式，将应用程序的运行情况以多种方式展现出来，而 REST 则提供 API 给用户，根据 API 开发监控应用程序运行的各阶段信息。

4.5.1 UI 监控

Spark 的 UI 监控分为实时 UI 监控和历史 UI 监控两种方式，默认情况下启用实时 UI 监控，历史 UI 监控需要手动启用。在实时 UI 监控页面中可以即时刷新查看作业的运行情况，而历史 UI 监控则是保存了应用程序的运行状态数据，根据用户需要可以查询这些应用程序历史运行情况。

1. 实时 UI 监控

实时 UI 监控分为 Master UI 监控和应用程序 UI 监控，其中 Master UI 监控在 Master 启动过程中启用，而应用程序 UI 监控在 SparkContext 启用。以 Standalone 为例，在 Master 启动的时候会启动 UI 监控和 REST 服务，Master UI 监控默认使用 8080 端口，访问地址为 http://host:8080，同时启动 Master 和应用程序的 Metrics 服务，具体代码在 Master 的 onStart 方法中。

```
override def onStart(): Unit = {
//启动 Master 的 UI 监控页面，其中 HTTP 服务由 Jetty 进行提供
  webUi = new MasterWebUI(this, webUiPort)
//默认使用 8080 端口，访问地址为 http://host:8080
  webUi.bind()
  masterWebUiUrl = "http://" + masterPublicAddress + ":" + webUi.boundPort
  checkForWorkerTimeOutTask  =  forwardMessageThread.scheduleAtFixedRate(new
    Runnable {
    override def run(): Unit = Utils.tryLogNonFatalError {
      self.send(CheckForWorkerTimeOut)
    }
  }, 0, WORKER_TIMEOUT_MS, TimeUnit.MILLISECONDS)

//启动 REST 服务，默认端口为 6066
  if (restServerEnabled) {
    val port = conf.getInt("spark.master.rest.port", 6066)
    restServer = Some(new StandaloneRestServer(address.host, port, conf, self,
      masterUrl))
  }
  restServerBoundPort = restServer.map(_.start())

//启动 Master 和应用程序的 Metrics 服务，把 Master UI 监控句柄注入到 Master 和应用程序的
//Metrics 服务中，这样 Master 监控信息会同时发送到 Master 和应用程序的 Metrics 中
  masterMetricsSystem.registerSource(masterSource)
  masterMetricsSystem.start()
  applicationMetricsSystem.start()
  masterMetricsSystem.getServletHandlers.foreach(webUi.attachHandler)
  applicationMetricsSystem.getServletHandlers.foreach(webUi.attachHandler)
......
}
```

在 Master UI 监控页面上有 4 部分内容，界面如图 4-14 所示。

- Master 概要信息：内容包括集群 Master 的访问地址和 REST 访问地址，集群的可用 Worker 节点数、CPU 核数、内存总量、正在运行/完成运行的应用数量和集群 Master 的状态等。

- 集群的 Worker 列表：运行在集群中的 Worker 节点列表，显示内容包括 Worker 的编号、访问地址（IP 地址和端口）、节点状态、CPU 核数和内存大小等信息，其中 CPU 核数和内存同时显示了总量和使用量。

图 4-14　Spark WebUI 监控页面

- 正在运行的应用程序：显示正在运行在集群中的应用程序列表，显示内容包括应用程序的编号、名称、使用的 CPU 核数、每个 Worker 节点平均使用内存大小、提交时间、提交用户、运行状态和运行耗时等信息。通过单击应用程序的编号或名称能够看到该应用程序作业和调度阶段的运行情况。
- 完成运行的应用程序：显示已经完成运行的应用程序，已经运行状态包括成功和失败两种情况，显示内容和正在运行的应用程序的显示内容相同。

而 SparkContext 启动时，启动应用程序的 UI 监控界面，默认端口为 4040，访问地址为 http://host:4040，其启动代码在 SparkContext 的初始化过程中。

```
_ui =
//默认情况下启动应用程序的 UI 监控，在监控过程中加入把作业处理监听器 JobProgressListener
//注入到消息总线 ListenerBus 中，用于监控作业处理状态
  if (conf.getBoolean("spark.ui.enabled", true)) {
    Some(SparkUI.createLiveUI(this, _conf, listenerBus, _jobProgressListener,
      _env.securityManager, appName, startTime = startTime))
```

```
        } else {None}
```

//如果端口被占用会逐步递增直至可用，默认端口为 4040，访问地址为 http://host:4040
ui.foreach(.bind())

应用程序 UI 监控一般包括作业、调度阶段、存储、运行环境、Executor 和 SQL 等信息，在 Spark Streaming 中会增加 Streaming 监控信息。在 Spark1.4 版本中，UI 监控加入了数据可视化功能，增加了事件时间轴、执行 DAG 和 Spark Streaming 统计 3 个视图，使得监控更为直观，可视化视图更加具体。

（1）作业监控页面。

在该监控页面中显示了作业的运行情况，内容包括作业的概要信息、事件时间轴视图、正在运行的作业和已成功运行的作业等信息。

在作业概要信息中，显示了作业运行的用户、总共运行的时间、采用的调度模式，以及正在运行/已成功运行作业的数量。正在运行/已成功运行作业列表显示的信息项包括作业编号、描述、提交时间、运行耗时、调度阶段总数/成功数、任务总数/成功数等信息，页面如图 4-15 所示。

图 4-15 Spark 作业监控页面

单击"Event Timeline"前面的三角图标，可以展开事件时间轴视图，在该视图中可以进行所有作业、指定作业和指定调度阶段三级钻取，图 4-16 中的时间轴显示了横跨整个应用程序中所有作业发生的事件。在图中，先进行 Executor 注册，这里使用了 4 个 Executor；注册完毕后进行运行应用程序，该应用程序运行中包括 4 个作业，其中 3 个成功，1 个失败；当所有作业运行完成后，回收资源移除该应用程序所有的 Executor。

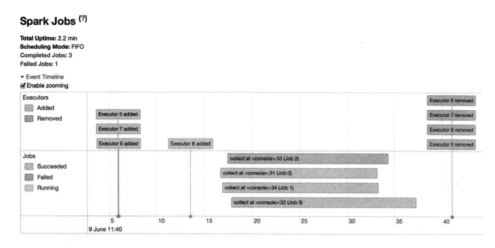

图 4-16 作业事件时间轴视图

在图 4-16 中，直接单击 Job1 查看该作业运行细节。在该作业中，并行运行 3 个调度阶段，这些调度阶段分别进行单词计数统计，它们之间没有依赖关系；当所有调度阶段执行完毕后，启动一个调度阶段，在该调度阶段中把前面 3 个调度阶段计数进行汇总。具体关系如图 4-17 所示。

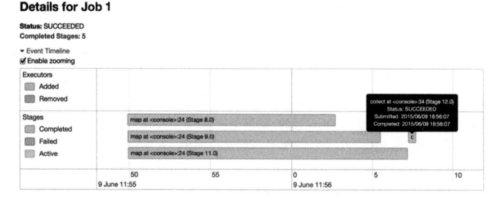

图 4-17 作业详细运行情况界面

在图 4-17 中，直接单击 Stage11 查看该调度阶段运行细节。在该调度阶段划分为 20 个任务（与数据分片关系密切），每段代表一个任务，这些任务分别在 4 个节点上进行运行，如图 4-18 所示（图片并没有完全显示）。从该图中可以看出，调度阶段的任务比较均匀地分散在节点中运行，而且大部分的执行时间用于计算（每个任务中占时间窗口最长的部分），而不是网络传输或者 I/O 开销。

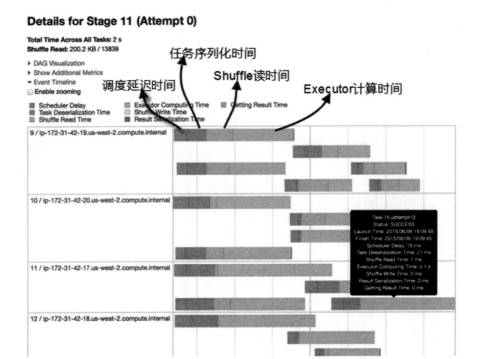

图 4-18　调度阶段详细运行情况的界面

值得注意的是，应用程序在运行中的 Executor 并不是一开始就分配好的，而是动态进行分配。由于在运行过程中，随着任务运行完毕，会回收闲置 Executor，调度器会把该 Executor 加入到下次分配任务进行调度，代码分析请参见 4.3.3 节的内容。这样不仅可以增加集群资源利用率，而且调度 Executor 越多任务越能够在更优的本地性位置进行运行，从而减少运行时间。

在作业监控页面提供了执行 DAG 视图，如图 4-19 所示，其中蓝色阴影框对应 Spark 操作，即用户调用的代码，每个框中点代表对应操作下创建的 RDD。在单词计数的例子中，作业包含两个调度阶段：Stage0 和 Stage1。在 Stage0 中，先由 SparkContext 的 textFile() 方法读取文件，然后进行 flatMap 操作，把每行进行分词成 word，其次对分词的结果进行 map 操作，形成（word，1）对，最后使用 reduceByKey 操作汇总单词数量。

通过执行 DAG 视图，可以发现对于每个调度阶段中操作是连续的，例如在单词计算的 Stage0 中，读取数据后每个 Executor 随即对相同任务的数据分块进行 flatMap 和 map 操作。其次在该图中，对于处理数据进行了缓存（用绿色突出表示），避免了持久化/读取操作，从而提高处理性能。

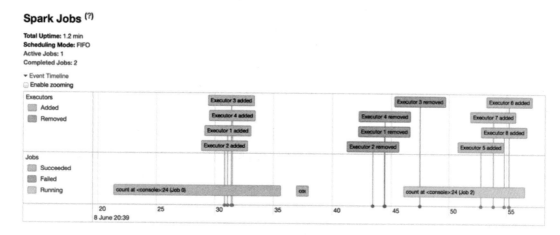

图 4-19　资源分配时间轴

与事件时间轴视图类似，执行 DAG 视图可以单击某个调度阶段，查看该调度阶段详细信息，如图 4-20 所示，在图中展示了该调度阶段所有 RDD 操作细节和依赖关系。

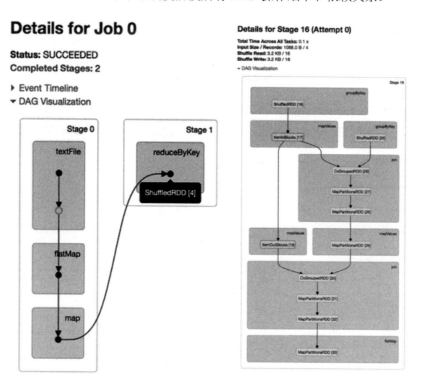

图 4-20　调度阶段依赖关系图

（2）调度阶段监控页面。

在该监控页面中显示了应用程序调度阶段的运行情况，内容包括调度阶段概要信息、正在运行的调度阶段、已成功运行的调度阶段和运行失败的调度阶段等 4 个部分。在调度阶段概要信息中，显示了该应用程序运行耗时、采用的调度模式，以及正在运行/已成功运行/运行失败的调度阶段的数量。

正在运行/已成功运行/运行失败的调度阶段列表显示的信息项包括调度阶段编号、名称、提交时间、运行耗时、运行任务统计、输入数据量、Shuffle 读数据量和 Shuffle 写数据量等信息，页面如图 4-21 所示。

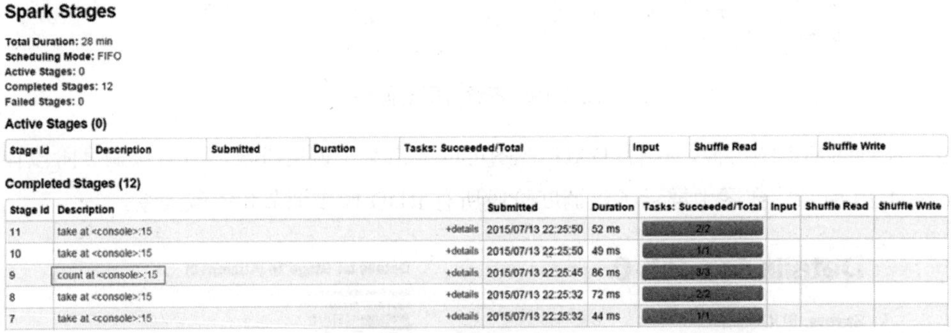

Spark Stages

Total Duration: 28 min
Scheduling Mode: FIFO
Active Stages: 0
Completed Stages: 12
Failed Stages: 0

Active Stages (0)

Stage Id	Description	Submitted	Duration	Tasks: Succeeded/Total	Input	Shuffle Read	Shuffle Write

Completed Stages (12)

Stage Id	Description	Submitted	Duration	Tasks: Succeeded/Total	Input	Shuffle Read	Shuffle Write
11	take at <console>:15 +details	2015/07/13 22:25:50	52 ms	2/2			
10	take at <console>:15 +details	2015/07/13 22:25:50	49 ms	1/1			
9	count at <console>:15 +details	2015/07/13 22:25:45	86 ms	3/3			
8	take at <console>:15 +details	2015/07/13 22:25:32	72 ms	2/2			
7	take at <console>:15 +details	2015/07/13 22:25:32	44 ms	1/1			

图 4-21　调度阶段监控页面

通过单击调度阶段的名称，可以看到调度阶段的详细信息，在页面中显示了该调度阶段中任务运行的总体信息、Executor 运行信息和任务的详细运行情况，如图 4-22 所示。

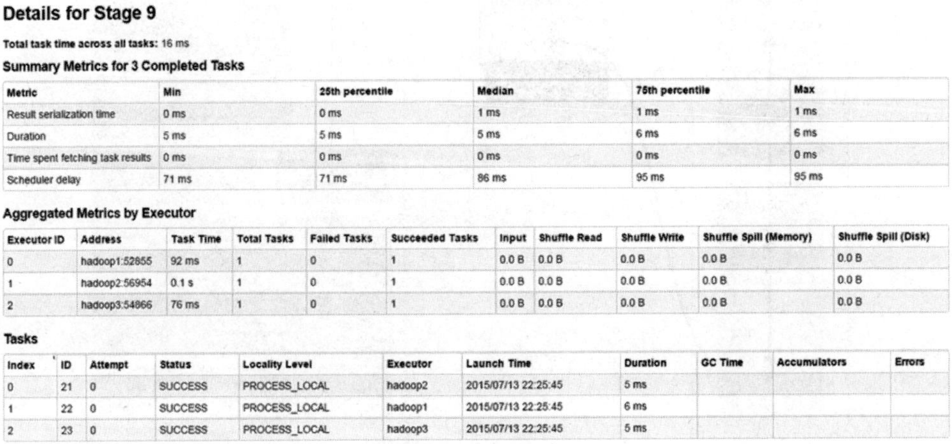

Details for Stage 9

Total task time across all tasks: 16 ms

Summary Metrics for 3 Completed Tasks

Metric	Min	25th percentile	Median	75th percentile	Max
Result serialization time	0 ms	0 ms	1 ms	1 ms	1 ms
Duration	5 ms	5 ms	5 ms	6 ms	6 ms
Time spent fetching task results	0 ms	0 ms	0 ms	0 ms	0 ms
Scheduler delay	71 ms	71 ms	86 ms	95 ms	95 ms

Aggregated Metrics by Executor

Executor ID	Address	Task Time	Total Tasks	Failed Tasks	Succeeded Tasks	Input	Shuffle Read	Shuffle Write	Shuffle Spill (Memory)	Shuffle Spill (Disk)
0	hadoop1:52855	92 ms	1	0	1	0.0 B	0.0 B	0.0 B	0.0 B	0.0 B
1	hadoop2:56954	0.1 s	1	0	1	0.0 B	0.0 B	0.0 B	0.0 B	0.0 B
2	hadoop3:54866	76 ms	1	0	1	0.0 B	0.0 B	0.0 B	0.0 B	0.0 B

Tasks

Index	ID	Attempt	Status	Locality Level	Executor	Launch Time	Duration	GC Time	Accumulators	Errors
0	21	0	SUCCESS	PROCESS_LOCAL	hadoop2	2015/07/13 22:25:45	5 ms			
1	22	0	SUCCESS	PROCESS_LOCAL	hadoop1	2015/07/13 22:25:45	6 ms			
2	23	0	SUCCESS	PROCESS_LOCAL	hadoop3	2015/07/13 22:25:45	5 ms			

图 4-22　某调度界面详细信息

- 任务运行的总体信息：显示对结果数据序列化时间、任务执行耗时、获取任务结果时间和调度延迟 4 项信息的最小、25%、50%、75%和最大的统计信息。
- Executor 运行信息：显示了每个 Executor 执行信息，内容包括 Executor 编号、Executor 地址、任务运行时间、运行任务个数、成功/失败任务个数、输入数据量、Shuffle 读数据量和 Shuffle 写数据量等。
- 任务的详细运行情况：包括任务的序号、编号、运行状态、数据本地性、执行的 Executor、开始执行时间、任务耗时、GC 时间和错误等信息。

（3）存储监控页面。

Spark 中如果 RDD 进行了缓存，则该情况可以通过存储监控页面进行查看，监控的内容包括缓存 RDD 名称、存储级别、缓存 Partition 数、缓存百分比、缓存内存数据大小、缓存 Alluxio 数据大小和存入磁盘数据大小，如图 4-23 所示。

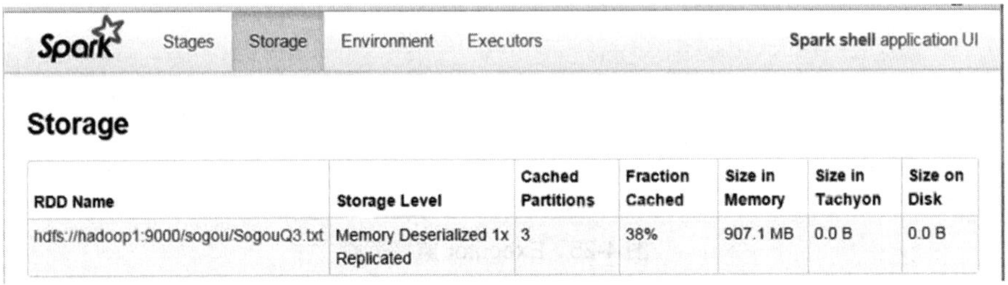

图 4-23　存储监控页面

（4）运行环境监控页面。

在该页面中显示了 Spark 的运行环境，内容包括运行时信息、系统参数和 Spark 参数等信息。通过该页面可以确认运行环境以及 Spark 设置的参数是否正确，该页面如图 4-24 所示。

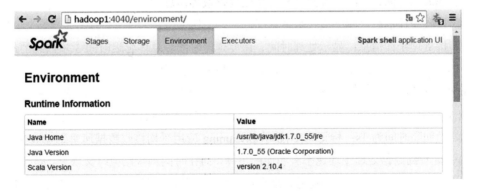

图 4-24　运行环境监控页面

（5）Executor 监控页面。

在该页面中先显示了 Spark 集群中运行的 Driver 和 Executor 列表，记录了 Driver 和 Executor 基本信息，如编号、地址信息和 RDD 数据块个数；另一方面也记录了在应用程序运行期间的参数，如内存使用情况、磁盘使用情况和正在运行/成功完成/失败任务数、输入数据量、Shuffle 读数据量和 Shuffle 写数据量。具体页面如图 4-25 所示。

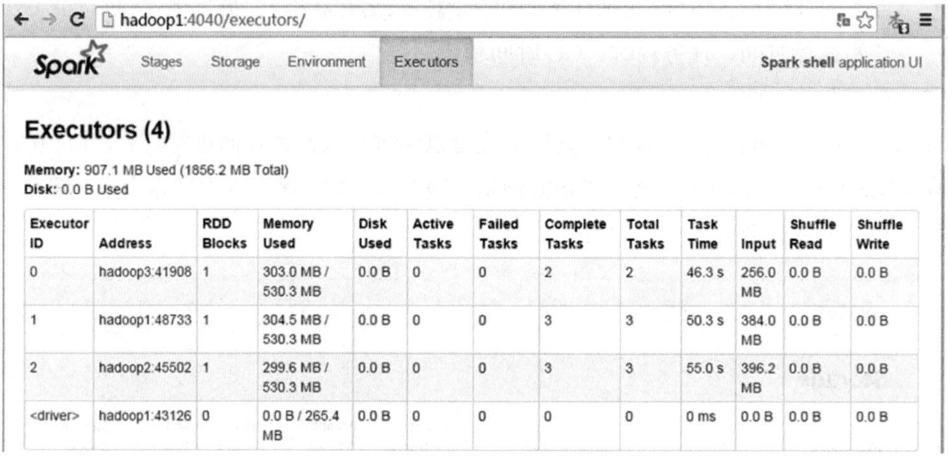

Executor ID	Address	RDD Blocks	Memory Used	Disk Used	Active Tasks	Failed Tasks	Complete Tasks	Total Tasks	Task Time	Input	Shuffle Read	Shuffle Write
0	hadoop3:41908	1	303.0 MB / 530.3 MB	0.0 B	0	0	2	2	46.3 s	256.0 MB	0.0 B	0.0 B
1	hadoop1:48733	1	304.5 MB / 530.3 MB	0.0 B	0	0	3	3	50.3 s	384.0 MB	0.0 B	0.0 B
2	hadoop2:45502	1	299.6 MB / 530.3 MB	0.0 B	0	0	3	3	55.0 s	396.2 MB	0.0 B	0.0 B
<driver>	hadoop1:43126	0	0.0 B / 265.4 MB	0.0 B	0	0	0	0	0 ms	0.0 B	0.0 B	0.0 B

图 4-25 Executor 监控页面

（6）Streaming 监控页面。

在该页面中显示了 Spark Streaming 处理情况，内容包括 Streaming 概要情况、接收统计信息和批处理统计信息。

- Streaming 概要情况：Streaming 开始时间、系统运行时间、网络接收数据情况、批处理间隔时间、已经处理批数和等待处理批数等信息。
- 接收统计信息：网络接收数据情况。
- 批处理统计信息：显示处理时间、调度延迟时间和总体延迟时间 3 项信息的最近一次任务、最小、25%、50%、75%和最大情况的统计信息。

Spark Streaming 统计可视化视图中，可以观察到数据接收的速率和每个批次处理的时间。标记为[A]显示了 Streaming 应用程序概要信息，在图 4-26 中显示该应用已经运行了 39 分 26 秒，其中批处理间隔时间为 1s；标记为[B]显示 Streaming 应用平均接收数据速率为 48.61events/s；标记为[C]处看到数据处理平均速率有明显的下降，而在时间轴快结束时又恢复，对于这种情况可单击“Input Rate”左边的三角符号，来显示每个源头各自的时间轴；标记[D]显示的是处理时间（Processing Time）情况，这些批次大约在平均 20ms 内被处理完成，和批处理间隔（在本例中是 1s）相比花费的处理时间更少，意味着调度延迟（一个批次等待之前批次处理完成的时间，标记为 [E]）几乎是零。由于调度延迟 Streaming 引用程序是否稳定的关键所在，所以 UI 的新功能使得对它的监控更加容易。

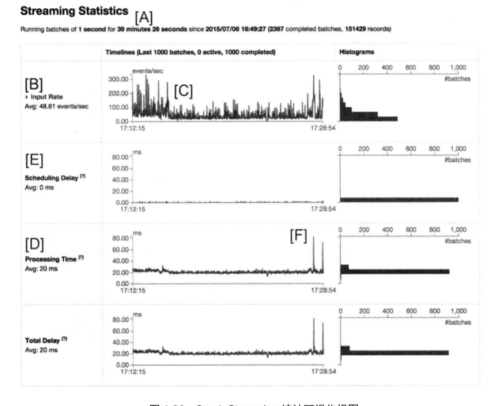

图 4-26　Spark Streaming 统计可视化视图

2．历史 UI 监控

默认情况下 Spark 没有打开历史 UI 监控功能，当 Spark 集群重启时，之前运行的应用程序状态信息不能够访问。此时需要打开 Spark 历史 UI 监控配置，配置项如下所示。

- 是否保存历史监控信息：配置项为 spark.eventLog.enabled，默认值为 false，启动保存历史监控信息需要修改该配置为 true。
- 保存历史监控信息日志目录：配置项为 spark.eventLog.dir，当设置完毕后，在 Spark 应用程序运行时，每个应用程序会根据名称创建对应的子目录，把其运行日志文件保存到该子目录中。

- Spark 历史监控服务器地址：配置项为 spark.yarn.historyServer.address，这个地址会在 Spark 应用程序完成后提交给 YARN RM，然后 RM 将信息从 RM UI 写到 History Server UI 上。

4.5.2 Metrics

Spark 内置了一个可配置的度量系统（Metrics），它是基于 Coda Hale 的 Metrics 库构建的，能够将 Spark 内部状态通过 HTTP、JMX、CSV 等形式呈现给用户。同时，用户也可以以插件的方式将自己实现的数据源（Metrics Source）和数据输出方式（Metrics Sink）添加到 Metrics 系统中，从而获取自己需要的数据。

在 Metrics 系统中有两个重要的概念：Metrics Source 和 Metrics Slink，通过这两个概念定义了 Metrics 系统的输入和输出。

- Metrics Source：定义了所需要采集的各种 Metrics，以及如何采集这些 Metrics。这些定义好的 Metrics 会注册到 Metrics 系统内部以备使用。
- Metrics Slink：定义了 Metrics 的输出行为，采集到的数据以何种方式呈现给用户。

图 4-27 显示了 Metrics 系统各模块之间的关系，Metrics Source 和 Metrics Slink 以插件的形式被 Metrics 系统所管理，通过 MetricsConfig 配置属性启动相应的插件。

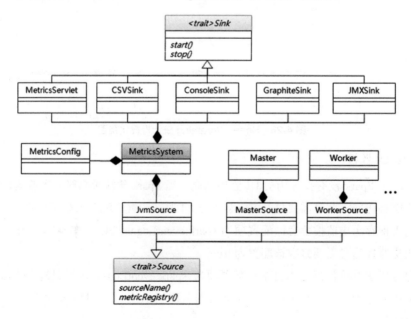

图 4-27　Metrics 系统各模块类图

Metrics 系统配置文件为$Spark/conf/metrics.properties，配置项格式如下，不符合规则配置项系统将忽略。

```
[instance].sink | source.[name].[options]=value
```

- instance：表示 Spark 内部监控的不同对象，在 Spark 定义了 Master、Worker、Executor、Driver 和 Application5 种对象。用户可以使用通配符*表示所有对象，不过配置项有更详细的描述，则通配符设置值将被覆盖，如下配置中：*.sink.console. period 和 *.sink.console. unit 将被 master.sink.console. period 和 master.sink.console. unit 所覆盖。

```
*.sink.console.period=10
*.sink.console.unit=seconds
master.sink.console.period=15
master.sink.console.unit=seconds
```

- sink | source：设置采集数据输出方式和采集数据源。
- name：该名称由用户定义，表示 sink 或 source 名称。在如下配置项中，配置了获取的度量数据以 CSV 的形式输出。

```
*.sink.csv.class=org.apache.spark.metrics.sink.CsvSink
*.sink.csv.period=1
*.sink.csv.unit=minutes
*.sink.csv.directory=/tmp/
```

- option：定义了配置项的名称和对应值，在下面的配置项中，配置不同对象数据源名称和对应值。

```
master.source.jvm.class=org.apache.spark.metrics.source.JvmSource
worker.source.jvm.class=org.apache.spark.metrics.source.JvmSource
driver.source.jvm.class=org.apache.spark.metrics.source.JvmSource
executor.source.jvm.class=org.apache.spark.metrics.source.JvmSource
```

1. 输入源（Metrics Source）介绍

在 Spark 中内置了 6 种 Metrics Source，这些数据源在 Master 启动或应用程序启动时会随之启动（参见 4.5.1 节中的 Master 监控 UI 启动），具体如下。

- MasterSource：主要统计连接的 Worker/可用的 Worker 数量、提交应用程序/等待应用程序数等 Master 信息。
- ApplicationSource：统计应用程序运行状态、运行时间和使用的 CPU 核数等信息。
- WorkerSource：统计 Worker 节点中运行 Executor 信息、使用/剩余 CPU 核数和使用/剩余内存大小等信息。
- ExecutorSource：Executor 中正在处理任务、已经完成任务、线程池总量和使用量等统计信息，如果使用 HDFS 文件系统，则统计其读数据、写数据等统计信息。

- DAGSchedulerSource：统计作业总数和正在运行作业数量，同时统计正在运行、等待和失败的调度阶段。
- BlockManagerSource：统计内存总量、已使用和剩余的内存大小以及使用磁盘空间大小。

除了内置的 Metrics Source 外，用户还可以通过自定义方式进行设置数据源。定义时使用 Metrics 系统通用的 Metrics Source，如 JvmSource，通过该配置项表示 Driver 对象将加载 JvmSource，配置方式如下：

```
driver.source.jvm.class=org.apache.spark.metrics.source.JvmSource
```

2. 输出方式（Metrics Sink）介绍

Spark 在 Metrics 系统中内置了 MetricsServlet，类似于 Metrics Source，在程序启动时自动加载 Metrics Sink。该 Metrics Sink 可以通过发送 HTTP 请求 "/metrics/json"，以 JSON 格式获取监控的 Metrics 数据，如发送 "/metrics/master/json" 和 "/metrics/applications/json" 可以分别获得 Master 对象和 Application 对象的度量数据。另外可以通过自定义方式配置 Metrics Sink，此时配置项值需要以 "class" 结尾，通过反射机制加载到 Metrics 系统中。

Spark 的 Metrics 系统提供如下几种可以配置的 Metrics Sink。

- ConsoleSink：把获取的 Metrics 数据输出到终端上。
- CsvSink：把获取的 Metrics 数据定期保存为 CSV 文件。
- GangliaSink：把获取的 Metrics 数据传输给 Ganglia 监控系统，用于图形等方式展现。
- GraphiteSink：把获取的 Metrics 数据传输给 Graphite 监控系统，用于图形等方式展现。
- JmxSink：用于图形等方式以 Java Bean 的形式输出，可以通过 JmxConsole 等工具进行查看。

下面配置项配置了使用 CSV 文件保存 Metrics 数据：

```
*.sink.csv.class=org.apache.spark.metrics.sink.CsvSink
```

4.5.3 REST

REST 是一种基于网络的软件架构设计风格，其英文全称是 Representational State Transfer，翻译成中文为 "表述性状态转移"。对 REST 更通俗的解释就是：它是软件架构的一种分类，把具有某一组特征的软件架构设计称为 REST 架构约束。基于 Web 的架构，实际上就是各种规范的集合，这些规范共同组成了 Web 架构。RESTful Web Services 使用 RESTful 架构风格构建的服务，由于其轻量的特性和在 HTTP 上直接传输数据的能力，在互联网服务部署技术的选择上，使用 RESTful 风格构建服务正在逐渐成为基于 SOAP 技术的有力挑战者。

Spark1.4 版本中引入了 REST API，API 返回的信息是 JSON 格式的，开发者们可以很方便地通过这个 API 来创建可视化的 Spark 监控工具。该个 API 支持正在运行的应用程序，通过 http://host:4040/api/v1 来获取一些信息，同时也支持历史服务器，访问地址为 http://host:18080/api/v1。现有 Spark 提供如下 REST API。

1．应用程序信息

- /applications：显示所有的应用程序列表。

2．作业信息

- /applications/[app-id]/jobs：获取给定应用程序的所有 Jobs。
- /applications/[app-id]/jobs/[job-id]：获取给定 Job 的信息。

3．调度阶段信息

- /applications/[app-id]/stages：获取给定应用程序的所有 stages。
- /applications/[app-id]/stages/[stage-id]：获取给定 stages 的所有 attempts。
- /applications/[app-id]/stages/[stage-id]/[stage-attempt-id]：获取给定 stages attempts 的信息。
- /applications/[app-id]/stages/[stage-id]/[stage-attempt-id]/taskSummary：获取给定 stage attempt 中所有 tasks 的 metrics 信息。
- /applications/[app-id]/stages/[stage-id]/[stage-attempt-id]/taskList：获取给定 stage attempt 的所有 tasks。

4．Executor 信息

- /applications/[app-id]/executors：获取给定应用程序的所有 Executors。

5．存储信息

- /applications/[app-id]/storage/rdd：获取给定应用程序的所有缓存的 RDDs。
- /applications/[app-id]/storage/rdd/[rdd-id]：获取给定 RDDs 的存储状态详情。

当应用程序运行在 YARN 模式，每个应用程序将有多个 attempts，需要将上面所有[app-id]改成[app-id]/[attempt-id]。上面所有的 URL 都是有版本的，这样将使得开发者们很容易在上面开发应用程序。

4.6 实例演示

4.6.1 计算年降水实例

在该实例中，对俄罗斯某地区的气象观测数据进行分析，通过分析获取该地区每年的降水量，并按照降水量降序进行排序保存到 HDFS 中。该数据集观测主要 4 个观测量：每天平均、最小、最大温度和每日总降水量，下载地址为 http://cdiac.ornl.gov/ftp/ndp040/中 dat 格式数据，数据最大跨度从 1881 年～2001 年。下面是某一个观测站记录数据片段：

```
31909 1992  6  6 0  6.8 0  8.2 0  10.0 0   0.0 3 0 0000
31909 1992  6  7 0  5.5 0  6.0 0   7.8 0  14.5 0 0 0000
31909 1992  6  8 0  4.9 0  5.9 0   7.9 0  25.7 0 0 0000
31909 1992  6  9 0  5.8 0  7.7 0  11.6 0   0.0 2 0 0000
31909 1992  6 10 0  5.1 0  7.1 0  12.0 0   0.0 2 0 0000
31909 1992  6 11 0  6.0 0  7.7 0  11.8 0   0.3 0 0 0000
31909 1992  6 12 0  7.1 0  8.9 0  11.0 0   7.2 0 0 0000
31909 1992  6 13 0  6.7 0  8.4 0  13.1 0   5.5 0 0 0000
31909 1992  6 14 0  7.5 0  9.4 0  12.7 0   2.0 0 0 0000
31909 1992  6 15 0  7.9 0  9.3 0  12.0 0   1.6 0 0 0000
31909 1992  6 16 0  8.6 0  9.6 0  10.9 0  13.7 0 0 0000
```

该数据集字段描述如表 4-1 所示，详细可以参见 data_format.html 页面。

表 4-1　气象观测数据集字段描述

序号	开始位置	字段长度	字段描述
1	2～6	5	WMO 气象站编码
2	9～12	4	年
3	15～16	2	月
4	19～20	2	日
5	22	1	空气温度质量标记
6	24～28	5	每日最低温度
7	30	1	每日最低温度质量标记，0 表示正常，1 表示存疑，9 表示异常值或无观察值
8	32～36	5	每日平均温度
9	38	1	每日平均温度质量标记，0 表示正常，1 表示存疑，9 表示异常值或无观察值
10	40～44	5	每日最高温度
11	46	1	每日最高温度质量标记，0 表示正常，1 表示存疑，9 表示异常值或无观察值
12	48～52	5	每日降水量

续表

序号	开始位置	字段长度	字段描述
13	54	1	每日降水量标记，0 表示降水量超过 0.1mm，1 表示数天累计统计量，2 表示无观察值，3 表示降水小于 0.1mm
14	56	1	每日降水量标记，0 表示正常，1 表示存疑，9 表示异常值或无观察值
15	58～61	4	4 位数据标记，AAAA 表示使用新版规范的数据集，O 可以存在数据标记任何位置，表示 4 个变量比较值不变；R 可以存在数据标记任何位置，表示 4 个变量比较值变化

1．启动运行环境

在该实例中需要启动 HDFS、Spark 集群和 Spark Shell，在 Spark 集群中有 3 个节点，每个节点中的 Executor 使用 1.5GB 内存，以下为执行命令：

```
$cd /app/spark/hadoop-2.7.2/sbin
$./start-dfs.sh

$cd /app/spark/spark-2.0.0/sbin
$./start-all.sh

$cd /app/spark/spark-2.0.0/bin
$./spark-shell --master spark://master:7077 --executor-memory 1536m
 --driver-memory 1024m
```

2．准备数据

先把下载的数据上传到/home/spark/work/ussr 目录下，然后使用 Hadoop 命令把这些数据文件放到 HDFS 的/ussr 目录下，具体命令如下：

```
$cd /home/spark/work
$hadoop fs -mkdir /ussr
$hadoop fs -put /ussr /
```

3．运行代码

下面为计算该地区每年降水量代码：

```scala
scala> val pre_data = sc.textFile("hdfs://master:9000/ussr/")
scala> val fields = pre_data.map(line => line.trim().replace("   "," ").replace
    ("  "," ").split(" ")).filter(_.length==15).filter(_(13)!="9").filter(_(11)!=
    "999.9")
scala> val pre_ann = fields.map(fields => (fields(1),fields(11).
    toDouble)).groupByKey().map{ x=>(x._1, x._2.reduce(_+_)) }
scala>val pre_sort = pre_ann.map(x=>(x._2*365,x._1)).sortByKey(false).
```

```
    map(x=>(x._2,x._1))
scala> pre_sort.saveAsTextFile("hdfs://master:9000/chapter4/out_pre")
```
在 Spark Shell 中计算年降水过程如图 4-28 所示。

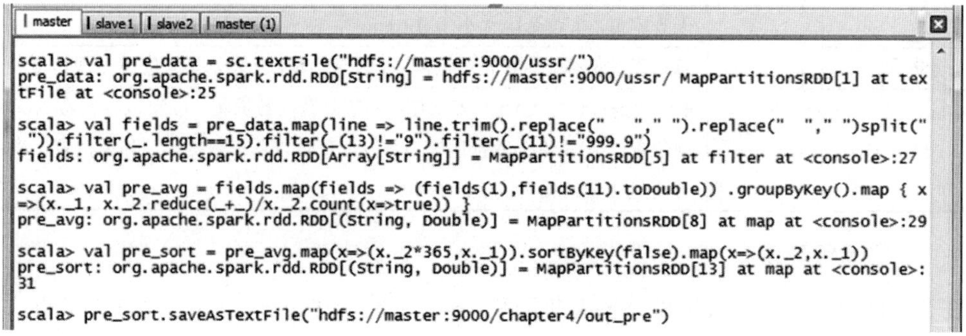

```
scala> val pre_data = sc.textFile("hdfs://master:9000/ussr/")
pre_data: org.apache.spark.rdd.RDD[String] = hdfs://master:9000/ussr/ MapPartitionsRDD[1] at tex
tFile at <console>:25

scala> val fields = pre_data.map(line => line.trim().replace("   "," ").replace("  "," ")split("
 ")).filter(_.length==15).filter(_(13)!="9").filter(_(11)!="999.9")
fields: org.apache.spark.rdd.RDD[Array[String]] = MapPartitionsRDD[5] at filter at <console>:27

scala> val pre_avg = fields.map(fields => (fields(1),fields(11).toDouble)) .groupByKey().map { x
=>(x._1, x._2.reduce(_+_)/x._2.count(x=>true)) }
pre_avg: org.apache.spark.rdd.RDD[(String, Double)] = MapPartitionsRDD[8] at map at <console>:29

scala> val pre_sort = pre_avg.map(x=>(x._2*365,x._1)).sortByKey(false).map(x=>(x._2,x._1))
pre_sort: org.apache.spark.rdd.RDD[(String, Double)] = MapPartitionsRDD[13] at map at <console>:
31

scala> pre_sort.saveAsTextFile("hdfs://master:9000/chapter4/out_pre")
```

图 4-28　在 Spark Shell 中计算年降水过程

处理过程分析如下：

（1）从 HDFS 读取数据，读取的是/ussr 目录下所有数据文件。

（2）对读取的数据进行清洗，需要注意的是，这些数据字段之间空白有占一个字符和两个字符，使用 replace 进行替换。另外，由于有些观察站数据缺失或异常，使用 filter 进行过滤。

（3）对清洗的数据获取年份和降水量字段进行 map 操作，再对该 map 按年份进行聚合求平均值。

（4）对平均值进行排序，排序方法是先把 map 的键值位置对换，对换时需要对值乘以天数 365，然后使用 sortByKey 进行排序，排序后再次把 map 的键值位置对换，就得到按照每年降水量降序排序。

（5）对排序的数据输出到 HDFS 的 out_pre 目录中。

4．查看结果

对保存到 HDFS 的结果进行合并，合并完查看结果，命令如下：

```
$hdfs dfs -getmerge /chapter4/out_pre ~/work/ussr.txt
$cat ~/work/ussr.txt
```

列出前十大降水量最大的年份：

```
(1966,513.4403000663264)
(1990,509.4891802572917)
(1981,507.928451106001)
(1997,506.01286414278667)
(1998,499.7649198582197)
(1978,498.46516285183947)
```

```
(2001,497.9254447462938)
(1970,494.44538460592867)
(1922,493.21714595277746)
(1977,492.7648129473363)
......
```

4.6.2 HA 配置实例

在 4.4.3 节中介绍了 Spark 高可用（HA）策略包括 ZooKeeper、文件系统和自定义方式 3 种方式，在下面将演示以 ZooKeeper 配置 Spark 高可用。在实例中，Spark 集群启动 3 个 Master。这 3 个 Master 连接到同一个 ZooKeeper 实例，利用 ZooKeeper 保存状态和选举机制，模拟主 Master 出现异常，其他两个 Standby 的 Master 选举进行接管，接管后恢复到旧的 Master 状态。

该实例先需要在 Spark 集群的每个节点中先启动 ZooKeeper，而在启动 Master 之前修改 Spark 的配置文件 spark-env.sh，在该文件中需要加入 ZooKeeper 配置信息，具体如表 4-2 所示。

表 4-2 ZooKeeper 配置项列表

属性项	描述
spark.deploy.recoveryMode	Spark 集群 Master 的恢复模式，默认为 NONE
spark.deploy.zookeeper.url	ZooKeeper 集群地址，可以使用 IP 地址或机器名加端口号，ZooKeeper 实例之间使用逗号进行分隔（例如 192.168.1.100:2181,192.168.1.101:2181 或者 master:2181,slave1:2181 等）
spark.deploy.zookeeper.dir	用于存放 ZooKeeper 恢复状态目录，默认为/spark

1. 修改配置

修改 Spark 配置文件 spark-env.sh，在该配置文件中加入 SPARK_DAEMON_JAVA_OPTS 配置项。在该实例中，ZooKeeper 集群地址为 master:2181,slave1:2181,slave2:2181，而存放 ZooKeeper 恢复状态目录/app/soft/zookeeper-3.4.8/recover。需要注意的是，在该配置文件中，Master 地址不能进行设置，否则发生故障时无法进行切换。

```
export SPARK_DAEMON_JAVA_OPTS="-Dspark.deploy.recoveryMode=ZOOKEEPER
  -Dspark.deploy.zookeeper.url=master:2181,slave1:2181,slave2:2181
  -Dspark.deploy.zookeeper.dir=/app/soft/zookeeper-3.4.8"
#export SPARK_MASTER_IP=master
export SPARK_MASTER_PORT=7077
export SPARK_EXECUTOR_INSTANCES=1
......
```

配置好把该文件复制到 Spark 的其他节点中，具体命令如下：

```
$cd /app/spark/spark-2.0.0/conf
```

```
$scp spark-env.sh spark@slave1:/app/spark/spark-2.0.0/conf
$scp spark-env.sh spark@slave2:/app/spark/spark-2.0.0/conf
```

2. 启动 ZooKeeper、Spark 和 Spark Shell

参见附录 C 安装 ZooKeeper，并使用如下命令启动 ZooKeeper：

```
$zkServer.sh start
$zkServer.sh status
```

使用如下命令启动 Spark 和 Spark Shell，如图 4-29 所示的启动后 master 节点为主 Master，状态为 ALIVE，Spark 集群 3 个节点已经被 Spark Shell 所使用。

```
$cd /app/spark/spark-2.0.0/sbin
$./start-all.sh
$cd /app/spark/spark-2.0.0/bin
$./spark-shell    --master    spark://master:7077,slave1:7077,slave2:7077
    --executor-memory 1536m --driver-memory 1024m
```

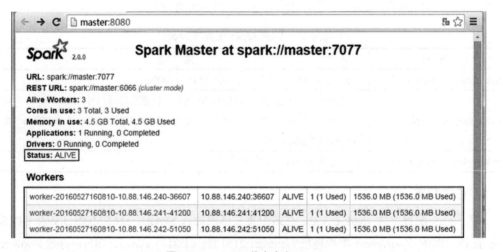

图 4-29　master 节点为主 Master

需要注意的是，由于集群中有多个 Master，所以在提交应用程序需指向一个 Master 列表，如 spark://master:7077,slave1:7077,slave2:7077，这样应用程序会轮询列表。

接着，在 slave1 和 slave2 中启动 Master：

```
$cd /app/spark/spark-2.0.0/sbin
$./start-master.sh
```

如图 4-30 所示的 slave1 和 slave2 节点中，Master 状态为 STANDBY。

图 4-30　slave1 节点状态为 STANDBY

3．模拟主 Master 故障

在 master 节点中使用如下命令停止 Master 进程：

```
$cd /app/spark/spark-2.0.0/sbin
$./stop-master.sh
```

等大概 1～2min 后，在 slave1 节点的监控 UI 上显示了已经接管 Spark 集群所有状态，从图 4-31 中可以看到 Spark Shell 使用了 3 个节点，每个节点使用一个 CPU 核和 1.5GB 内存。

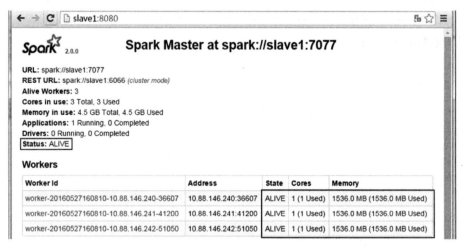

图 4-31　slave1 节点为主 Master 接管 Spark 集群

4.7 小结

在消息通信原理小节中,介绍了 Spark 的消息通信框架,在该框架中定义通信框架的接口,这些接口调用 Netty 具体方法进行实现。

在作业执行原理小节中,首先介绍了作业执行相关术语,包括作业、调度阶段、任务、DAGScheduler 和 TaskScheduler 等,然后对执行过程进行了概述,最后按照六个步骤介绍了作业执行的全过程,这些步骤包括提交作业、划分调度阶段、提交调度阶段、提交任务、执行任务和获取执行结果。

在调度算法章节中,介绍了三种粒度的调度算法。在应用程序间可以认为执行的是有条件的 FIFO 策略,在作业及调度阶段间提供了 FIFO(先进先出)模式和 FAIR 模式,而任务之间是由数据本地性和延迟执行等因素共同决定。

在容错及 HA 章节中介绍了 Spark 三种异常:Executor 异常、Worker 异常和 Master 异常,以及这些异常的恢复方式。在监控管理中介绍了 Spark 提供了 UI 监控、Spark Metrics 和 REST 三种方式监控应用程序运行状态,其中 UI 监控包括实时 UI 监控和历史 UI 监控方式,通过这些方式可以有效进行集群和运行状态的监控。

本章最后介绍了计算年降水实例和 HA 配置实例,在计算年降水实例介绍了 Spark 应用程序运行的过程,并使用 UI 监控查看运行过程,在 HA 配置实例中介绍了 Spark 集群中如何进行设置及演示 HA 恢复过程。

第 5 章

Spark 存储原理

5.1 存储分析

在本节我们将对 Spark 的存储进行全面的分析，Spark 存储介质包括内存和磁盘等。通过本节的分析，我们可以了解到 Spark 在不同运行场景下读写数据过程。

5.1.1 整体架构

Spark 的存储采取了主从模式，即 Master/Slave 模式，整个存储模块使用了前面介绍的 RPC 的消息通信方式。其中，Master 负责整个应用程序运行期间的数据块元数据的管理和维护，而 Slave 一方面负责将本地数据块的状态信息上报给 Master，另一方面接收从 Master 传过来的执行命令，如获取数据块状态、删除 RDD/数据块等命令。在每个 Slave 存在数据传输通道，根据需要在 Slave 之间进行远程数据的读取和写入。Spark 的存储整体架构如图 5-1 所示。

根据 Spark 存储整体架构图，下面将根据数据生命周期过程中进行的消息通信。

（1）在应用程序启动时，SparkContext 会创建 Driver 端的 SparkEnv，在该 SparkEnv 中实例化 BlockManager 和 BlockManagerMaster，在 BlockManagerMaster 内部创建消息通信的终端点 BlockManagerMasterEndpoint。

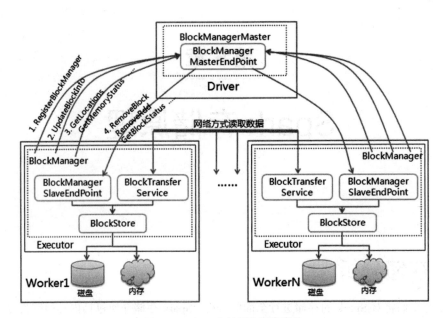

图 5-1 Spark 存储消息通信架构

在 Executor 启动时也会创建其 SparkEnv，在该 SparkEnv 中实例化 BlockManager 和负责网络数据传输服务的 BlockTransferService。在 BlockManager 初始化过程中，一方面会加入 BlockManagerMasterEndpoint 终端点的引用，另一方面会创建 Executor 消息通信的 BlockManagerSlaveEndpoint 终端点，并把该终端点的引用注册到 Driver 中，这样 Driver 和 Executor 相互持有通信终端点的引用，可以在应用程序执行过程中进行消息通信。实例化 BlockTransferService 过程中，使用 Netty 的数据传输服务方式（在 Spark2.0 版本之前，提供 Netty 和 NIO 进行选择）。由于该数据传输服务隐藏了集群间不同节点间的消息传输操作，可类似于本地数据操作方式进行数据读写，大大简化了网络间数据传输的复杂程度。在 SparkEnv 类中创建 BlockTransferService、BlockManager 和 BlockManagerMaster 代码如下：

```
//创建远程数据传输服务，使用Netty方式
val blockTransferService =
  new NettyBlockTransferService(conf, securityManager, hostname, numUsableCores)

//创建BlockManagerMaster，如果是Driver端在BlockManagerMaster内部，则创建终端点
//BlockManagerMasterEndpoint；如果是Executor，则创建BlockManagerMasterEndpoint
//的引用
val blockManagerMaster = new BlockManagerMaster(registerOrLookupEndpoint(
  BlockManagerMaster.DRIVER_ENDPOINT_NAME,
  new BlockManagerMasterEndpoint(rpcEnv, isLocal, conf, listenerBus)),conf,
    isDriver)
```

```
//创建 BlockManager,如果是 Driver 端包含 BlockManagerMaster,如果是 Executor 包含的是
//BlockManagerMaster 的引用,另外 BlockManager 包含了远程数据传输服务,当 BlockManager
//调用 initialize()方法初始化时真正生效
val blockManager = new BlockManager(executorId, rpcEnv, blockManagerMaster,
  serializerManager, conf, memoryManager, mapOutputTracker, shuffleManager,
  blockTransferService, securityManager, numUsableCores)
```

其中 BlockManager 初始化的代码如下,如果是 Executor 创建其消息通信的终端点 BlockManagerSlaveEndpoint,并向 Driver 端发送 RegisterBlockManager 消息,把该 Executor 的 BlockManager 和其所包含的 BlockManagerSlaveEndpoint 引用注册到 BlockManagerMaster 中。

```
def initialize(appId: String): Unit = {
  //在 Executor 中启动远程数据传输服务,根据配置启动传输服务器 BlockTransferServer,
  //该服务器启动后等待其他节点发送请求消息
  blockTransferService.init(this)
  shuffleClient.init(appId)

  //获取 BlockManager 编号
  blockManagerId = BlockManagerId(
    executorId, blockTransferService.hostName, blockTransferService.port)

  //获取 Shuffle 服务编号,如果启动外部 Shuffle 服务,则加入外部 Shuffle 服务端口信息,
  //否则使用 BlockManager 编号
  shuffleServerId = if (externalShuffleServiceEnabled) {
    logInfo(s"external shuffle service port = $externalShuffleServicePort")
    BlockManagerId(executorId, blockTransferService.hostName,
      externalShuffleServicePort)
  } else {
    blockManagerId
  }

  //把 Executor 的 BlockManager 注册到 BlockManagerMaster 中,其中包括其终端点
  //BlockManagerSlaveEndpoint 的引用,Master 端持有该引用可以向 Executor 发送消息
  master.registerBlockManager(blockManagerId, maxMemory, slaveEndpoint)

  //如果外部 Shuffle 服务启动并且为 Executor 节点,则注册该外部 Shuffle 服务
  if (externalShuffleServiceEnabled && !blockManagerId.isDriver) {
    registerWithExternalShuffleServer()
  }
}
```

(2)当写入、更新或删除数据完毕后,发送数据块的最新状态消息 UpdateBlockInfo 给 BlockManagerMasterEndpoint 终端点,由其更新数据块的元数据。该终端点的元数据存放在

BlockManagerMasterEndpoint 的 3 个 HashMap 中，分别如下：

```
//该 HashMap 中存放了 BlockManagerId 与 BlockManagerInfo 的对应,其中 BlockManagerInfo
//包含了 Executor 内存使用情况、数据块的使用情况、已被缓存的数据块和 Executor 终端点的引用,
//通过该引用可以向该 Executor 发送消息
private   val   blockManagerInfo   =   new   mutable.HashMap[BlockManagerId,
  BlockManagerInfo]

//该 HashMap 中存放了 ExecutorID 和 BlockManagerId 对应列表
private val blockManagerIdByExecutor=new mutable.HashMap[String,BlockManagerId]

//该 HashMap 存放了 BlockId 和 BlockManagerId 序列所对应的列表,原因在于一个数据块可能存储
//有多个副本,保存在多个 Executor 中
private val blockLocation=new JHashMap[BlockId,mutable.HashSet[BlockManagerId]]
```

在更新数据块的元数据时，更新 blockManagerInfo 和 blockLocations 两个列表。

- 在处理 blockManagerInfo 时，传入的 blockManagerId、blockId 和 storageLevel 等参数，通过这些参数判断数据的操作是插入、更新还是删除操作。当插入或删除数据块时，会增加或更新 BlockManagerInfo 该数据块信息。如果是删除数据块，则会在 BlockManagerInfo 移除该数据块信息，这些操作结束后还会对该 Executor 的内存使用信息进行更新。

- 在处理 blockLocations 时，根据 blockId 判断 blockLocations 中是否包含该数据库。如果包含该数据块，则根据数据块的操作，当进行数据更新时，更新该数据块所在的 BlockManagerId 信息，当进行数据删除时，则移除该 BlockManagerId 信息，在删除的过程中判断数据块对应的 Executor 是否为空，如果为空表示在集群中删除了该数据块，则在 blockLocations 删除该数据块信息。如果不包含该数据块，则进行的是数据添加操作，把该数据块对应的信息加入到 blockLocations 列表中。

（3）应用程序数据存储后，在获取远程节点数据、获取 RDD 执行的首选位置等操作时需要根据数据块的编号查询数据块所处的位置，此时发送 GetLocations 或 GetLocationsMultipleBlockIds 等消息给 BlockManagerMasterEndpoint 终端点,通过对元数据的查询获取数据块的位置信息。例如 BlockManagerMasterEndpoint 中的 getLocations 方法，该方法获取 blockLocations 中键值为数据块编号 BlockId 的所有 BlockManagerId 序列，代码如下所示：

```
private def getLocations(blockId: BlockId) : Seq[BlockManagerId] = {
   //根据 blockId 判断是否包含该数据块，如果包含，则返回其所对应的 BlockManagerId 序列
   if (blockLocations.containsKey(blockId)) blockLocations.get(blockId).toSeq
     else Seq.empty
}
```

（4）Spark 提供删除 RDD、数据块和广播变量等方式。当数据需要删除时，提交删除消息给 BlockManagerSlaveEndpoint 终端点，在该终端点发起删除操作，删除操作一方面需要删除 Driver 端元数据信息，另一方面需要发送消息通知 Executor，删除对应的物理数据。下面以 RDD 的 unpersistRDD 方法描述其删除过程，类调用关系如图 5-2 所示。

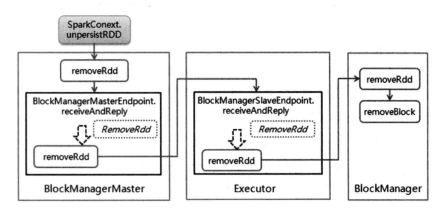

图 5-2　RDD 的 unpersistRDD 类调用关系图

首先，在 SparkConext 中调用 unpersistRDD 方法，在该方法中发送 removeRdd 消息给 BlockManagerMasterEndpoint 终端点；然后，该终端点接收到消息时，从 blockLocations 列表中找出该 RDD 对应的数据存在 BlockManagerId 列表，查询完毕后，更新 blockLocations 和 blockManagerInfo 两个数据块元数据列表；最后，把获取的 BlockManagerId 列表，发送消息给所在 BlockManagerSlaveEndpoint 终端点，通知其删除该 Executor 上的 RDD，删除时调用 BlockManager 的 removeRdd 方法，删除在 Executor 上 RDD 所对应的数据块。其中在 BlockManagerMasterEndpoint 终端点的 removeRdd 代码如下：

```
private def removeRdd(rddId: Int): Future[Seq[Int]] = {
  //在 blockLocations 和 blockManagerInfo 中删除该 RDD 的数据元信息
  //首先，根据 RDD 编号获取该 RDD 存储的数据块信息
  val blocks = blockLocations.keys.flatMap(_.asRDDId).filter(_.rddId == rddId)
  blocks.foreach { blockId =>
    //然后，根据数据块的信息找出这些数据块所在 BlockManagerId 列表，遍历这些列表并删除
    //BlockManager 包含该数据块的元数据，同时删除 blockLocations 对应该数据块的元数据
    val bms: mutable.HashSet[BlockManagerId] = blockLocations.get(blockId)
    bms.foreach(bm=>blockManagerInfo.get(bm).foreach(_.removeBlock (blockId)))
    blockLocations.remove(blockId)
  }

  //最后，发送 RemoveRdd 消息给 Executor，通知其删除 RDD
```

```
val removeMsg = RemoveRdd(rddId)
Future.sequence(blockManagerInfo.values.map { bm =>bm.slaveEndpoint.
    ask[Int] (removeMsg)}.toSeq
)
}
```

除了前面操作中 BlockManagerMasterEndpoint 和 BlockManagerSlaveEndpoint 进行的消息通信外，它们之间还进行数据的状态信息、获取 BlockManager 副本等信息的交互，这些方法可以查看这两个类的 receiveAndReply 方法中的处理消息。

在实际研究存储实现之前，我们看一下 Spark 存储模块类之间的关系，如图 5-3 所示。在整个模块中 BlockManager 是其核心，它不仅提供存储模块处理各种存储方式的读写方法，而且为 Shuffle 模块提供数据处理等操作接口。

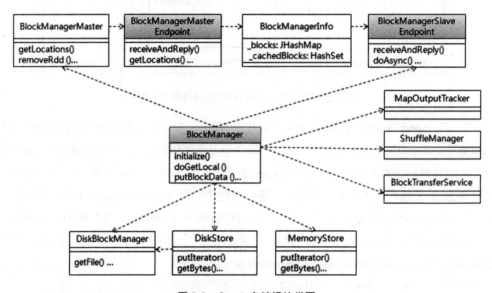

图 5-3 Spark 存储模块类图

BlockManager 存在于 Driver 端和每个 Executor 中，在 Driver 端的 BlockManager 保存了数据的元数据信息，而在 Executor 的 BlockManager 根据接收到消息进行操作：

- 当 Executor 的 BlockManager 接收到读取数据的时,根据数据块所在的节点是否本地使用 BlockManager 不同的方法进行处理。如果在本地，则直接调用 MemoryStore 和 DiskStore 中的取方法 getValues/getBytes 进行读取；如果是在远程，则调用 BlockTransferService 的服务进行获取远程节点上的数据。
- 当 Executor 的 BlockManager 接收到写入数据时，如果不需要创建副本，则调用

BlockStore 的接口方法进行处理（这点和数据读取不同，数据读取是直接调用不同存储的读取方法），根据数据写入的存储类型，决定调用对应的写入方法。

5.1.2 存储级别

Spark 虽是基于内存的计算，但 RDD 的数据集不仅可以存储在内存中，还可以使用 persist 或 cache 方法显式地将 RDD 的数据集缓存到内存或者磁盘中。下面我们看一下 RDD 中 persist 的源代码。

```
private def persist(newLevel: StorageLevel, allowOverride: Boolean): this.type = {
  //如果RDD指定了非NONE的存储级别，该存储级别将不能修改
  if (storageLevel != StorageLevel.NONE && newLevel != storageLevel
       && !allowOverride) {
    throw new UnsupportedOperationException(
      "Cannot change storage level of an RDD after it was already assigned a level")
  }

  //当RDD原来的存储级别为NONE时,可以对RDD进行持久化处理,在处理前需要先清除SparkConext
  //中原来RDD相关存储元数据，然后加入该RDD持久化信息
  if (storageLevel == StorageLevel.NONE) {
    sc.cleaner.foreach(_.registerRDDForCleanup(this))
    sc.persistRDD(this)
  }

  //当RDD原来的存储级别为NONE时，把RDD存储级别修改为传入新值
  storageLevel = newLevel
  this
}
```

RDD 第一次被计算时，persist 方法会根据参数 StorageLevel 的设置采取特定的缓存策略，当 RDD 原本存储级别为 NONE 或者新传递进来的存储级别值与原来的存储级别相等时才进行操作。由于 persist 操作是控制操作的一种，它只是改变了原 RDD 的元数据信息，并没有进行数据的存储操作，真正进行是在 RDD 的 iterator 方法中。对于 cache 方法而言，它只是 persist 方法的一个特例，即 persist 方法的参数为 MEMORY_ONLY 的情况。

在 StorageLevel 类中，根据 useDisk、useMemory、useOffHeap、deserialized、replication5 个参数的组合，Spark 提供了 12 种存储级别的缓存策略，这可以将 RDD 持久化到内存、磁盘和外部存储系统，或者是以序列化的方式持久化到内存中，甚至可以在集群的不同节点之间存储多份副本。代码如下：

```
class StorageLevel private(
```

```
        private var _useDisk: Boolean,
        private var _useMemory: Boolean,
        private var _useOffHeap: Boolean,
        private var _deserialized: Boolean,
        private var _replication: Int = 1)
```

这些缓存策略都被包含在类 StorageLevel 中

```
object StorageLevel {
  val NONE = new StorageLevel(false, false, false, false)
  val DISK_ONLY = new StorageLevel(true, false, false, false)
  val DISK_ONLY_2 = new StorageLevel(true, false, false, false, 2)
  val MEMORY_ONLY = new StorageLevel(false, true, false, true)
  val MEMORY_ONLY_2 = new StorageLevel(false, true, false, true, 2)
  val MEMORY_ONLY_SER = new StorageLevel(false, true, false, false)
  val MEMORY_ONLY_SER_2 = new StorageLevel(false, true, false, false, 2)
  val MEMORY_AND_DISK = new StorageLevel(true, true, false, true)
  val MEMORY_AND_DISK_2 = new StorageLevel(true, true, false, true, 2)
  val MEMORY_AND_DISK_SER = new StorageLevel(true, true, false, false)
  val MEMORY_AND_DISK_SER_2 = new StorageLevel(true, true, false, false, 2)
  val OFF_HEAP = new StorageLevel(false, false, true, false)
```

可选用的存储级别如表 5-1 所示。

表 5-1 Spark 存储级别

存储级别	描述
NONE	不进行数据存储
MEMORY_ONLY	将 RDD 作为反序列化的对象存储 JVM 中。如果 RDD 不能被内存装下，一些分区将不会被缓存，并且在需要的时候被重新计算。这是默认的级别
MEMORY_AND_DISK	将 RDD 作为反序列化的对象存储在 JVM 中。如果 RDD 不能被内存装下，超出的分区将被保存在硬盘上，并且在需要时被读取
MEMORY_ONLY_SER	将 RDD 作为序列化的对象进行存储（每一分区占用一个字节数组）
MEMORY_AND_DISK_SER	与 MEMORY_ONLY_SER 相似，但是把超出内存的分区将存储在硬盘上，而不是在每次需要的时候重新计算
DISK_ONLY	只将 RDD 分区存储在硬盘上
DISK_ONLY_2 等带 2 的	与上述的存储级别一样，但是将每一个分区都复制到两个集群节点上
OFF_HEAP	可以将 RDD 存储到分布式内存文件系统中，如 Alluxio

5.1.3 RDD 存储调用

在分析存储模块原理时，需要明确 RDD 与数据块 Block 之间的关系。在前面的章节中我们了解到 RDD 包含多个 Partition，每个 Partition 对应一个数据块 Block，那么每个 RDD 中包含一个或多个数据块 Block，每个 Block 拥有唯一的编号 BlockId，对应数据块编号规则为："rdd_" + rddId + "_" + splitIndex，其中 splitIndex 为该数据块对应 Partition 的序列号。

在 5.1.2 节中知道在 persist 方法中并没有发生数据存储操作动作，实际发生数据操作是任务运行过程中，RDD 调用 iterator 方法时发生的。在调用过程中，先根据数据块 Block 编号在判断是否已经按照指定的存储级别进行存储，如果存在该数据块 Block，则从本地或远程节点读取数据；如果不存在该数据块 Block，则调用 RDD 的计算方法得出结果，并把结果按照指定的存储级别进行存储。RDD 的 iterator 方法代码如下：

```
final def iterator(split: Partition, context: TaskContext): Iterator[T] = {
  if (storageLevel != StorageLevel.NONE) {
    //如果存在存储级别，尝试读取内存的数据进行迭代计算
    getOrCompute(split, context)
  } else {
    //如果不存在存储级别，则直接读取数据进行迭代计算或读取检查点结果进行迭代计算
    computeOrReadCheckpoint(split, context)
  }
}
```

其中调用的 getOrCompute 方法是存储逻辑的核心，代码如下：

```
private[spark] def getOrCompute(partition: Partition, context: TaskContext):
    Iterator[T] = {
  //通过 RDD 的编号和 Partition 序号获取数据块 Block 的编号
  val blockId = RDDBlockId(id, partition.index)
  var readCachedBlock = true

  //由于该方法由 Executor 调用，可使用 SparkEnv 代替 sc.env
  //根据数据块 Block 编号先读取数据，然后再更新数据，这里是读写数据的入口点
  SparkEnv.get.blockManager.getOrElseUpdate(blockId,storageLevel,elementClassTag,
    () => {
    //如果数据块不在内存，则尝试读取检查点结果进行迭代计算
    readCachedBlock = false
    computeOrReadCheckpoint(partition, context)
  }) match {
    //对 getOrElseUpdate 返回结果进行处理，该结果表示处理成功，记录结果度量信息
    case Left(blockResult) =>
      if (readCachedBlock) {
        val existingMetrics = context.taskMetrics().registerInputMetrics
            (blockResult.readMethod)
        existingMetrics.incBytesReadInternal(blockResult.bytes)
        new InterruptibleIterator[T](context, blockResult.data.asInstanceOf
            [Iterator[T]]) {
```

```
          override def next(): T = {
           existingMetrics.incRecordsReadInternal(1)
            delegate.next()
          }
        }
      } else {
        new InterruptibleIterator(context, blockResult.data.asInstanceOf
          [Iterator[T]])
      }

    //对getOrElseUpdate返回结果进行处理，该结果表示保存失败，例如数据太大无法放到内存中
    //并且也无法保存在磁盘中，把该结果返回给调用者，由其决定如何处理
    case Right(iter) =>
      new InterruptibleIterator(context, iter.asInstanceOf[Iterator[T]])
  }
}
```

在 getOrCompute 调用 getOrElseUpdate 方法，该方法是存储读写数据的入口点：

```
def getOrElseUpdate[T](blockId: BlockId, level: StorageLevel, classTag: ClassTag
  [T],makeIterator:()=>Iterator[T]): Either[BlockResult, Iterator[T]] = {
 //读取数据块入口，尝试从本地或者远程读取数据
 get(blockId) match {
   case Some(block) => return Left(block)
   case _ =>
 }

 //写数据入口
 doPutIterator(blockId, makeIterator, level, classTag, keepReadLock = true) match {
   case None =>
     val blockResult = getLocalValues(blockId).getOrElse {
       releaseLock(blockId)
       throw new SparkException(s"get() failed for block $blockId even though
         we held a lock")
     }
     releaseLock(blockId)
     Left(blockResult)
   case Some(iter) =>Right(iter)
  }
}
```

5.1.4 读数据过程

BlockManager 的 get 方法是读数据的入口点，在读取时分为本地读取和远程节点读取两个

步骤。本地读取使用 getLocalValues 方法，在该方法中根据不同的存储级别直接调用不同存储实现的方法；而远程节点读取使用 getRemoteValues 方法，在 getRemoteValues 方法中调用了 getRemoteBytes 方法，在方法中调用远程数据传输服务类 BlockTransferService 的 fetchBlockSync 进行处理，使用 Netty 的 fetchBlocks 方法获取数据。整个数据读取类调用如图 5-4 所示。

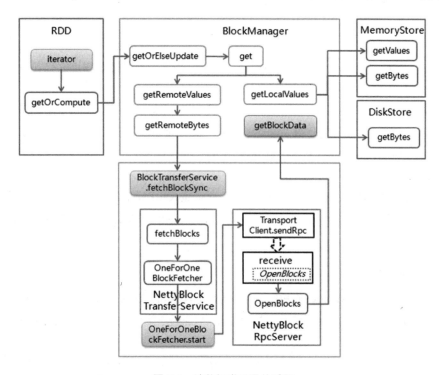

图 5-4 读数据类调用关系图

在本地读取的时候，根据不同的存储级别可以分为内存和磁盘两种读取方式（在 Spark 2.0 版本之前支持外部存储系统读取，如 Alluxio，在该版本及之后不支持）。

1．内存读取

在 getLocalValues 方法中，读取内存中的数据根据返回的是封装成 BlockResult 类型还是数据流，分别调用 MemoryStore 的 getValues 和 getBytes 两种方法，代码如下：

```
//使用内存存储级别，并且数据存储在内存情况
if (level.useMemory && memoryStore.contains(blockId)) {
  val iter: Iterator[Any] = if (level.deserialized) {
    //如果存储时使用反序列化，则直接读取内存中的数据
    memoryStore.getValues(blockId).get
```

```
    } else {
      //如果存储时未使用反序列化,则内存中的数据后做反序列化处理
      serializerManager.dataDeserializeStream(
        blockId, memoryStore.getBytes(blockId).get.toInputStream())(info.classTag)
    }
    //数据读取完毕后,返回数据及数据块大小、读取方法等信息
    val ci = CompletionIterator[Any, Iterator[Any]](iter, releaseLock(blockId))
    Some(new BlockResult(ci, DataReadMethod.Memory, info.size))
}
```

在 MemoryStore 的 getValues 和 getBytes 方法中,最终都是通过数据块编号获取内存中的数据,其代码为:

```
val entry = entries.synchronized { entries.get(blockId) }
```

你一定不相信自己的眼睛,怎么会这么简单?这需要介绍 Spark 内存存储其实是使用了一个大的 LinkedHashMap 来保存数据,其定义如下:

```
private val entries = new LinkedHashMap[BlockId, MemoryEntry[_]](32, 0.75f, true)
```

相对于 HashMap,LinkedHashMap 保存了记录的插入顺序,在遍历 LinkedHashMap 元素时,先得到的记录是先插入的。由于内存大小的限制和 LinkedHashMap 特性,先保存的数据会先被清除,该策略类似于 FIFO。

2. 磁盘读取

磁盘读取在 getLocalValues 方法中,调用的是 DiskStore 的 getBytes 方法,在读取磁盘中的数据后需要把这些数据缓存到内存中,其中 BlockManager 类中的 getLocalValues 方法代码片段如下:

```
    else if (level.useDisk && diskStore.contains(blockId)) {
      val iterToReturn: Iterator[Any] = {
        //从磁盘中获取数据,由于保存到磁盘的数据是序列化的,读取得到的数据也是序列化
        val diskBytes = diskStore.getBytes(blockId)
        if (level.deserialized) {
          //如果存储级别需反序列化,则把读取的数据反序列化,然后存储到内存中
          val diskValues = serializerManager.dataDeserializeStream(blockId,
            diskBytes.toInputStream(dispose = true))(info.classTag)
          maybeCacheDiskValuesInMemory(info, blockId, level, diskValues)
        } else {
          //如果存储级别不需要反序列化,则直接把这些序列化数据存储到内存中
          val stream = maybeCacheDiskBytesInMemory(info, blockId, level, diskBytes)
            .map {_.toInputStream(dispose = false)}
            .getOrElse { diskBytes.toInputStream(dispose = true) }

          //返回的数据需进行反序列化处理
```

```
      serializerManager.dataDeserializeStream(blockId, stream)(info.classTag)
    }
  }
  //数据读取完毕后，返回数据及数据块大小、读取方法等信息
  val ci = CompletionIterator[Any, Iterator[Any]](iterToReturn,
      releaseLock(blockId))
  Some(new BlockResult(ci, DataReadMethod.Disk, info.size))
}
```

在 Spark 中由 spark.local.dir 设置磁盘存储的一级目录，默认情况下设置 1 个一级目录，在每个一级目录下最多创建 64 个二级子目录。一级目录命名为 spark-UUID.randomUUID，其中，UUID.randomUUID 为 16 位的 UUID，二级目录以数据命名，范围是 00~63，目录中文件的名字是数据块的名称 blockId.name，磁盘的目录形式如图 5-5 所示。二级子目录在启动时并没有创建，而是当进行数据操作时才创建。

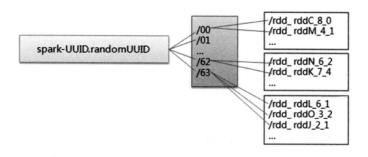

图 5-5　磁盘的目录组织形式

在 DiskStore 中的 getBytes 方法中，调用 DiskBlockManager 的 getFile 方法获取数据块所在文件的句柄。该文件名为数据块的名称，文件所在的一级目录和二级子目录索引值通过文件名的哈希值取模获取。其代码如下：

```
def getFile(filename: String): File = {
  //根据文件名的哈希值获取一级目录和二级子目录索引值，其中一级目录索引值为哈希值与一级目录
  //个数的模，而二级子目录索引值为哈希值与二级目录个数的模
  val hash = Utils.nonNegativeHash(filename)
  val dirId = hash % localDirs.length
  val subDirId = (hash / localDirs.length) % subDirsPerLocalDir

  //先通过一级目录和二级子目录索引值获取该目录，然后判断该子目录是否存在
  val subDir = subDirs(dirId).synchronized {
    val old = subDirs(dirId)(subDirId)
    if (old != null) {
      old
```

```
    } else {
      //如果不存在该目录创建该目录,范围为 00～63
      val newDir = new File(localDirs(dirId), "%02x".format(subDirId))
      if (!newDir.exists() && !newDir.mkdir()) {
        throw new IOException(s"Failed to create local dir in $newDir.")
      }
      //判断该文件是否存在,如果不存在,则创建
      subDirs(dirId)(subDirId) = newDir
      newDir
    }
  }
  //通过文件的路径获取文件的句柄并返回
  new File(subDir, filename)
}
```

获取文件句柄后,读取整个文件内容,以 RandomAccessFile 的只读方式打开该文件。该文件的打开方式支持从偏移位置开始读取指定大小,如果该文件足够小直接读取,而较大文件则指定通道的文件区域直接映射到内存中进行读取。其代码如下:

```
def getBytes(blockId: BlockId): ChunkedByteBuffer = {
  //获取数据块所在文件的句柄
  val file = diskManager.getFile(blockId.name)

  //以只读的方式打开文件,并创建读取文件的通道
  val channel = new RandomAccessFile(file, "r").getChannel
  Utils.tryWithSafeFinally {
    //对于小文件,直接读取
    if (file.length < minMemoryMapBytes) {
      val buf = ByteBuffer.allocate(file.length.toInt)
      channel.position(0)
      while (buf.remaining() != 0) {
        if (channel.read(buf) == -1) {
          throw new IOException("Reached EOF before filling buffer\n" +
            s"offset=0\nfile=${file.getAbsolutePath}\nbuf.remaining=
              ${buf.remaining}")
        }
      }
      buf.flip()
      new ChunkedByteBuffer(buf)
    } else {
      //将指定通道的文件区域直接映射到内存中
      new ChunkedByteBuffer(channel.map(MapMode.READ_ONLY, 0, file.length))
    }
  } {
```

```
      channel.close()
    }
}
```

在远程节点读取数据的时候，Spark 只提供了 Netty 远程读取方式，鉴于 Netty 远程读取数据方式的高效、稳定，在 Spark1.6 版本起将不提供 NIO 方式。下面我们分析 Netty 远程数据读取过程。在 Spark 中主要由下面两个类处理 Netty 远程数据读取。

- NettyBlockTransferService：该类向 Shuffle、存储模块提供了数据存取的接口，接收到数据存取的命令时，通过 Netty 的 RPC 架构发送消息给指定节点，请求进行数据存取操作。
- NettyBlockRpcServer：当 Executor 启动时，同时会启动 RCP 监听器，当监听到消息时把消息传递到该类进行处理，消息内容包括读取数据 OpenBlocks 和写入数据 UploadBlock 两种。

使用 Netty 处理远程数据读取流程如下：

（1）Spark 远程读取数据入口为 getRemoteValues，然后调用 getRemoteBytes 方法，在该方法中调用 getLocations 方法先向 BlockManagerMasterEndpoint 终端点发送 GetLocations 消息，请求据数据块所在的位置信息。当 Driver 的终端点接收到请求消息时，根据数据块的编号获取该数据块所在的位置列表，根据是否是本地节点数据对位置列表进行排序。其中 BlockManager 类中的 getLocalValues 方法代码片段如下：

```
private def getLocations(blockId: BlockId): Seq[BlockManagerId] = {
  //向 BlockManagerMasterEndpoint 终端点发送消息，获取数据块所在节点信息
  val locs = Random.shuffle(master.getLocations(blockId))

  //从获取的节点信息中，优先读取本地节点数据
  val (preferredLocs, otherLocs) = locs.partition { loc => blockManagerId.host
    == loc.host }
  preferredLocs ++ otherLocs
}
```

获取数据块的位置列表后，在 BlockManager.getRemoteBytes 方法中调用 BlockTransferService 提供的 fetchBlockSync 方法进行读取远程数据，具体代码如下：

```
def getRemoteBytes(blockId: BlockId): Option[ChunkedByteBuffer] = {
  var runningFailureCount = 0
  var totalFailureCount = 0

  //获取数据块位置列表
  val locations = getLocations(blockId)
  val maxFetchFailures = locations.size
```

```
    var locationIterator = locations.iterator
    while (locationIterator.hasNext) {
      val loc = locationIterator.next()

      //通过BlockTransferService提供的fetchBlockSync方法获取远程数据
      val data = try {
        blockTransferService.fetchBlockSync(
          loc.host, loc.port, loc.executorId, blockId.toString).nioByteBuffer()
      } catch { …… }

      //获取到数据后，返回该数据块
      if (data != null) {
        return Some(new ChunkedByteBuffer(data))
      }
      logDebug(s"The value of block $blockId is null")
    }
    logDebug(s"Block $blockId not found")
    None
}
```

（2）调用远程数据传输服务 BlockTransferService 的 fetchBlockSync 方法后，在该方法中继续调用 fetchBlocks 方法。该方法是一个抽象方法，实际上调用的是 Netty 远程数据服务 NettyBlockTransferService 类中的 fetchBlocks 方法。在 fetchBlocks 方法中，根据远程节点的地址和端口创建通信客户端 TransportClient，通过该 RPC 客户端向指定节点发送读取数据消息。

```
override def fetchBlocks(host: String, port: Int, execId: String, blockIds:
    Array[String], listener: BlockFetchingListener): Unit = {
  try {
    val blockFetchStarter = new RetryingBlockFetcher.BlockFetchStarter {
      override def createAndStart(blockIds: Array[String], listener:
          BlockFetchingListener) {
        //根据远程节点的节点和端口创建通信客户端
        val client = clientFactory.createClient(host, port)

        //通过该客户端向指定节点发送获取数据消息
        new OneForOneBlockFetcher(client, appId, execId, blockIds.toArray,
            listener).start()
      }
    }
    ……
}
```

其中发送读取消息是在 OneForOneBlockFetcher 类中实现，在该类中的构造函数定义了该消息 this.openMessage = new OpenBlocks(appId, execId, blockIds)，然后在该类的 start 方法中向 RPC

客户端发送消息：
```
public void start() {
  ......
  //通过客户端发送读取数据块消息
  client.sendRpc(openMessage.toByteBuffer(), new RpcResponseCallback() {
    @Override
    public void onSuccess(ByteBuffer response) {...... }

    @Override
    public void onFailure(Throwable e) {......}
  });
}
```

（3）当远程节点的 RPC 服务端接收到客户端发送消息时，在 NettyBlockRpcServer 类中对消息进行匹配。如果是请求读取消息时，则调用 BlockManager 的 getBlockData 方法读取该节点上的数据，读取的数据块封装为 ManagedBuffer 序列缓存在内存中，然后使用 Netty 提供的传输通道，把数据传递到请求节点上，完成远程传输任务。

```
override def receive(client: TransportClient, messageBytes: Array[Byte],
  responseContext: RpcResponseCallback): Unit = {
  val message = BlockTransferMessage.Decoder.fromByteArray(messageBytes)

  message match {
    case openBlocks: OpenBlocks =>
      //调用 blockManager 的 getBlockData 读取该节点上的数据，读取的数据块封装为
      //ManagedBuffer 序列缓存在内存中
      val blocks: Seq[ManagedBuffer] = openBlocks.blockIds.map
        (BlockId.apply).map(blockManager.getBlockData)

      //注册 ManagedBuffer 序列，利用 Netty 传输通道进行传输数据
      val streamId=streamManager.registerStream(appId, blocks.iterator.asJava)
      responseContext.onSuccess(new StreamHandle(streamId, blocks.size).toByteArray)
      ......
  }
}
```

5.1.5　写数据过程

通过前面的分析，我们了解到 BlockManager 的 doPutIterator 方法是写数据的入口点。在该方法中，根据数据是否缓存到内存中进行处理。如果不缓存到内存中，则调用 BlockManager 的 putIterator 方法直接存储磁盘；如果缓存到内存中，则先判断数据存储级别是否进行了反序列化。如果设置反序列化，则说明获取的数据为值类型，调用 putIteratorAsValues 方法把数据存

入内存；如果没有设置反序列化，则获取的数据为字节类型，调用 putIteratorAsBytes 方法把数据存入内存中。在把数据存入内存过程中，需要判断在内存中展开该数据大小是否足够，当足够时调用 BlockManager 的 putArray 方法写入内存，否则把数据写入到磁盘。

在写入数据完成时，一方面把数据块的元数据发送给 Driver 端的 BlockManagerMasterEndpoint 终端点，请求其更新数据元数据，另一方面判断是否需要创建数据副本，如果需要则调用 replicate 方法，把数据写到远程节点上，类似于读取远程节点数据，Spark 提供 Netty 方式写数据。整个数据写入类调用关系如图 5-6 所示。

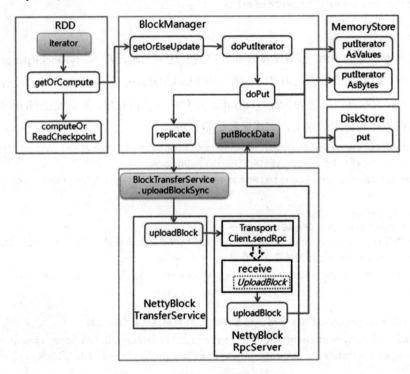

图 5-6　写入数据类调用关系图

通过上面的方法调用图，可以知道在 BlockManager 的 doPutIterator 方法中根据存储级别和数据类型确定调用的方法，当存储级别为内存时，调用 MemoryStore 的写入方法；当存储级别为硬盘时，调用 DiskStore 的写入方法。BlockManager.doPutIterator 代码如下所示：

```
private def doPutIterator[T](blockId: BlockId, iterator: () => Iterator[T],
    level: StorageLevel, classTag: ClassTag[T], tellMaster: Boolean = true,
    keepReadLock: Boolean = false): Option[PartiallyUnrolledIterator[T]] = {
  //辅助类，用于获取数据块信息，并对写数据结果进行处理
```

```scala
doPut(blockId, level, classTag, tellMaster = tellMaster, keepReadLock =
  keepReadLock) { info =>
  val startTimeMs = System.currentTimeMillis
  var iteratorFromFailedMemoryStorePut: Option[PartiallyUnrolledIterator[T]]
      = None
  var size = 0L

  //把数据写放到内存中
  if (level.useMemory) {
    if (level.deserialized) {
      //如果设置反序列化，则说明获取的数据为值类型，调用 putIteratorAsValues 方法
      //把数据存入内存
      memoryStore.putIteratorAsValues(blockId, iterator(), classTag) match {
        //数据写入内存成功，返回数据块大小
        case Right(s) =>size = s
        //数据写入内存失败，如果存储级别设置写入磁盘，则写到磁盘中，否则返回结果
        case Left(iter) =>
          if (level.useDisk) {
            diskStore.put(blockId) { fileOutputStream =>
              serializerManager.dataSerializeStream(blockId, fileOutputStream,
                  iter)(classTag)
            }
            size = diskStore.getSize(blockId)
          } else {
            iteratorFromFailedMemoryStorePut = Some(iter)
          }
      }
    } else {
      //如果没有设置反序列化，则获取的数据为字节类型，调用 putIteratorAsBytes 方法
      //把数据存入内存中
      memoryStore.putIteratorAsBytes(blockId, iterator(), classTag,
        level.memoryMode) match {
        //数据写入内存成功，返回数据块大小
        case Right(s) =>size = s
        //数据写入内存失败，如果存储级别设置写入磁盘，则写到磁盘中，否则返回结果
        case Left(partiallySerializedValues) =>
          if (level.useDisk) {
            diskStore.put(blockId) { fileOutputStream =>
              partiallySerializedValues.finishWritingToStream(fileOutputStream)
            }
            size = diskStore.getSize(blockId)
          } else {
            iteratorFromFailedMemoryStorePut=Some(partiallySerializedValues.
              valuesIterator)
```

```
            }
          }
        }
      }
      //调用 DiskStore 的 put 方法把数据写放到磁盘中
      else if (level.useDisk) {
        diskStore.put(blockId) { fileOutputStream =>
          serializerManager.dataSerializeStream(blockId, fileOutputStream,
            iterator())(classTag)
        }
        size = diskStore.getSize(blockId)
      }

      val putBlockStatus = getCurrentBlockStatus(blockId, info)
      val blockWasSuccessfullyStored = putBlockStatus.storageLevel.isValid
      if (blockWasSuccessfullyStored) {
        //如果成功写入，则把该数据块的元数据发送给 Driver 端
        info.size = size
        if (tellMaster) {
          reportBlockStatus(blockId, info, putBlockStatus)
        }
        Option(TaskContext.get()).foreach { c =>
          c.taskMetrics().incUpdatedBlockStatuses(Seq((blockId, putBlockStatus)))
        }
        //如果需要创建副本，则根据数据块编号获取数据复制到其他节点上
        if (level.replication > 1) {
          val remoteStartTime = System.currentTimeMillis
          val bytesToReplicate = doGetLocalBytes(blockId, info)
          try {
            replicate(blockId, bytesToReplicate, level, classTag)
          } finally {
            bytesToReplicate.dispose()
          }
        }
      }
      iteratorFromFailedMemoryStorePut
    }
  }
```

在 Spark 中写入数据分为内存和磁盘两种方式，其对应写入过程如下：

1. 写入内存

在分析数据写入内存前，我们先看一下 Spark 使用的内存结构，如图 5-7 所示。在该图中，

内存大致分为两部分：图中下半部分为已经使用的内存，在这些内存中存放在 entries 中，该 entries 由不同数据块生成的 MemoryEntry 构成；图中上半部分为可用内存，这些内存用于尝试展开数据块，这些展开数据块的线程并不是一下子把数据展开到内存中（需要注意的是，展开动作是在内存中占位置，并没有真正写入），而是采取"步步为营"的策略，在每个步中都会先检查内存大小是否足够，如果内存大小不足，则尝试把内存中的数据写入到磁盘中，需要释放空间用来存放新写入的数据。而在读数据时了解内存中的数据是以 LinkedHashMap 保存，由于 LinkedHashMap 保存了记录的插入顺序，在计算释放空间时会以类似 FIFO 的顺序进行计算。当计算释放空间足够时，则把内存中释放的数据写入到磁盘并返回内存足够的结果，而当计算出释放所有空间都不足时（但属于同一个 RDD 中的数据块不能被释放），则返回内存不足的结果。Spark 的内存管理相对比较简单，内存的替换算法类似于 FIFO。当机器有足够的内存时，可以明显减少内存释放次数，提高系统处理速度。

如果数据展开成功，则需要把这些数据写入内存中，在写入之前会再次判断内存中展开数据与数据块估计存储在内存中的大小。如果小于或者等于，则需要多分配它们之间的差值，然后尝试写入到内存中；如果大于，则直接把数据写入到内存中，如图5-7所示。

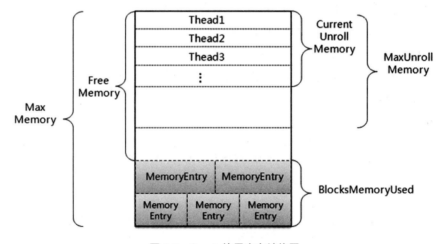

图 5-7　Spark 使用内存结构图

在内存处理类 MemoryStore 中，存在两种写入方法，分别为 putIteratorAsValues 和 putIteratorAsBytes。这两个方法区别在于写入内存的数据类型不同，putIteratorAsValues 针对的是值类型的数据写入，而 putIteratorAsBytes 针对的是字节码数据的写入。这两个方法写入内存过程基本类似，下面以 putIteratorAsValues 讲解写入过程。

（1）在数据块展开前，为该展开线程获取初始化内存,该内存大小为 unrollMemoryThreshold,

获取完毕后返回是否成功的结果 keepUnrolling。

（2）如果 Iterator[T] 存在元素且 keepUnrolling 为真，则继续向前遍历 Iterator[T]，内存展开元素的数量 elementsUnrolled 自增 1。如果遍历 Iterator[T] 到头或者 keepUnrolling 为假，则跳到步骤（4）。

（3）当每 memoryCheckPeriod 即 16 次展开动作后，进行一次检查展开的内存大小是否超过当前分配的内存。如果没有超过则继续展开，如果不足则根据增长因子计算需要增加的内存大小，然后根据该大小申请，申请增加的内存大小：当前展开大小*内存增长因子-当前分配的内存大小。如果申请成功，则把内存大小加入到已使用内存中，而该展开线程获取的内存大小为：当前展开大小*内存增长因子。

（4）判断数据块是否在内存中成功展开，如果展开失败，则记录内存不足并退出；如果展开成功，则继续进行下一步骤。

（5）先估算该数据块在内存中存储的大小，然后比较数据块展开的内存和数据块在内存中存储的大小，如果数据块展开的内存<=数据块存储的大小，说明展开内存的大小不足以存储数据块，需要申请它们之间的差值，如果申请成功，则调用 transferUnrollToStorage 方法处理；数据块展开的内存>数据块存储的大小，说明展开内存的大小足以存储数据块，那么先释放多余的内存，然后调用 transferUnrollToStorage 方法处理。

（6）在 transferUnrollToStorage 方法中释放该数据块在内存展开的空间，然后判断内存是否足够用于写入数据。如果有足够的内存，则把数据块放到内存的 entries 中，否则返回内存不足，写入内存失败的消息。

其中 putIteratorAsValues 详细代码如下：

```
private[storage] def putIteratorAsValues[T](blockId:BlockId,values:
  Iterator[T] ,classTag: ClassTag[T]): Either[PartiallyUnrolledIterator[T],
  Long] = {
//在内存展开元素的数量
var elementsUnrolled = 0
//是否存在足够的内存用于继续展开该数据块，true 表示有，false 表示内存不足
var keepUnrolling = true
//每个展开线程初始化内存大小，这里设置为 unrollMemoryThreshold,
//该变量使用了 spark.storage.unrollMemoryThreshold 配置
val initialMemoryThreshold = unrollMemoryThreshold
//数据块在内存展开时，设置每经过几次展开动作去检查是否申请内存，默认为 16
val memoryCheckPeriod = 16
//当前线程保留用于处理展开操作保留的内存大小，初始值为 initialMemoryThreshold
var memoryThreshold = initialMemoryThreshold
//内存增长因子，每次请求的内存大小为：该因子乘以 vector 大小，减去 memoryThreshold
```

```scala
val memoryGrowthFactor = 1.5
//展开该数据块已使用内存大小
var unrollMemoryUsedByThisBlock = 0L
//用于跟踪该数据块展开所使用的内存大小
var vector = new SizeTrackingVector[T]()(classTag)

//在数据块展开前，根据设置为该线程尝试获取初始化内存
keepUnrolling = reserveUnrollMemoryForThisTask(blockId,
  initialMemoryThreshold, MemoryMode.ON_HEAP)

//如果获取失败记录日志，如果获取成功，则把该大小加入到已使用内存中
if (!keepUnrolling) {
  logWarning(s"Failed to reserve initial memory threshold of " +
    s"${Utils.bytesToString(initialMemoryThreshold)} for computing block
      $blockId in memory.")
} else {
  unrollMemoryUsedByThisBlock += initialMemoryThreshold
}

//在内存中安全展开该数据块，定期判断是否超过分配内存大小
while (values.hasNext && keepUnrolling) {
  vector += values.next()
  //每memoryCheckPeriod 即16 次展开动作，进行一次检查展开的内存大小是否超过
  //当前分配的内存大小
  if (elementsUnrolled % memoryCheckPeriod == 0) {
    //展开所需的内存与该线程分配到的内存大小比较，如果内存充足，则继续展开；
    //如果不足，则根据增长因子计算需要增加的内存大小，然后根据该大小申请
    val currentSize = vector.estimateSize()
    if (currentSize >= memoryThreshold) {
      //获取申请增加的内存大小：当前展开大小*内存增长因子-当前分配的内存大小
      val amountToRequest = (currentSize * memoryGrowthFactor -
        memoryThreshold).toLong

      //申请需要增加的内存，如果申请成功，则把内存大小加入到已使用内存中，
      //而该展开线程获取的内存大小为：当前展开大小*内存增长因子
      keepUnrolling =reserveUnrollMemoryForThisTask(blockId,
        amountToRequest, M emoryMode.ON_HEAP)
      if (keepUnrolling) {unrollMemoryUsedByThisBlock += amountToRequest}
      memoryThreshold += amountToRequest
    }
  }
  elementsUnrolled += 1
}
```

```scala
if (keepUnrolling) {
  //成功地在内存中展开数据块,估算该数据块在内存中存储的大小
  val arrayValues = vector.toArray
  vector = null
  val entry =new DeserializedMemoryEntry[T](arrayValues, SizeEstimator.
     estimate(arrayValues),classTag)
  val size = entry.size

  //定义内部方法,在该方法中释放该数据块在内存展开的空间,然后判断内存是否足够用于写入数据
  def transferUnrollToStorage(amount: Long): Unit = {
    memoryManager.synchronized {
      releaseUnrollMemoryForThisTask(MemoryMode.ON_HEAP, amount)
      val success = memoryManager.acquireStorageMemory(blockId, amount,
        MemoryMode.ON_HEAP)
    }
  }
  //判断内存是否有足够的空间保存该数据块
  val enoughStorageMemory = {
    //判断数据块展开的内存大小和数据块存储的大小
    if (unrollMemoryUsedByThisBlock <= size) {
      //数据块展开的内存<=数据块存储的大小,说明展开内存的大小不足以存储数据块,
      //需要申请它们之间的差值,如果申请成功,则调用transferUnrollToStorage方法处理
      val acquiredExtra =memoryManager.acquireStorageMemory(
        blockId, size - unrollMemoryUsedByThisBlock, MemoryMode.ON_HEAP)
      if (acquiredExtra) {
        transferUnrollToStorage(unrollMemoryUsedByThisBlock)
      }
      acquiredExtra
    } else {
      //数据块展开的内存>数据块存储的大小,说明展开内存的大小足以存储数据块,
      //那么先释放多余的内存,然后调用transferUnrollToStorage方法处理
      val excessUnrollMemory = unrollMemoryUsedByThisBlock - size
      releaseUnrollMemoryForThisTask(MemoryMode.ON_HEAP, excessUnrollMemory)
      transferUnrollToStorage(size)
      true
    }
  }
  if (enoughStorageMemory) {
    //如果有足够的内存,则把数据块放到内存的entries中,并返回占用内存大小
    entries.synchronized {entries.put(blockId, entry)}
    Right(size)
  } else {
    //如果内存不足,则返回该数据块在内存部分展开消息及大小等信息
    Left(new PartiallyUnrolledIterator(this,unrollMemoryUsedByThisBlock,
```

```
        unrolled = arrayValues.toIterator,rest = Iterator.empty))
    }
  } else {
    //内存不足无法展开
    logUnrollFailureMessage(blockId, vector.estimateSize())
    Left(new PartiallyUnrolledIterator(
      this,unrollMemoryUsedByThisBlock,unrolled=vector.iterator,rest=values))
  }
}
```

2. 写入磁盘

Spark 写入磁盘的方法调用了 DiskStore 的 put 方法，该方法提供了写入文件的回调方法 writeFunc。在该方法中先获取写入文件句柄，然后把数据序列化为数据流，最后根据回调方法把数据写入文件中。其处理代码如下：

```
def put(blockId: BlockId)(writeFunc: FileOutputStream => Unit): Unit = {
  if (contains(blockId)) {
    throw new IllegalStateException(s"Block $blockId is already present in the
      disk store")
  }
  logDebug(s"Attempting to put block $blockId")
  val startTime = System.currentTimeMillis

  //获取需要写入文件句柄，参见外部存储系统的读过程
  val file = diskManager.getFile(blockId)
  val fileOutputStream = new FileOutputStream(file)
  var threwException: Boolean = true
  try {
    //使用回调方法，写入前需要把值类型数据序列化成数据流
    writeFunc(fileOutputStream)
    threwException = false
  } finally {
    try {
      Closeables.close(fileOutputStream, threwException)
    } finally {
      if (threwException) { remove(blockId) }
    }
  }
  val finishTime = System.currentTimeMillis
}
```

5.2 Shuffle 分析

5.2.1 Shuffle 简介

在 Hadoop 的 MapReduce 框架中，Shuffle 是连接 Map 和 Reduce 之间的桥梁，Map 的输出要用到 Reduce 中必须经过 Shuffle 这个环节。由于 Shuffle 阶段涉及磁盘的读写和网络传输，因此 Shuffle 的性能高低直接影响整个程序的性能和吞吐量。图 5-8 描述了 Hadoop 的 MapReduce 整个流程，其中 Shuffle 阶段是介于 Map 阶段和 Reduce 阶段之间。

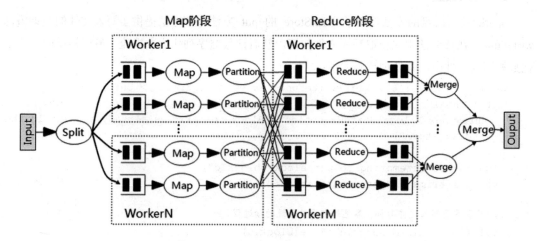

图 5-8 Hadoop 的 MapReduce 过程

Shuffle 中文意思是"洗牌、混洗"的意思，在 MapReduce 过程中需要各节点上的同一类数据汇集到某一节点进行计算，把这些分布在不同节点的数据按照一定的规则聚集到一起的过程称为 Shuffle。Spark 作为 MapReduce 框架的一种实现，自然也实现了 Shuffle 的逻辑，与 Hadoop 遇到的情况类似，在 Shuffle 过程中存在如下问题：

- 数据量非常大，达到 TB 甚至 PB 级别。这些数据分散到数百甚至数千的集群中运行，如何管理为后续任务创建数据众多的文件，以及处理大小超过内存的数据量？
- 如果对结果进行序列化和反序列化，以及在传输之前如何进行压缩处理？

5.2.2 Shuffle 的写操作

Spark 在 Shuffle 的处理方式也是一个迭代的过程，从最开始避免 Hadoop 多余的排序（即在 Reduce 之前获取的数据经过排序），提供了基于哈希的 Shuffle 写操作，但是这种方式在 Map

和 Reduce 数量较大的情况会导致写文件数量大和缓存开销过大的问题。为解决该问题，在 Spark1.2 版本中把默认的 Shuffle 写替换为基于排序的 Shuffle 写，该操作中会把所有的结果写到一个文件中，同时生成一个索引文件进行定位。下面将介绍两种写操作过程。

1. 基于哈希的 Shuffle 写操作

参见图 5-8，在 Hadoop 中 Reduce 所处理的数据都是经过排序的，但在实际处理中很多场景并不需要排序，因此在 Spark1.0 之间实现的是基于哈希的 Shuffle 写操作机制。在该机制中每一个 Mapper 会根据 Reduce 的数量创建出相应的 bucket，bucket 的数据是 M*R，其中 M 是 Map 的个数，R 是 Reduce 的个数；Mapper 生成的结果会根据设置的 Partition 算法填充到每个 bucket 中。这里的 bucket 是一个抽象的概念，在该机制中每个 bucket 对应一个文件；当 Reduce 启动时，会根据任务的编号和所依赖的 Mapper 的编号从远端或者是本地取得相应的 bucket 作为 Reduce 任务的输入进行处理，其处理流程如图 5-9 所示。

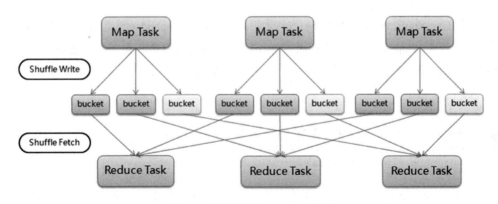

图 5-9　基于哈希的 Shuffle 写操作

相对于传统的 MapReduce，Spark 假定大多数情况下 Shuffle 的数据排序是不必要的，比如 Word Count，强制进行排序只能使性能变差，因此 Spark 并不在 Reduce 端做 merge sort，而是使用聚合（aggregator）。聚合实际上是一个 HashMap，它以当前任务输出结果的 Key 为键，以任意要 combine 类型为值，当在 Word Count 的 Reduce 进行单词计数时，它会将 Shuffle 读到的每一个键值对更新或者插入到 HashMap 中。这样就不需要预先把所有的键值对进行 merge sort，而是来一个处理一个，省下了外部排序这个步骤。

下面从源码角度分析如何基于哈希的 Shuffle 写操作。

（1）在 ShuffleMapTask 的 runTask 方法中，ShuffleManager 调用 getWriter 方法得到 ShuffleWriter 对象，由于 ShuffleWriter 本身是 trait，这里采用的是 ShuffleWriter 的默认值

HashShuffleWriter，通过 HashShuffleWriter 的 writer 方法把 RDD 的计算结果持久化。持久化完毕后，将元数据信息写入到 MapStatus 中，后续的任务可以通过该 MapStatus 得到处理结果信息。

```
override def runTask(context: TaskContext): MapStatus = {
  ……
  var writer: ShuffleWriter[Any, Any] = null
  try {
    //从 SparkEnv 获取 ShuffleManager，在系统启动时，会根据设置进行初始化
    val manager = SparkEnv.get.shuffleManager
    writer=manager.getWriter[Any,Any](dep.shuffleHandle,partitionId,context)

    //调用 RDD 进行计算，通过 HashShuffleWriter 的 writer 方法把 RDD 的计算结果持久化
    writer.write(rdd.iterator(partition, context).asInstanceOf[Iterator[_ <:
        Product2[Any, Any]]])
    writer.stop(success = true).get
  } catch { …… }
}
```

（2）在 HashShuffleWriter 的 writer 方法中，通过 ShuffleDependency 是否定义了 Aggregator 判断是否需要在 Map 端对数据进行聚合操作，如果需要则对数据进行聚合处理。然后调用 ShuffleWriterGroup 的 writers 方法得到一个 DiskBlockObjectWriter 对象，调用该对象的 writer 方法写入。

```
//写入前定义 blockManager 和 shuffle
private val blockManager = SparkEnv.get.blockManager
private val shuffle = shuffleBlockResolver.forMapTask(dep.shuffleId, mapId,
   numOutputSplits, dep.serializer, writeMetrics)

override def write(records: Iterator[Product2[K, V]]): Unit = {
  //判断是否需要聚合，如果需要，则对数据按照键值进行聚合
  val iter = if (dep.aggregator.isDefined) {
    if (dep.mapSideCombine) {
      dep.aggregator.get.combineValuesByKey(records, context)
    } else {
      records
    }
  } else {
    records
  }

  //对原始结果或者是聚合后的结果调用 ShuffleWriterGroup 的 writers 方法得到
  //一个 DiskBlockObjectWriter 对象，调用该对象的 writer 方法写入
  for (elem <- iter) {
    val bucketId = dep.partitioner.getPartition(elem._1)
```

```
      shuffle.writers(bucketId).write(elem._1, elem._2)
    }
  }
```

（3）在步骤（2）中生成 writer 中定义了 shuffle，该 shuffle 在 FileShuffleBlockResolver 由 forMapTask 方法定义。需要注意的是，writers 是通过 ShuffleDependency 来获取后续 Partion 的数量，这个数量和后续任务数相对应，这样在运行的时候，每个 Partition 对应一个任务，从而形成流水线高效并发地处理数据。具体的实现代码如下：

```
private val shuffleStates = new ConcurrentHashMap[ShuffleId, ShuffleState]
def forMapTask(shuffleId: Int, mapId: Int, numReducers: Int, serializer:
  Serializer, writeMetrics: ShuffleWriteMetrics): ShuffleWriterGroup = {
  new ShuffleWriterGroup {
    //在这里使用 Java 的 ConcurrentHashMap,而不使用 Scala 的 Map 并使用 getOrElseUpdate
    //方法，是因为其不是原子操作
    private val shuffleState: ShuffleState = {
      shuffleStates.putIfAbsent(shuffleId, new ShuffleState(numReducers))
      shuffleStates.get(shuffleId)
    }
    val openStartTime = System.nanoTime

    //获取序列化实例，该实例在 SparkEnv 实例化，由 spark.serializer 进行配置
    val serializerInstance = serializer.newInstance()
    val writers: Array[DiskBlockObjectWriter] = {
      Array.tabulate[DiskBlockObjectWriter](numReducers) { bucketId =>
        //获取 blocket 编号，规则为"shuffle_" + shuffleId + "_" + mapId + "_" + reduceId
        val blockId = ShuffleBlockId(shuffleId, mapId, bucketId)

        //在 DiskBlockManager.getFile 方法中，根据该编号获取该数据文件
        val blockFile = blockManager.diskBlockManager.getFile(blockId)

        //返回临时文件路径，该路径为：blockFile 绝对路径+UUID，然后根据该临时文件路径生成
        //磁盘写入器实例 writers
        val tmp = Utils.tempFileWith(blockFile)
        blockManager.getDiskWriter(blockId, tmp, serializerInstance, bufferSize,
          writeMetrics)
      }
    }
    writeMetrics.incWriteTime(System.nanoTime - openStartTime)

    //释放写入器实例方法
    override def releaseWriters(success: Boolean) {
      shuffleState.completedMapTasks.add(mapId)
    }
```

        }
    }

2．基于排序的 Shuffle 写操作

虽然基于哈希的 Shuffle 写操作能够较好地完成 Shuffle 数据写入，但是存在两大问题：

- 每个 Shuffle Map Task 为后续的任务创建一个单独的文件，因此在运行过程中所创建的文件数是 S*F，其中 S 为当前 Shuffle Map Task 的任务数，而 F 为后续任务的任务数。假定当前任务数 S 为 1000，后续任务 F 数为 1000，那么会产生 1M 个文件，这对于文件系统是一个非常大的负担，同时在 Shuffle 数据量不大而文件非常多的情况下，随机写会严重降低 I/O 的性能。

- 虽然 Shuffle 写时数据不需要存储在内存再写到磁盘中，但是 DiskBlockObjectWriter 所带来的开销也是一个不容小视的内存开销。假定当前任务数 S 为 1000，后续任务 F 数为 1000，那么会产生 1M 个文件，而每一个 Writer Handler 默认需要 100KB 的内存，那么缓存的开销就需要 100GB 的内存！当然实际情况 Shuffle 写是分时运行的，其内存所需是 C*F*100KB，其中 C 是 Spark 集群中运行的核数，F 为后续任务数，如果后续任务很大的话，缓存所占用的内存也是一笔不小的开销。

为了缓解 Shuffle 过程中产生文件过多的文件和 Writer Handlerde 缓存开销过大的问题，在 Spark1.1 版本中借鉴了 Hadoop 在 Shuffle 中的处理方式，引入基于排序的 Shuffle 写操作机制。在该机制中，每个 Shuffle Map Task 不会为后续的每个任务创建单独的文件，而是会将所有的结果写到同一个文件中，对应的生成一个 Index 文件进行索引。通过这种机制避免了大量文件的产生，一方面可以减轻文件系统管理众多文件的压力，另一方面可以减少 Writer Handlerde 缓存所占用的内存大小，节省了内存同时避免了 GC 的风险和频率。

在前面我们知道基于哈希的 Shuffle 写操作输出结果是放在 HashMap 中，没有经过排序，但是对于一些如 groupByKey 的操作，如果使用 HashMap，则需要将所有的键值对放在该 HashMap 中并将值合并成一个数组。可以想象为了能够存放所有数据，必须确保每一个 Partition 足够小到内存能够存放，这对于内存来说是一个很大的挑战。为了减少内存的使用，可以将 Aggregator 的操作从内存转移到磁盘中，在结束的时候再将这些不同的文件进行归并排序，从而减少内存的使用量。

基于排序的 Shuffle 写操作解决了文件创建数目过多的问题，也化解了在需要排序时内存占用过大的问题，其实现架构如图 5-10 所示。

前面我们分析了 HashShuffleWriter 的 writer 方法，在这里我们看一下 SortShuffleWriter 的 write 方法是如何实现的？在该方法中，先判断 Shuffle Map Task 输出结果在 Map 端是否需要合

并（Combine），如果需要合并，则外部排序中进行聚合并排序；如果不需要，则外部排序中不进行聚合和排序，例如 sortByKey 操作在 Reduce 端会进行聚合并排序。确认外部排序方式后，在外部排序中将使用 PartitionedAppendOnlyMap 来存放数据，当排序中的 Map 占用的内存已经超越了使用的阈值，则将 Map 中的内容溢写到磁盘中，每一次溢写产生一个不同的文件。当所有数据处理完毕后，在外部排序中有可能一部分计算结果在内存中，另一部分计算结果溢写到一或多个文件之中，这时通过 merge 操作将内存和 spill 文件中的内容合并整到一个文件里。

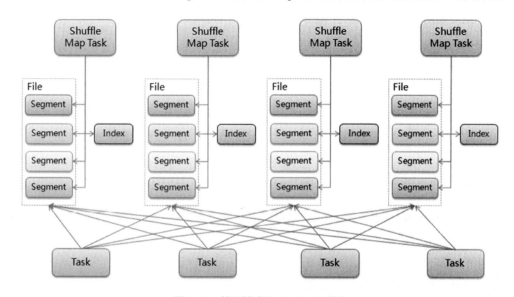

图 5-10　基于排序的 Shuffle 写操作

（1）在 SortShuffleWriter 中 write 代码如下：

```
override def write(records: Iterator[Product2[K, V]]): Unit = {
  //获取 Shuffle Map Task 输出结果的排序方式
  sorter = if (dep.mapSideCombine) {
    //当输出结果需要 Combine，那么外部排序算法中进行聚合
    require(dep.aggregator.isDefined, "Map-side combine without Aggregator
      specified!")
    new ExternalSorter[K, V, C](dep.aggregator, Some(dep.partitioner),
    dep.keyOrdering, dep.serializer)
  } else {
    //其他情况下，外部排序算法不进行聚合
    new ExternalSorter[K, V, V](aggregator = None, Some(dep.partitioner),
      ordering = None, dep.serializer)
  }
  //根据获取的排序方式，对数据进行排序并写入到内存缓冲区中。如果排序中的 Map 占用的内存
```

```
//已经超越了使用的阈值，则将 Map 中的内容溢写到磁盘中，每一次溢写产生一个不同的文件
sorter.insertAll(records)

//通过 Shuffle 编号和 Map 编号获取该数据文件
val output= shuffleBlockResolver.getDataFile(dep.shuffleId, mapId)
val tmp = Utils.tempFileWith(output)

//通过 Shuffle 编号和 Map 编号获取 ShuffleBlock 编号
val blockId = ShuffleBlockId(dep.shuffleId, mapId,
    IndexShuffleBlockResolver.NOOP_REDUCE_ID)

//在外部排序中有可能一部分计算结果在内存中，另一部分计算结果溢写到一或多个文件之中，
//这时通过 merge 操作将内存和 spill 文件中的内容合并整到一个文件里
val partitionLengths = sorter.writePartitionedFile(blockId, tmp)

//创建索引文件，将每个 partition 的在数据文件中的起始位置和结束位置写入到索引文件
shuffleBlockResolver.writeIndexFileAndCommit(dep.shuffleId, mapId,
    partitionLengths, tmp)
//将元数据信息写入到 MapStatus 中，后续的任务可以通过该 MapStatus 得到处理结果信息
mapStatus = MapStatus(blockManager.shuffleServerId, partitionLengths)
}
```

（2）在 ExternalSorter 的 insertAll 方法中，先判断是否需要进行聚合（Aggregation），如果需要，则根据键值进行合并（Combine），然后把这些数据写入到内存缓冲区中，如果排序中的 Map 占用的内存已经超越了使用的阈值，则将 Map 中的内容溢写到磁盘中，每一次溢写产生一个不同的文件。如果不需要聚合，则直接把数据到内存缓存区。

```
override def insertAll(records: Iterator[Product2[K, V]]): Unit = {
    //获取外部排序中是否需要进行聚合（Aggregation）
    val shouldCombine = aggregator.isDefined
    if (shouldCombine) {
        //如果需要聚合，则使用 AppendOnlyMap 根据键值进行合并
        val mergeValue = aggregator.get.mergeValue
        val createCombiner = aggregator.get.createCombiner
        var kv: Product2[K, V] = null
        val update = (hadValue: Boolean, oldValue: C) => {
            if (hadValue) mergeValue(oldValue, kv._2) else createCombiner(kv._2)
        }
        while (records.hasNext) {
            addElementsRead()
            kv = records.next()
            map.changeValue((getPartition(kv._1), kv._1), update)
            //对数据进行排序并写入到内存缓冲区中，如果排序中的 Map 占用的内存已经超越了使用的阈值，
            //则将 Map 中的内容溢写到磁盘中，每一次溢写产生一个不同的文件
```

```
      maybeSpillCollection(usingMap = true)
    }
  } else {
    //不需要进行聚合（Aggregation），对数据进行排序并写入到内存缓冲区中
    while (records.hasNext) {
      addElementsRead()
      val kv = records.next()
      buffer.insert(getPartition(kv._1), kv._1, kv._2.asInstanceOf[C])
      maybeSpillCollection(usingMap = false)
    }
  }
}
```

5.2.3 Shuffle 的读操作

在前面一节我们了解到了 Shuffle 写的过程，在下游调度阶段执行时，需要读取这些数据，而在读取前需要解决两个问题。

- Shuffle 写有基于哈希和排序两种方式，它们对应读取方式如何？
- 如何确认下游任务读取数据的位置信息，位置信息包括所在节点、Executor 编号和读取数据块序列等？

带着这两个问题先梳理一下 Shuffle 读的流程，如图 5-11 所示。

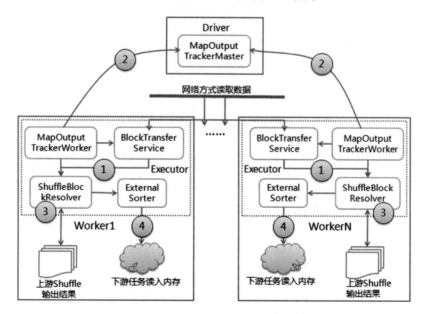

图 5-11　Shuffle 读的流程图

（1）在 SparkEnv 启动时，会对 ShuffleManager、BlockManager 和 MapOutputTracker 等实例化。ShuffleManager 配置项有 HashShuffleManager、SortShuffleManager 和自定义的 ShuffleManager 等3种选项，前两种在 Shuffle 读中均实例化 BlockStoreShuffleReader（参见后面 Shuffle 相关类图），但是在 HashShuffleManager 中所持有的是 FileShuffleBlockResolver 实例，SortShuffleManager 中所持有的 IndexShuffleBlockResolver 实例。对于第一个问题在该步骤就能够知道答案：将根据不同的写入方式将采取相同的读取方式，读取的数据放在哈希列表中便于后续处理。

（2）在 BlockStoreShuffleReader 的 read 方法中，调用 MapOutputTracker 的 getMapSizesByExecutorId 方法，由 Executor 的 MapOutputTracker 发送获取结果状态的 GetMapOutputStatuses 消息给 Driver 端的 MapOutputTrackerMaster，请求获取上游 Shuffle 输出结果对应的 MapStatus，在该 MapStatus 存放了结果数据的位置信息，这个信息也就是在 4.2.7 节中介绍的 ShuffleMapTask 执行结果元信息。到这里我们就可以知道第二个问题的答案，通过请求 Driver 端的 MapOutputTrackerMaster，可以得到上游 Shuffle 结果的位置信息。

（3）知道 Shuffle 结果的位置信息后，对这些位置进行筛选，判断当前运行的数据是从本地还是从远程节点获取。如果是本地获取，直接调用 BlockManager 的 getBlockData 方法，在读取数据的时候会根据写入方式采取不同 ShuffleBlockResolver 读取；如果是在远程节点上，需要通过 Netty 网络方式读取数据。在远程读取的过程中使用多线程的方式进行读取，一般来说，会启动5个线程到5个节点进行读取所有，每次请求的数据大小不会超过系统设置的 1/5，该大小由 spark.reducer.maxSizeInFlight 配置项进行设置，默认情况该配置为 48MB。

（4）读取数据后，判断 ShuffleDependency 是否定义聚合（Aggregation），如果需要，则根据键值进行聚合。需要注意的是，如果在上游 ShuffleMapTask 已经做了合并，则在合并数据的基础上做键值聚合。待数据处理完毕后，使用外部排序（ExternalSorter）对数据进行排序并放入存储中，至此完成 Shuffle 读数据操作。

下面将从代码分析 Shuffle 读的实现，其类调用关系如图 5-12 所示。

（1）Shuffle 读的起始点是由 ShuffledRDD.computer 发起的，在该方法中会调用 ShuffleManager 的 getReader 方法，在前面我们已经知道，基于哈希和排序 Shuffle 读都是使用了 HashShuffleReader 的 read 方式。

```
override def compute(split: Partition, context: TaskContext): Iterator[(K, C)]
    = {
  val dep = dependencies.head.asInstanceOf[ShuffleDependency[K, V, C]]
  //根据配置 ShuffleReader,基于哈希和排序的 Shuffle 读都使用了 HashShuffleReader
  SparkEnv.get.shuffleManager.getReader(dep.shuffleHandle, split.index,
```

```
            split.index + 1, context).read().asInstanceOf[Iterator[(K, C)]]
}
```

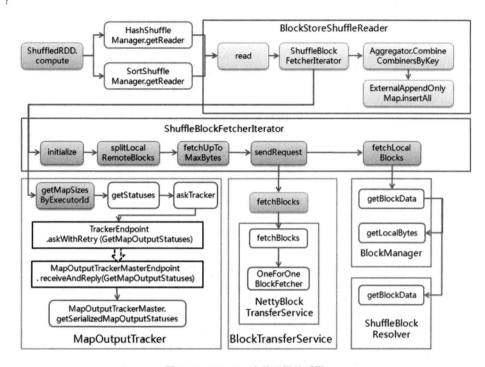

图 5-12　Shuffle 读类调用关系图

（2）在 HashShuffleReader 的 read 方式中先实例化 ShuffleBlockFetcherIterator，在该实例化过程中，通过 MapOutputTracker 的 getMapSizesByExecutorId 获取上游 ShffleMapTask 输出数据的元数据。先尝试在本地的 mapStatuses 获取，如果获取不到，则通过 RPC 通信框架发送消息给 MapOutputTrackerMaster，请求获取该 ShffleMapTask 输出数据的元数据，获取这些元数据转换成 Seq[(BlockManagerId, Seq[(BlockId, Long)])]的序列。在这个序列中的元素包括两部分信息，BlockManagerId 可以定位数据所处的 Excutor，而 Seq[(BlockId, Long)]可以定位 Excutor 的数据块编号和获取数据的大小。

```
override def read(): Iterator[Product2[K, C]] = {
  val blockFetcherItr = new ShuffleBlockFetcherIterator(
    context,blockManager.shuffleClient,blockManager,
    //通过消息发送获取 ShuffleMapTask 存储数据位置的元数据
    mapOutputTracker.getMapSizesByExecutorId(handle.shuffleId, startPartition),
    //远程获取数据时，每次传输数据设置大小
    SparkEnv.get.conf.getSizeAsMb("spark.reducer.maxSizeInFlight", "48m") *
      1024 * 1024)
```

......
}
```

在 MapOutputTracker 的 getMapSizesByExecutorId 方法处理代码如下：

```scala
def getMapSizesByExecutorId(shuffleId: Int, startPartition: Int, endPartition:
  Int): Seq[(BlockManagerId, Seq[(BlockId, Long)])] = {
//通过 shuffleId 获取上游 ShffleMapTask 输出数据的元数据
logDebug(s"Fetching outputs for shuffle $shuffleId, partitions
  $startPartition-$endPartition")
val statuses = getStatuses(shuffleId)

//使用同步方式把获取的 MapStatuses 序列号为
//Seq[(BlockManagerId, Seq[(BlockId, Long)])]格式
statuses.synchronized {
  return  MapOutputTracker.convertMapStatuses(shuffleId,  startPartition,
    endPartition, statuses)
}
}
```

获取上游 ShffleMapTask 输出数据的元数据是在 getStatuses 方法中，在该方法中通过同步的方式尝试在本地 mapStatus 中读取，如果成功获取，则返回这些信息；如果失败，则通过 RPC 通信框架发送请求到 MapOutputTrackerMaster 进行获取，具体代码如下：

```scala
private def getStatuses(shuffleId: Int): Array[MapStatus] = {
//根据 ShuffleMapTask 编号尝试从本地获取输出结果的元数据 MapStatus，
//如果不能获取这些信息，则向 MapOutputTrackerMaster 请求获取
val statuses = mapStatuses.get(shuffleId).orNull
if (statuses == null) {
  logInfo("Don't have map outputs for shuffle " + shuffleId + ", fetching them")
  val startTime = System.currentTimeMillis
  var fetchedStatuses: Array[MapStatus] = null
  fetching.synchronized {
    //其他人也在读取该信息，等待其他人读取完毕后再进行读取
    while (fetching.contains(shuffleId)) {
      try {fetching.wait()} catch {case e: InterruptedException =>}
    }

    //使用同步操作读取指定 Shuffle 编号的数据，该操作要么成功读取数据，要么其他人同时在
    //读取，此时把读取 Shuffle 编号加入到 fetching 读取列表中，在后续中读取
    fetchedStatuses = mapStatuses.get(shuffleId).orNull
    if (fetchedStatuses == null) {
      fetching += shuffleId
    }
  }
```
```

```
    if (fetchedStatuses == null) {
      try {
        //发送消息给MapOutputTrackerMaster，获取该ShffleMapTask输出的元数据
        val fetchedBytes=askTracker[Array[Byte]](GetMapOutputStatuses (shuffleId))

        //对获取的元数据进行反序列化
        fetchedStatuses = MapOutputTracker.deserializeMapStatuses(fetchedBytes)
        logInfo("Got the output locations")
        mapStatuses.put(shuffleId, fetchedStatuses)
      } finally {
        fetching.synchronized {
          fetching -= shuffleId
          fetching.notifyAll()
        }
……
}
```

（3）获取读取数据的位置信息后，返回到 ShuffleBlockFetcherIterator 的 initialize 方法，该方法是 Shufffle 读的核心代码所在。在该方法中先通过调用 splitLocalRemoteBlocks 方法对获取的数据位置信息进行区分，判断数据所处的位置是本地节点还是远程节点。如果是远程节点使用 fetchUpToMaxBytes 方法，从远程节点获取数据；如果是本地节点使用 fetchLocalBlocks 方法获取数据。

```
    private[this] def initialize(): Unit = {
      context.addTaskCompletionListener(_ => cleanup())
      //对获取数据位置的元数据进行区分，区分为本地节点还是远程节点
      val remoteRequests = splitLocalRemoteBlocks()
      fetchRequests ++= Utils.randomize(remoteRequests)

      //对于远程节点的数据，使用Netty网络方式读取
      fetchUpToMaxBytes()
      val numFetches = remoteRequests.size - fetchRequests.size
      logInfo("Started " + numFetches + " remote fetches in" +
        Utils.getUsedTimeMs(startTime))

      //对于本地节点的数据，需要注意的是，不同的写入方式，所采用的读取方式也不同，基于哈希的
      //Shuffle 写使用的是 FileShuffleBlockResolver 的 getBlockData 方法获取数据，而基于
      //排序的 Shuffle 写使用的是 IndexShuffleBlockResolver 的 getBlockData 方法获取数据
      fetchLocalBlocks()
      logDebug("Got local blocks in " + Utils.getUsedTimeMs(startTime))
    }
在 splitLocalRemoteBlocks 方法中划分数据读取方式：
    private[this] def splitLocalRemoteBlocks(): ArrayBuffer[FetchRequest] = {
```

```scala
        //设置每次请求的大小不超过maxBytesInFlight的1/5,该阈值由
        //spark.reducer.maxSizeInFlight配置项进行设置,默认情况该配置为48MB
        val targetRequestSize = math.max(maxBytesInFlight / 5, 1L)

        val remoteRequests = new ArrayBuffer[FetchRequest]
        var totalBlocks = 0
        for ((address, blockInfos) <- blocksByAddress) {
          totalBlocks += blockInfos.size
          if (address.executorId == blockManager.blockManagerId.executorId) {
            //当数据和所在BlockManager在一个节点时,把该信息加入到localBlocks列表中,
            //通过需要过滤大小为0的数据块
            localBlocks ++= blockInfos.filter(_._2 != 0).map(_._1)
            numBlocksToFetch += localBlocks.size
          } else {
            val iterator = blockInfos.iterator
            var curRequestSize = 0L
            var curBlocks = new ArrayBuffer[(BlockId, Long)]
            while (iterator.hasNext) {
              val (blockId, size) = iterator.next()
              if (size > 0) {
                //对于不空数据块,把其信息加入到列表中
                curBlocks += ((blockId, size))
                remoteBlocks += blockId
                numBlocksToFetch += 1
                curRequestSize += size
              } else if (size < 0) {
                throw new BlockException(blockId, "Negative block size " + size)
              }

              //按照不大于maxBytesInFlight的标准,把这些需要处理数据组合在一起
              if (curRequestSize >= targetRequestSize) {
                remoteRequests += new FetchRequest(address, curBlocks)
                curBlocks = new ArrayBuffer[(BlockId, Long)]
                curRequestSize = 0
              }
            }
            //剩余的处理数据组成一次请求
            if (curBlocks.nonEmpty) {
              remoteRequests += new FetchRequest(address, curBlocks)
            }
          }
        }
        remoteRequests
      }
```

(4) 数据读取完毕后，回到 BlockStoreShuffleReader 的 read 方法，判断是否定义聚合，如果需要，则根据键值调用 Aggregator 的 combineCombinersByKey 方法进行聚合。聚合完毕，使用外部排序（ExternalSorter）对数据进行排序并放入内存中。

(5) 读取数据后，判断 ShuffleDependency 是否定义聚合（Aggregation），如果需要，则根据键值进行聚合，需要注意的是，如果在上游 ShuffleMapTask 已经做了合并，则在合并数据的基础上做键值聚合。待数据处理完毕后，使用 ExternalSorter（外部排序）的 insertAll 方法对数据进行排序，该操作和 Shuffle 写排序是类似的。

```
override def read(): Iterator[Product2[K, C]] = {
  ……
  val aggregatedIter: Iterator[Product2[K, C]] = if (dep.aggregator.isDefined) {
    if (dep.mapSideCombine) {
      //对于上游 ShuffleMapTask 已经合并的，对合并的结果数据进行聚合
      val combinedKeyValuesIterator = interruptibleIter.asInstanceOf
        [Iterator[(K, C)]]
      dep.aggregator.get.combineCombinersByKey(combinedKeyValuesIterator,
        context)
    } else {
      //对未合并的数据进行聚合处理
      val keyValuesIterator = interruptibleIter.asInstanceOf[Iterator[(K,
        Nothing)]]
      dep.aggregator.get.combineValuesByKey(keyValuesIterator, context)
    }
  } else {
    interruptibleIter.asInstanceOf[Iterator[Product2[K, C]]]
  }

  dep.keyOrdering match {
    //对于需要排序，使用 ExternalSorter 进行排序，根据获取的排序方式，对数据进行排序
    //并写入到内存缓冲区中。如果排序中的 Map 占用的内存已经超越了使用的阈值，则将 Map 中
    //的内容溢写到磁盘中
    case Some(keyOrd: Ordering[K]) =>
      val sorter = new ExternalSorter[K, C, C](ordering = Some(keyOrd), serializer
        = Some(ser))
      sorter.insertAll(aggregatedIter)
      context.taskMetrics().incMemoryBytesSpilled(sorter.memoryBytesSpilled)
      context.taskMetrics().incDiskBytesSpilled(sorter.diskBytesSpilled)
      context.taskMetrics().incPeakExecutionMemory(sorter.peakMemoryUsedBytes)
      CompletionIterator[Product2[K, C], Iterator[Product2[K,
        C]]](sorter.iterator, sorter.stop())
    case None =>
      aggregatedIter
```

        }
    }

5.3 序列化和压缩

在分布式计算中，序列化和压缩是提升性能的两个重要手段。Spark 通过序列化将链式分布的数据转化为连续分布的数据，这样就能够进行分布式的进程间数据通信或者在内存进行数据压缩等操作，通过压缩能够减少内存占用以及 IO 和网络数据传输开销，提升 Spark 整体的应用性能。

5.3.1 序列化

在 Spark 中内置了两个数据序列化类：JavaSerializer 和 KryoSerializer，这两个继承于抽象类 Serializer，而在 Spark SQL 中 SparkSqlSerializer 继承于 KryoSerializer，它们之间关系如 5-13 Spark 序列化类图所示。

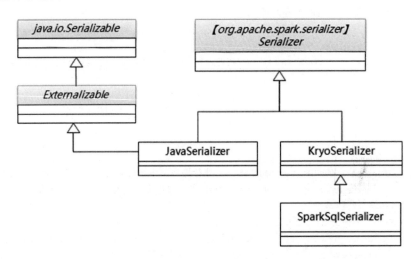

图 5-13　Spark 序列化类图

默认情况下 Spark 默认使用的 JavaSerializer 序列方法，它使用的是 Java 的 ObjectOutputStream 序列化框架。JavaSerializer 继承于 java.io.Serializable，虽然其灵活，但是它的性能不佳，而且生成的序列结果也较大，因此 Spark 提供性能更佳（一般比 Java 序列化快一个数量级）、压缩效率更高的 KryoSerializer 方法。不过使用中需要注意，KryoSerializer 并不支持所有的系列化对象，而且要求用户注册类。如果这两种序列化方法不满足要求，也可以通过集成 Serializer 类自定义新的序列化方法。

Spark 初始序列化是在 SparkEnv 类进行创建，在该类中根据 spark.serializer 配置初始化序列化实例，然后把该实例作为参数初始化 SerializerManager 实例，而 SerializerManager 作为参数初始化 BlockManager，代码如下所示：

```
val serializer = instantiateClassFromConf[Serializer](
  "spark.serializer", "org.apache.spark.serializer.JavaSerializer")
val serializerManager = new SerializerManager(serializer, conf)
val closureSerializer = new JavaSerializer(conf)
```

需要注意的是，这里可配的序列化的对象是 Shuffle 数据以及 RDD 缓存等场合，对于 Spark 任务的序列化是通过 spark.closure.serializer 来配置，目前只支持 JavaSerializer。

5.3.2 压缩

Spark 内置提供了三种压缩方法，分别是：LZ4、LZF 和 Snappy，这三个方法均继承于特质类 CompressionCodec，并实现了其压缩和解压两个方法，它们之间关系如 5-14 Spark 压缩类图所示。

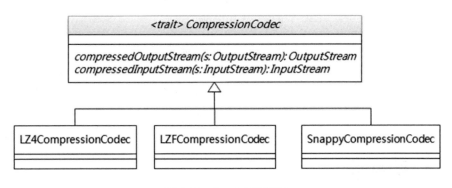

图 5-14　Spark 压缩类图

这三个压缩算法采用了第三方库实现的，Snappy 提供了更高的压缩速度，LZF 提供了更高的压缩比，LZ4 提供了压缩速度和压缩比俱佳的性能：

- Snappy：使用了 org.xerial.snappy 库，Snappy 算法的前身是 Zippy，被 Google 用于 MapReduce、BigTable 等许多内部项目；
- LZF：使用了 com.ning.compress.lzf 库，其中 Ning-compress 是一个对数据进行 LZF 格式压缩和解压缩的库；

- LZ4：使用了 net.jpountz.lz4 库，它是一个高效的无损压缩算法，可扩展支持多核 CPU。

Spark 中相关压缩配置项如表 5-4 所示。

表 5-4　Spark 压缩配置项

配置项	默认值	说明
spark.io.compression.codec	LZ4	RDD 缓存和 Shuffle 数据压缩所采用的算法，具体采用什么算法需要权衡 CPU、网络、磁盘的能力和负载。在 Shuffle 数据时，可能采取基于哈希的写操作，如果 Reduce 分区数量巨大，需要同时打开大量的压缩数据流用于写文件，会需要大量的内存。对于 RDD 缓存而言，绝大多数场合都是内存操作或者本地 IO，所以 CPU 负载的问题可能比 IO 的问题更加突出。LZF 和 Snappy 相比较，前者压缩率比较高（通常要高 20%左右），但是 CPU 代价较大，LZ4 提供了压缩速度和压缩比俱佳的性能。
spark.rdd.compress	false	该参数决定了 RDD 缓存过程中，RDD 数据在序列化之后是否进一步进行压缩再储存到内存或磁盘上，该参数主要考虑磁盘的 IO 带宽。默认情况下该参数 false，也就是不使用压缩。
spark.broadcast.compress	true	由于广播变量需要通过网络发送，所以通过压缩减小传播数据大小，以减少网络传输开销和内存占用是必要的。默认情况下该参数 true，也就是使用压缩。
spark.io.compression.snappy.blockSize	32k	设置使用 Snappy 压缩算法的块大小
spark.io.compression.lz4.blockSize	32k	设置使用 LZ4 压缩算法的块大小

5.4　共享变量

通常情况下，当一个函数传递给远程集群节点上运行的 Spark 操作时（如 Map、Reduce），该函数中所有的变量都会在各节点中创建副本，在各节点中的变量相互隔离并由所在节点的函数进行调用，并且这些变量的更新都不会传递回 Driver 程序。在任务间进行通用、可读写的共享变量是低效的，然而 Spark 还是提供了两种类型的共享变量：广播变量和累加器。

5.4.1　广播变量

广播变量允许开发人员在每个节点缓存只读的变量，而不是在任务之间传递这些变量。例如，使用广播变量能够高效地在集群每个节点创建大数据集的副本。同时，Spark 还使用高效的广播算法分发这些变量，从而减少通信的开销。

Spark 应用程序作业的执行由一系列调度阶段构成，而这些调度阶段通过 Shuffle 进行分隔。Spark 能够在每个调度阶段自动广播任务所需通用的数据，这些数据在广播时需进行序列化缓存，并在任务运行前需进行反序列化。这就意味着当多个调度阶段的任务需要相同的数据，显式地创建广播变量才有用。

可以通过调用 SparkContext.broadcast(v) 创建一个广播变量 v，该广播变量封装在 v 变量中，可使用获取该变量 value 的方法进行访问。代码如下所示

```
scala> val broadcastVar = sc.broadcast(Array(1, 2, 3))
broadcastVar: org.apache.spark.broadcast.Broadcast[Array[Int]] = Broadcast(0)

scala> broadcastVar.value
res0: Array[Int] = Array(1, 2, 3)
```

当广播变量创建后，在集群中所有函数将以变量 v 代表该播变量，并且该变量 v 一次性分发到各节点上。另外，为了确保所有的节点获得相同的变量，对象 v 广播后只读不能够被修改。

5.4.2 累加器

累加器是 Spark 中仅有通过关联操作进行累加的变量，因此能够有效地支持并行计算，它们能够用于计数（如 MapReduce）和求和。Spark 原生支持数值类型的累加器，不过开发人员能够定义新的类型。如果在创建累加器时指定了名称，可以通过 Spark 的 UI 监控界面中进行查看，这种方式能够帮助理解作业所构成的调度阶段执行过程。

通过调用 SparkContext.accumulator(v) 方法初始化累加器变量 v，在集群中的任务能够使用加法或者"+="操作符进行累加操作（在 Scala 和 Python 中）。然而，它们不能在应用程序中读取这些值，只能由 Driver 程序通过读方法获取这些累加器的值。

下面代码演示如何把一个数组的元素追加到累加器中：

```
scala> val accum = sc.accumulator(0, "My Accumulator")
accum: spark.Accumulator[Int] = 0

scala> sc.parallelize(Array(1, 2, 3, 4)).foreach(x => accum += x)
...
10/09/29 18:41:08 INFO SparkContext: Tasks finished in 0.317106 s

scala> accum.value
res2: Int = 10
```

尽管上面的例子使用 Spark 原生所支持的累加器 Int 类型，但是开发人员能够通过继承 AccumulatorParam 类来创建自定义的累加类型。AccumulatorParam 接口提供了两个方法：zero 方法为自定义类型设置"0 值"和 addInPlace 方法将两个变量进行求和。例如，下面将对 Vector 类所提供的向量 vector 进行求和，代码如下：

```
object VectorAccumulatorParam extends AccumulatorParam[Vector] {
  def zero(initialValue: Vector): Vector = {
    Vector.zeros(initialValue.size)
  }
  def addInPlace(v1: Vector, v2: Vector): Vector = {
    v1 += v2
  }
}

//可以创建向量的累加器变量
val vecAccum = sc.accumulator(new Vector(...))(VectorAccumulatorParam)
```

在 Scala 中，尽管结果的类型和累加元素的数据类型可能存在不一致的情况，Spark 提供更通用的接口来累加数据（例如，通过创建一个列表来容纳累加的元素），另外 SparkContext.accumulableCollection 提供了通用的方法来累加 Scala 集合类型。

累加器只能由 Spark 内部进行更新，并保证每个任务在累加器的更新操作仅执行一次，也就是说，重启任务也不应该更新。在转换操作中，用户必须意识到任务和作业的调度过程重新执行会造成累加器多次更新。

累加器同样具有 Spark 懒加载的求值模型。如果它们在 RDD 的操作中进行更新，它们的值只在 RDD 进行行动操作时才进行更新。因此，当执行如 Map 懒加载操作时，累加器并没有立即更新。以下代码片段演示了该特性：

```
//此时 accum 的值仍然是 0，因为没有动作操作引起 map 的计算
val accum = sc.accumulator(0)
data.map { x => accum += x; f(x) }
```

5.5 实例演示

在该实例中，我们将计算所有年份的最高温度，在此过程中观察基于哈希的 Shuffle 写操作和基于排序的 Shuffle 写操作两种情况下中间文件的生成。为了便于观测，将使用第 4 章中计算气象数据的部分数据，Shuffle 中间文件个数据可以使用 find 进行查找，并与 UI 监控得到各调度阶段和每个调度阶段中任务数计算进行比对。

1. 启动运行环境

参见 4.6.1 节实例需要启动 HDFS、Spark 集群和 Spark Shell，在该 Spark 集群中有 3 个节点，每个节点中的 Executor 使用一个核数和 1.5GB 内存。

2. 运行代码

计算该地区所有年份的最高温度代码如下：

```
scala> val tem_data = sc.textFile("hdfs://master:9000/ussr/f234*.dat")
scala> val fields = tem_data.map(line => line.trim().replace("  "," ").replace("
    "," ")split(" ")).filter(_.length==15).filter(_(10)!="9").filter(_(9)!=
    "999.9")
scala> val map_temp = fields.map(fields => (fields(1),fields(9).toDouble))
scala> val max_temp = map_temp.map(x=>(x._2,x._1)).sortByKey(false).map(x=>
    (x._2,x._1)).collect
```

具体过程分为 4 个步骤：

（1）从 HDFS 读取数据，为方便观察方便，选取了/ussr 目录下匹配 f23*.dat 的数据。

（2）对读取的数据进行清洗，需要注意的是，这些数据字段之间空白有占一个字符和两个字符，使用 replace 进行替换。另外，由于有些观察站数据缺失或异常，使用 filter 进行过滤。

（3）对清洗的数据获取年份和每日最高温度字段进行 Map 操作。

（4）对每日最高温度进行排序，排序方法是先把 map 的键和值位置对换，然后使用 sortByKey 进行排序，排序后再次把 map 的键和值位置对换，就得到所有年份最高温度降序排序，如图 5-15 所示。

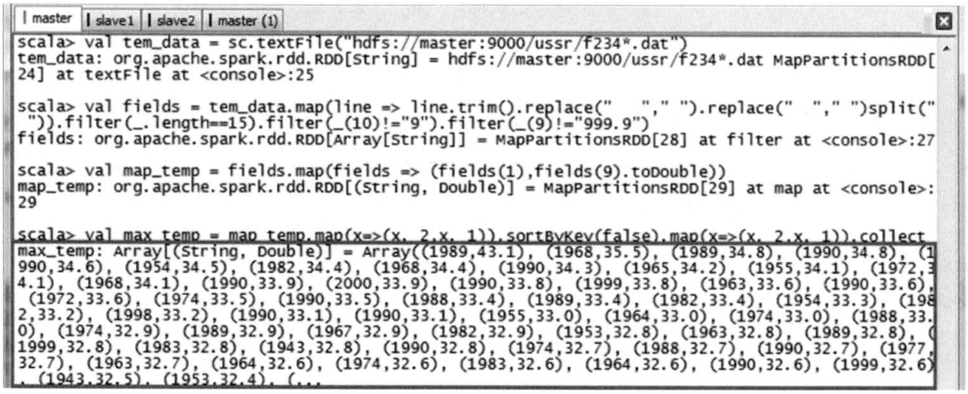

图 5-15 计算所有年份的最高温度结果

3. 基于排序的 Shuffle 写操作结果

在前面步骤执行过程中，新开一个终端使用如下命令查看在 Shuffle 阶段生成的数据：
`$find /tmp/spark* -type f ! -name "*.class"`

在该命令中，查找在 /tmp 目录下以 spark 字符开头的所有文件，但不包含 class 文件。由于 Spark 默认情况下是基于排序的 Shuffle 写操作，在 master、slave1 和 slave2 3 个节点中各生成了一个 data 文件和一个 index 文件，这些文件如下所述：

master 节点生成的 Shuffle 文件：
```
/tmp/spark-23f99…/executor-f46a3f…/blockmgr-02606111…/36/shuffle_0_2_0.data
/tmp/spark-23f99…/executor-f46a3f…/blockmgr-02606111…/32/shuffle_0_2_0.index
```

slave1 节点生成的 Shuffle 文件：
```
/tmp/spark-81203…/executor-1c609f…/blockmgr-d38ba…/0c/shuffle_0_0_0.data
/tmp/spark-81203…/executor-1c609…/blockmgr-d38ba…/30/shuffle_0_0_0.index
```

slave2 节点生成的 Shuffle 文件：
```
/tmp/spark-b429d…/executor-24e2f…/blockmgr-17699…/15/shuffle_0_1_0.data
/tmp/spark-b429d…/executor-24e2f…/blockmgr-17699…/0f/shuffle_0_1_0.index
```

通过查看 UI 监控页面，该应用程序生成了两个作业，其中在第二个作业中包含了两个调度阶段，在调度阶段 1 中进行了 Shuffle 写操作，写的数据为 289.3KB 数据，如图 5-16 所示。

Stage Id	Description	Submitted	Duration	Tasks: Succeeded/Total	Input	Output	Shuffle Read	Shuffle Write
2	collect at <console>:31 +details	2016/05/27 23:01:07	6 s	3/3			289.3 KB	
1	map at <console>:31 +details	2016/05/27 23:01:05	2 s	3/3	4.6 MB			289.3 KB

图 5-16　计算所有年份的最高温度 Shuffle 读写情况

通过观察调度阶段 1 的详细监控页面，我们可以观察到各个节点 Shuffle 写的情况，例如：master 写 66.5KB，数据条数为 15336 条。我们也观察到该调度阶段包含 3 个任务，每个任务在各个节点对应生成一个 data 文件和一个 index 文件，如图 5-17 所示。

Index	ID	Attempt	Status	Locality Level	Executor ID / Host	Launch Time	Duration	GC Time	Input Size / Records	Write Time	Shuffle Write Size / Records	Errors
0	3	0	SUCCESS	NODE_LOCAL	1 / slave1	2016/05/27 23:01:04	2 s	0.2 s	2.4 MB (hadoop) / 40177	15 ms	136.4 KB / 31872	
1	4	0	SUCCESS	NODE_LOCAL	2 / slave2	2016/05/27 23:01:04	2 s	0.1 s	1325.8 KB (hadoop) / 21850	27 ms	86.4 KB / 20053	
2	5	0	SUCCESS	NODE_LOCAL	0 / master	2016/05/27 23:01:04	2 s	15 ms	943.8 KB (hadoop) / 15341	35 ms	66.5 KB / 15336	

图 5-17　计算所有年份的最高温度调度阶段 1 的详细监控页面

调度阶段 2 即进行 Shuffle 读的操作，在该调度阶段有 3 个任务，分别读取了前一调度阶段写的数据，每个任务会分别从 Shuffle 写的文件读取数据，两者的大小也就前面显示的289.3KB。

4．基于哈希的 Shuffle 写操作结果

在$SPARK/conf 目录中修改 spark-default.conf，如果没有该文件先复制一份。在该配置文件中，加入如下配置项，即使用了基于哈希的 Shuffle 写：

```
spark.shuffle.manager=hash
```

再次执行计算所有年份最高温度代码，通过观察，在 master、slave1 和 slave2 3 个节点分别生成了 3 个文件。

master 节点生成的 Shuffle 文件：

```
/tmp/spark-97f3f…/executor-905f3fd1-…/blockmgr-7af93f50-…/0d/shuffle_0_1_0
/tmp/spark-97f3f…/executor-905f3fd1-…/blockmgr-7af93f50-…/0e/shuffle_0_1_1
/tmp/spark-97f3f…/executor-905f3fd1-…/blockmgr-7af93f50-…/0f/shuffle_0_1_2
```

slave1 节点生成的 Shuffle 文件：

```
/tmp/spark-d1a26…/executor-0ddfc4fa-…/blockmgr-b4b4ffbf-…/0d/shuffle_0_0_1
/tmp/spark-d1a26…/executor-0ddfc4fa-…/blockmgr-b4b4ffbf-…/0c/shuffle_0_0_0
/tmp/spark-d1a26…/executor-0ddfc4fa-…/blockmgr-b4b4ffbf-…/0e/shuffle_0_0_2
```

slave2 节点生成的 Shuffle 文件：

```
/tmp/spark-bd9ca…/executor-603495e3-…/blockmgr-749455cd-…/0e/shuffle_0_2_0
/tmp/spark-bd9ca…/executor-603495e3-…/blockmgr-749455cd-…/0f/shuffle_0_2_1
/tmp/spark-bd9ca…/executor-603495e3-…/blockmgr-749455cd-…/10/shuffle_0_2_2
```

前面我们知道 Shuffle 文件命名规则为："shuffle_" + shuffleId + "_" + mapId + "_" + reduceId，在同一个调度阶段中，shuffleId 相同，在每个节点中差异在 reduceId，而不同节点差异在 mapId。例如 shuffle_0_1_2 表示该 Shuffle 中，第 2 个 map 任务生成的文件，被后续的第 3 个 reduce 任务读取。

5．结论

通过两种 Shuffle 写的操作结果，我们可以得出如下结论，这些结论与 5.2.2 节中分析的情况保持一致。

- 基于哈希的 Shuffle 写操作，Shuffle 运行过程中所创建的文件数是 S*F，其中 S 为当前 Shuffle Map Task 的任务数，而 F 为后续任务的任务数。
- 基于排序的 Shuffle 写操作，Shuffle 运行过程中 Map Task 不会为后续的每个任务创建单独的文件，而是会将所有的结果写到同一个文件中，对应的生成一个 Index 文件进行索引。

5.6 小结

在本章节中先介绍了 Spark 存储原理，Spark 存储方式包括内存和磁盘，存储采取了主从模式（Master/Slave 模式），Master 负责整个应用程序运行期间的数据块元数据的管理和维护，而 Slave 一方面负责将本地数据块的状态信息上报给 Master，并接收从 Master 传过来的执行命令。Spark 存储过程分为读数据和写数据两个过程，读数据分为本地读取和远程节点读取两种方式。

然后，介绍了 Shuffle 操作，在 Spark 中 Shuffle 分为写和读两种操作。对于 Shuffle 写操作分为基于哈希的 Shuffle 写操作和基于排序的 Shuffle 写操作两种类型，基于哈希的 Shuffle 写操作在 Map 和 Reduce 数量较大的情况会导致写文件数量大和缓存开销过大，基于排序的 Shuffle 写操作通过把所有的结果写到一个文件中，同时生成一个索引文件进行定位进行解决。

接着，介绍了 Spark 中的序列化和压缩，Spark 通过序列化将不连续的数据转化为连续分布的数据，并对这些数据进行压缩，通过该操作能够减少内存占用以及 IO 和网络数据传输开销。同时也介绍了 Spark 的两种共享变量：广播变量和累加器，广播变量能够把数据高效地传递到集群的节点内存中，在节点中所运行的作业和任务能够进行调用，累加器则有效地支持并行计算，能够用于计数和求和，它不仅支持原生的数值类型，而且还通过扩展支持新的类型。

最后，观察基于哈希的 Shuffle 写操作和基于排序的 Shuffle 写操作两种情况下中间文件的生成，能够看到在两种情况下生成文件的类型和个数均不相同。

第 6 章

Spark 运行架构

Spark 注重建立良好的生态系统，它不仅支持多种外部文件存储系统，而且提供了多种运行模式。部署在单台机器上时，既可以用本地（Local）模式运行，也可以使用伪分布式模式来运行；当以分布式集群部署的时候，可以根据自己集群的实际情况选择独立（Standalone）运行模式（Spark 自带的模式）、YARN 运行模式，还是 Mesos 运行模式。Spark 各种运行模式虽然在启动方式、运行位置、调度策略和异常处理上各有不同，但它们的目的基本都是一致的，就是在合适的位置安全可靠地根据用户的配置运行/管理作业和任务。

6.1 运行架构总体介绍

6.1.1 总体介绍

Spark 虽然支持多种运行模式，但 Spark 应用程序的运行架构基本由三部分组成，包括 SparkContext（驱动程序）、ClusterManager（集群资源管理器）和 Executor（任务执行进程）组成。

其中，SparkContext 用于负责与 ClusterManager 通信，进行资源的申请、任务的分配和监控等，负责作业执行的全生命周期管理。ClusterManager 提供了资源的分配和管理，在不同的运行模式下，担任的角色有所不同，在本地运行、Spark Standalone 等运行模式中由 Master 提供，在 YARN 运行模式作用由 Resource Manager 担任，在 Mesos 运行模式下由 Mesos Manager 进行负责。当 SparkContext 对运行的作业划分并分配资源后，会把任务发送 Executor 进行运行。

每个应用程序获取专属的 Executor 进程，这些进程在应用程序运行的过程中一直驻留，并以多线程方式运行任务。这种隔离机制有以下两种优势，一是从调度角度来看，每个 Driver 端调度它自己的任务，二是从运行角度来看，来自不同应用程序的任务运行在不同的 JVM 中。当然，这也意味着 Spark 应用程序不能跨应用程序共享数据，除非将数据写入到外部存储系统。

对于潜在的集群资源管理器来说，Spark 是不可知的，也就是说 Spark 与资源管理器无关，只要能够获取 Executor 进程，并能保持相互之间的通信就可以了，即使是在支持其他应用程序的集群资源管理器（例如 YARN、Mesos 等）上运行也相对简单。

6.1.2 重要类介绍

在 4.2.1 节中我们知道，TaskScheduler 是最重要 Spark 调度器之一。它负责具体任务的调度执行，而 SchedulerBackend 则负责应用程序运行期间与底层资源调度系统交互。应用程序运行过程是：首先，在 SparkContext 启动时，调用 TaskScheduler.start 方法启动 TaskScheduler 调度器；然后，当 DAGScheduler 调度阶段和任务拆分完毕时，调用 TaskScheduler.submitTasks 方法提交任务，SchedulerBackend 接到执行任务时，通过 reviveOffers 方法分配运行资源并启动运行节点的 Executor；最后，由 TaskScheduler 接收任务运行状态，如果任务运行完毕，则继续分配，直至应用程序所有任务运行完毕。

- **TaskScheduler**：该类是高层调度器 DAGScheduler 与任务执行 SchedulerBackend 的桥梁。TaskScheduler 是特质类（trait），定义了任务调度相关的实现方法，TaskSchedulerImpl 是其最重要的子类，它实现了 TaskScheduler 所有的接口方法，而 TaskScheduler 的孙子类 YarnScheduler 和 YarnClusterScheduler 只是重写其中一两个方法，如图 6-1 所示。

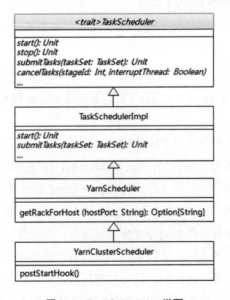

图 6-1　TaskScheduler 类图

- **SchedulerBackend**：该类为特质类（trait），子类根据不同运行模式分为本地运行模式的 LocalBackend、粗粒度运行模式的 CoarseGrainedSchedulerBackend 和细粒度 Mesos 运行模式的 MesosSchedulerBackend 等。粗粒度运行模式包括独立运行模式（Standalone）的 SparkDeploySchedulerBackend、YARN 运行模式的 YarnSchedulerBackend 和粗粒度 Mesos 运行模式的 CoarseMesosSchedulerBackend 等。其中，YARN 运行模式根据 SparkContext 运行位置的不同分为 Yarn-Client 运行模式的 YarnClientSchedulerBackend 和 Yarn-Cluster 运行模式的 YarnClusterSchedulerBackend。其类图关系如图 6-2 所示。

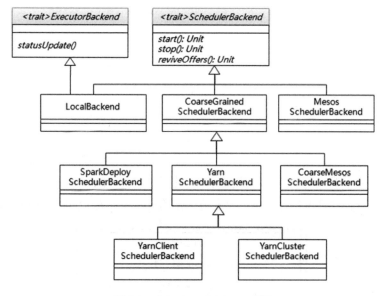

图 6-2　SchedulerBackend 类图

6.2　本地（Local）运行模式

6.2.1　运行模式介绍

在本地运行模式中，Spark 所有进程都运行在一台机器的 JVM 中。该运行模式一般用于测试等用途。在运行中，如果在命令语句中不加任何配置，Spark 默认设置为 Local 模式，本地模式的标准写法是 local[N]，这里的 N 表示的是打开 N 个线程进行多线程运行。下面演示的是使用 4 个线程来运行 LocalPi。

```
$./bin/run-example org.apache.spark.examples.localPi local[4]
```

除了在命令终端提交任务的部署工具后面的附加参数添加运行模式，我们还可以直接在编写的代码的具体实现中添加运行模式。通过在 SparkConf 对象的 setMaster 方法里添加运行模式来指定。如下面程序，我们指定应用程序采用 Local 模式来实现：

```
object sparkPi {
  def main(args: Array[String]) {
    val conf = new SparkConf().setMaster("local").setAppName("Spark Pi")
    val spark = new SparkContext(conf)
    val silces = if (args.length > 0) args(0).toInt else 2
    val n = 100000 * slices
    val count = spark.parallelize(1 to n, slices.map { i =>
      val x = random * 2 - 1
      val y = random * 2 -1
      if(x*x + y*y < 1) 1 else 0
    }.reduce(_ + _)
    println("Pi is roughly " + 4.0 * count / n)
    spark.stop()
  }
}
```

这里需要注意的是，在用户自己编写的程序中设置的运行模式的优先级要大于在 Spark 应用程序配置文件中参数设置的值。但是我们为了使得应用程序能够更加灵活地部署在各种模式下，不建议把运行模式硬编码在代码中。本地运行模式的运行流程如图 6-3 所示。

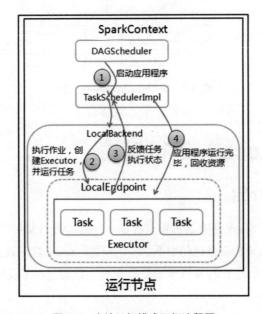

图 6-3　本地运行模式运行流程图

（1）启动应用程序，在 SparkContext 启动过程中，初始化 DAGScheduler 和 TaskSchedulerImpl 两个调度器，同时初始化 LocalBackend 以及 LocalEndpoint 本地终端点。

（2）对作业进行划分调度阶段后，任务集按照拆分顺序发送任务到 LocalEndpoint 本地终端点，本地终端点接收到任务集时在本地启动 Executor，启动完毕后在该 Executor 中执行接收到任务集。如果设置了多线程方式，则启动多个线程并行处理任务。

（3）Executor 执行任务状态通过 LocalEndpoint 本地终端点反馈给上层作业调度器，上层作业调度器根据接收到的信息更新任务状态，同时根据任务状态调整任务集状态。如果任务集运行状态为完成，则发送下一任务集。

（4）当应用程序完毕后进行回收资源，上层作业调度器注销在 LocalBackend 中运行的 Executor，注销完毕后释放 DAGScheduler、TaskSchedulerImpl 和 LocalBackend 等进程。

6.2.2 实现原理

图 6-4 是本地运行模式作业运行类调用关系图，相比独立运行模式作业运行类调用图，少了 CoarscGrainedExecutorBackend 类。其原因在于 LocalBackend 类继承于 SchedulerBackend 和 ExecutorBackend，在 LocalBackend 的 start 方法中启动本地终端点 LocalEndpoint，同时启动了 Executor。

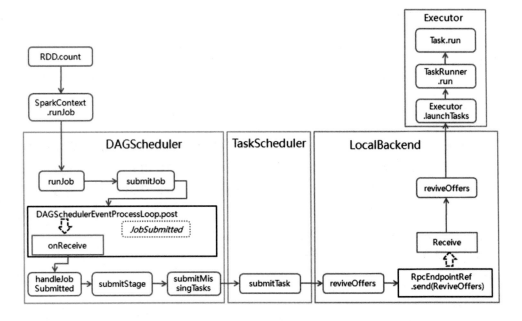

图 6-4 本地运行模式类调用关系图

前面对本地运行模式的运行流程进行了分析,下面将根据运行流程的步骤分析其代码实现原理,具体如下:

(1) 在 SparkContext 对象启动时,会在内部初始化 TaskScheduler 和 SchedulerBackend,在其 createTaskScheduler 方法中根据模式匹配初始化 TaskSchedulerImpl 和 LocalBackend。启动本地运行模式分为单线程和多线程运行处理模式。当匹配字符串中未指定运行线程数时,以单线程处理模式运行,运行时启动一个线程处理任务;当匹配字符串中以 local[N] 指定运行线程数时,启动 N 个线程同时处理任务。其匹配代码如下:

```
//未指定运行线程数时以单线程处理模式运行,运行时启动给一个线程处理任务
case "local" =>
val scheduler=new TaskSchedulerImpl(sc,MAX_LOCAL_TASK_FAILURES,isLocal = true)
//启动单线程处理任务
val backend = new LocalBackend(sc.getConf, scheduler, 1)
scheduler.initialize(backend)
(backend, scheduler)

//以 local[N]指定运行线程数时,启动 N 个线程同时处理任务
val LOCAL_N_REGEX = """local\[([0-9]+|\*)\]""".r
case LOCAL_N_REGEX(threads) =>
  //获取运行节点可用 CPU 核数,当匹配字符串为 local[*]时,启动 CPU 核数的进程数量
  def localCpuCount: Int = Runtime.getRuntime.availableProcessors()
  val threadCount = if (threads == "*") localCpuCount else threads.toInt
  if (threadCount <= 0) {
    throw new SparkException(s"Asked to run locally with $threadCount threads")
  }
  val scheduler=new TaskSchedulerImpl(sc,MAX_LOCAL_TASK_FAILURES,isLocal=true)
  val backend = new LocalBackend(sc.getConf, scheduler, threadCount)
  scheduler.initialize(backend)
  (backend, scheduler)
```

(2) 参见本地模式作业运行类调用关系图,在 DAGScheduler 类中对作业进行划分调度阶段,然后使用 submitStage 方法进行提交调度阶段。在该方法中使用 submitMissingTasks 提交调度阶段的任务集,提交的任务集在 TaskSchedulerImpl.submitTasks 方法中发送消息给 LocalEndpoint 本地终端点启动 Executor,启动完毕后在该 Executor 中执行任务集。在 LocalEndpoint 类中 Executor 启动和执行任务集代码如下:

```
//启动 Executor,启动 isLocal 为真表示本地启动
private val executor = new Executor(localExecutorId, localExecutorHostname,
    SparkEnv.get, userClassPath, isLocal = true)
def reviveOffers() {
  val offers = Seq(new WorkerOffer(localExecutorId, localExecutorHostname,
```

```
    freeCores))
//根据设置线程数启动相应的线程处理任务
for (task <- scheduler.resourceOffers(offers).flatten) {
  freeCores -= scheduler.CPUS_PER_TASK
  executor.launchTask(executorBackend, taskId = task.taskId, attemptNumber =
    task.attemptNumber,task.name, task.serializedTask)
  }
}
```

（3）Executor 执行任务状态通过 LocalEndpoint 本地终端点进行更新，该终端点调用 TaskSchedulerImpl.statusUpdate 方法实现，然后继续调用 reviveOffers 运行任务集中其他的任务，当任务集运行完毕后更新任务集状态，高层调度器会发送下一个任务集。在本地终端点进行任务更新代码如下：

```
case StatusUpdate(taskId, state, serializedData) =>
  scheduler.statusUpdate(taskId, state, serializedData)
  if (TaskState.isFinished(state)) {
    freeCores += scheduler.CPUS_PER_TASK
    reviveOffers()
  }
```

（4）当所有调度阶段执行完毕时（参见 4.2.7 节的内容）应用程序也就运行完成，此时需要进行回收资源，回收顺序为先注销在 LocalBackend 中运行的 Executor，然后释放 TaskSchedulerImpl 实例，再释放 DAGScheduler，最终注销 SparkContext。

6.3 伪分布（Local-Cluster）运行模式

6.3.1 运行模式介绍

伪分布运行模式顾名思义是在一台机器中模拟集群运行，相对独立运行模式中 Master、Worker 和 SpartContext 在不同节点，伪分布运行模式中这些进程都是在一台机器上。伪分布运行模式既可以在$SPARK_HOME/conf/slaves 配置 Woker 节点为本地机器名来实现，也可以在脚本中通过执行 local-cluster 匹配字符串来执行。下面演示在一个节点中模拟启动 3 个 Worker 进程，每个 Worker 进程启动两个 CPU 核和 1024MB 内存来运行 LocalPi。

```
$./bin/run-example org.apache.spark.examples.localPi local-cluster[3,2,1024]
```

伪分布运行模式运行流程和独立运行模式相同，区别在于伪分布运行模式运行在一个节点中。其运行流程图如图 6-5 所示。

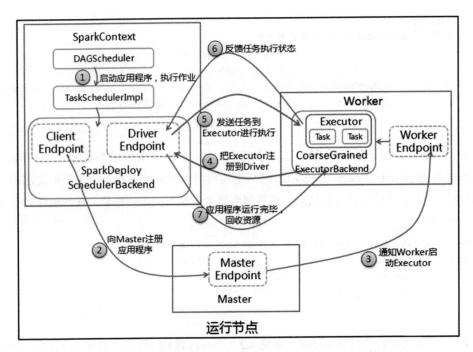

图 6-5 伪分布运行模式运行流程

6.3.2 实现原理

类似于独立运行模式，在 SparkContext 对象启动时，在 createTaskScheduler 方法中根据模式匹配初始化 TaskSchedulerImpl 和 SparkDeploySchedulerBackend。具体匹配代码如下：

```
//伪分布运行模式匹配字符串，需以 local-cluster 开头并包括 Worker 进程数、
//每个 Worker 进程启动 CPU 核数和使用的内存数
val LOCAL_CLUSTER_REGEX = """local-cluster\[\s*([0-9]+)\s*,\s*([0-9]+)\s*,
 \s*([0-9]+)\s*]""".r
case LOCAL_CLUSTER_REGEX(numSlaves, coresPerSlave, memoryPerSlave) =>
  val memoryPerSlaveInt = memoryPerSlave.toInt
  if (sc.executorMemory > memoryPerSlaveInt) {
    throw new SparkException("Asked to launch cluster with %d MB RAM / worker but
      requested %d MB/worker".format(memoryPerSlaveInt, sc.executorMemory))
  }
  val scheduler = new TaskSchedulerImpl(sc)

//通过 Worker 进程数、每个 Worker 进程启动 CPU 核数和使用的内存数模拟启动伪集群
val localCluster=new LocalSparkCluster(numSlaves.toInt, coresPerSlave.toInt,
  memoryPerSlaveInt, sc.conf)
```

```
val masterUrls = localCluster.start()
val backend = new SparkDeploySchedulerBackend(scheduler, sc, masterUrls)
scheduler.initialize(backend)

//通过回调函数设置伪集群关闭方法
backend.shutdownCallback = (backend: SparkDeploySchedulerBackend) => {
  localCluster.stop()
}
(backend, scheduler)
```

启动 Master 和 Worker 进程是在 LocalSparkCluster.start 方法中，先设置 REST 和 Shuffle 服务为 false，然后启动 Master 进程，默认情况下为 spark://localHost:7077,在启动 Worker 进程，Worker 进程数、每个 Worker 进程启动 CPU 核数和使用的内存数由传入的参数确定，启动完毕后返回 master 对象。LocalSparkCluster.start 具体代码如下：

```
def start(): Array[String] = {
  val _conf = conf.clone()
    .setIfMissing("spark.master.rest.enabled", "false")
    .set("spark.shuffle.service.enabled", "false")

  //启动 Master 进程，获取 Master 访问路径
  val (rpcEnv, webUiPort, _) = Master.startRpcEnvAndEndpoint(localHostname, 0,
    0, _conf)
  masterWebUIPort = webUiPort
  masterRpcEnvs += rpcEnv
  val masterUrl = "spark://" + Utils.localHostNameForURI() + ":" +
    rpcEnv.address.port
  val masters = Array(masterUrl)

  //根据参数启动对应 Worker 进程数、每个 Worker 进程启动 CPU 核数和使用的内存数
  for (workerNum <- 1 to numWorkers) {
    val workerEnv = Worker.startRpcEnvAndEndpoint(localHostname, 0, 0,
      coresPerWorker, memoryPerWorker, masters, null, Some(workerNum), _conf)
    workerRpcEnvs += workerEnv
  }

  masters
}
```

伪分布运行模式作业运行类调用关系图和独立运行模式一致，参见图 4-10。

6.4 独立（Standalone）运行模式

6.4.1 运行模式介绍

独立运行模式是 Spark 自身实现的资源调度框架，由客户端、Master 节点和 Worker 节点组成，其中 SparkContext 既可以运行在 Master 节点上，也可以运行在本地客户端。当用 Spark-Shell 交互式工具提交作业或者直接使用 run-example 脚本来运行示例时，SparkContext 在 Master 节点上运行；当使用 Spark-Submit 工具提交作业或者在 Eclipse、IDEA 等开发平台上运行 Spark 作业时，SparkContext 是运行在本地客户端。Worker 节点可以通过 ExecutorRunner 运行在当前节点上的 CoarseGrainedExecutorBackend 进程，每个 Worker 节点上存在一个或多个 CoarseGrainedExecutorBackend 进程，每个进程包含一个 Executor 对象。该对象持有一个线程池，每个线程可以执行一个任务。独立运行模式运行流程如图 6-6 所示。

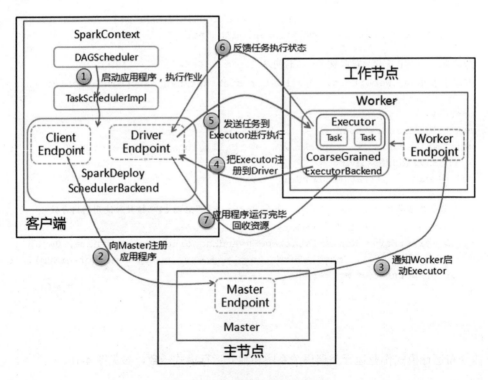

图 6-6 独立运行模式运行流程

（1）启动应用程序，在 SparkContext 启动过程中，先初始化 DAGScheduler 和

TaskSchedulerImpl 两个调度器，同时初始化 SparkDeploySchedulerBackend，并在其内部启动终端点 DriverEndpoint 和 ClientEndpoint。

（2）终端点 ClientEndpoint 向 Master 注册应用程序，Master 收到注册消息把该应用加入到等待运行应用列表中，等待由 Mater 分派给该应用程序 Worker。

（3）当应用程序获取到 Worker 时，Master 会通知 Worker 中的终端点 WokerEndpoint 创建 CoarseGrainedExecutorBackend 进程，在该进程中创建执行容器 Executor。

（4）Executor 创建完毕后发送消息给 Master 和终端点 DriverEndpoint，告知 Executor 已经创建完毕，在 SparkContext 成功注册后，等待接收从 Driver 终端点发送执行任务的消息。

（5）SparkContext 分配任务集给 CoarseGrainedExecutorBackend 执行，任务执行是在 Executor 按照一定调度策略进行的。

（6）CoarseGrainedExecutorBackend 在任务处理过程中，把处理任务的状态发送给 SparkContext 的终端点 DriverEndpoint，SparkContext 根据任务执行不同的结果进行处理。如果任务集处理完毕后，则会继续发送其他的任务集。

（7）应用程序运行完成后，SparkContext 会进行资源回收，先销毁在各 Worker 的 CoarseGrainedExecutorBackend 进程，然后注销其自身。

6.4.2 实现原理

在独立运行模式中 SparkContext 对象启动时，在 createTaskScheduler 方法中根据模式匹配初始化 TaskSchedulerImpl 和 SparkDeploySchedulerBackend。具体匹配代码如下：

```
//独立运行模式匹配字符串，以 spark 开头的字符串
val SPARK_REGEX = """spark://(.*)""".r
case SPARK_REGEX(sparkUrl) =>
  val scheduler = new TaskSchedulerImpl(sc)
  val masterUrls = sparkUrl.split(",").map("spark://" + _)
  val backend = new SparkDeploySchedulerBackend(scheduler, sc, masterUrls)
  scheduler.initialize(backend)
  //初始化 TaskSchedulerImpl 和 SparkDeploySchedulerBackend
  (backend, scheduler)
```

独立运行模式运行原理参见 4.2 节的内容，其中独立运行模式中作业调用类图如图 4-10 所示。

6.5　YARN 运行模式

YARN 是一种统一资源管理机制，在其上面可以运行多套计算框架。目前的大数据技术世界，大多数公司除了使用 Spark 来进行数据计算，由于历史原因或者单方面业务处理的性能考虑而使用着其他的计算框架，比如 MapReduce、Storm 等计算框架。Spark 基于此种情况开发了 Spark on YARN 的运行模式，由于借助了 YARN 良好的弹性资源管理机制，不仅部署应用程序更加方便，而且用户在 YARN 集群中运行的服务和 Application 的资源也完全隔离，更具实践应用价值的是 YARN 可以通过队列的方式，管理同时运行在集群中的多个服务。

YARN 运行模式根据 Driver 在集群中的位置分为两种模式：一种是 YARN-Client 模式，另一种是 YARN-Cluster（或称为 YARN-Standalone 模式）。

6.5.1　YARN 运行框架

任何框架与 YARN 的结合，都必须遵循 YARN 的开发模式。在分析 Spark on YARN 的实现细节之前，有必要先分析一下 YARN 框架的一些基本原理。

YARN 框架的基本运行流程如图 6-7 所示。

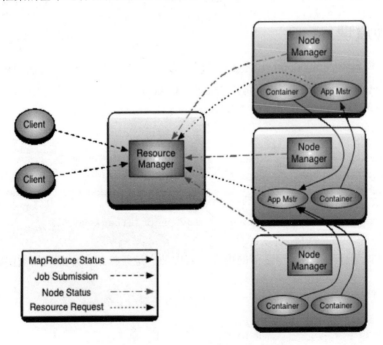

图 6-7　YARN 框架的基本运行流程图

ResourceManager 与 NodeManagers 共同组成了整个数据计算框架。其中，ResourceManager 负责将集群的资源分配给各个应用使用，而资源分配和调度的基本单位是 Container，其中封装了机器资源，如内存、CPU、磁盘和网络等，每个任务会被分配一个 Container，该任务只能在该 Container 中执行，并使用该 Container 封装的资源。NodeManager 是一个个的计算节点，主要负责启动 Application 所需的 Container，监控资源（内存、CPU、磁盘和网络等）的使用情况并将之汇报给 ResourceManager。ApplicationMaster 与具体的 Application 相关，主要负责同 ResourceManager 协商以获取合适的 Container，并跟踪这些 Container 的状态和监控其进度。

6.5.2　YARN-Client 运行模式介绍

YARN-Client 的工作流程分为以下几个步骤，处理流程如图 6-8 所示。

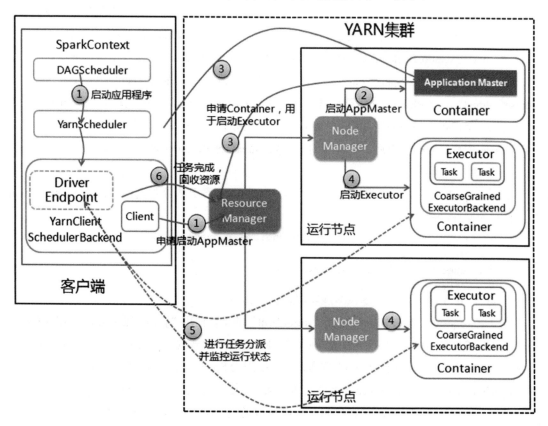

图 6-8　YARN-Client 的工作流程

（1）启动应用程序，在 SparkContext 启动过程中，初始化 DAGScheduler 调度器，使用反射方法初始化 YarnScheduler 和 YarnClientShcedulerBackend。YarnClientShcedulerBackend 在内部启动终端点 DriverEndpoint 和 Client，然后 Client 向 YARN 集群的 ResourceManager 申请启动 Application Master。

（2）ResourceManager 收到请求后，在集群中选择一个 NodeManager，为该应用程序分配第一个 Container，要求在这个 Container 中启动应用程序的 Application Master，与 YARN-Cluster 有区别的是，在该 Application Master 不运行 SparkContext，只与 SparkContext 进行联系进行资源的分派。

（3）客户端中的 SparkContext 启动完毕后，与 Application Master 建立通信，向 Resource Manager 注册，根据任务信息向 Resource Manager 申请资源（Container）。

（4）一旦 Application Master 申请到资源（也就是 Container）后，便与对应的 NodeManager 通信，要求它在获得的 Container 中启动 CoarseGrainedExecutorBackend，CoarseGrainedExecutorBackend 启动后会向客户端中的 SparkContext 注册并申请任务集。

（5）客户端中的 SparkContext 分配任务集给 CoarseGrainedExecutorBackend 执行，CoarseGrainedExecutorBackend 运行任务并向终端点 DriverEndpoint 汇报运行的状态和进度，让客户端随时掌握各个任务的运行状态，从而可以在任务失败时重新启动任务。

（6）应用程序运行完成后，客户端的 SparkContext 向 ResourceManager 申请注销并关闭自身。

下面用 spark-submit 脚本提交应用程序，可以在附加参数列表添加上面列出的 YARN-Client 模式特有的属性，命令如下：

```
$./bin/spark-submit --class org.apache.spark.examples.SparkPi \
  --master yarn-client \
  --num-executors 3 --driver-memory 4g \
  --executor-memory 2g \
  --executor-cores 1 lib\spark-examples*.jar
```

其中，"--master"参数如果不指定"yarn-client"的话，使用"yarn"也是运行 YARN-Client 模式。与 YARN-Cluster 模式不同的是，使用 YARN-Client 模式提交应用程序，当运行结束之后可以直接在客户本地看到控制台打印的结果，这是因为 SparkContext 直接运行在客户端中。

YARN-Client 模式中，Driver 在客户端本地运行，这种模式可以使得 Spark Application 和客户端进行交互，因为 Driver 在客户端，所以可以通过 webUI 访问 Driver 的状态，默认是 http://sparkmaster:4040 访问，而 YARN 通过 http:// hadoopmaster:8088 访问。

6.5.3 YARN-Client 运行模式实现原理

前面对 YARN-Client 运行模式的运行流程进行了分析，下面将根据运行流程的步骤分析其代码实现原理，具体如下：

（1）在 SparkContext 启动时，初始化 DAGScheduler 调度器，然后在 createTaskScheduler 方法中匹配为 YARN-Client 运行模式时，通过反射的方法初始化 YarnScheduler 和 YarnClientSchedulerBackend 两个对象，其中 YarnClientSchedulerBackend 类是 CoarseGrainedSchedulerBackend 类的子类，YarnScheduler 是 TaskSchedulerImpl 的子类，仅仅重写了 TaskSchedulerImpl 中的 getRackForHost 方法。在 SparkContext 的 createTaskScheduler 方法中，匹配模式代码及初始化如下：

```
case "yarn" if deployMode == "client" =>
 //任务调度器加载 YarnScheduler 类，使用反射方法创建 TaskSchedulerImpl 实例
 val scheduler = try {
   val clazz = Utils.classForName("org.apache.spark.scheduler.cluster.
     YarnScheduler")
   val cons = clazz.getConstructor(classOf[SparkContext])
   cons.newInstance(sc).asInstanceOf[TaskSchedulerImpl]
 } catch {……}

 //调度后台进程加载 YarnClientSchedulerBackend 类，
 //使用反射方法创建 CoarseGrainedSchedulerBackend 实例
 val backend = try {
   val clazz = Utils.classForName("org.apache.spark.scheduler.cluster.
     YarnClientSchedulerBackend")
   val cons = clazz.getConstructor(classOf[TaskSchedulerImpl],
     classOf[SparkContext])
   cons.newInstance(scheduler,sc).asInstanceOf[CoarseGrainedSchedulerBackend]
 } catch {……}
 scheduler.initialize(backend)
 (backend, scheduler)
```

在 YarnClientShcedulerBackend.start 方法中先在内部启动 Client，然后调用父类 start 方法启动 DriverEndpoint 终端点，并通过在 submitApplication 方法中申请启动 ApplicationMaster。其过程可以参考 YARN-Client 运行模式应用程序启动类图如图 6-9 所示。

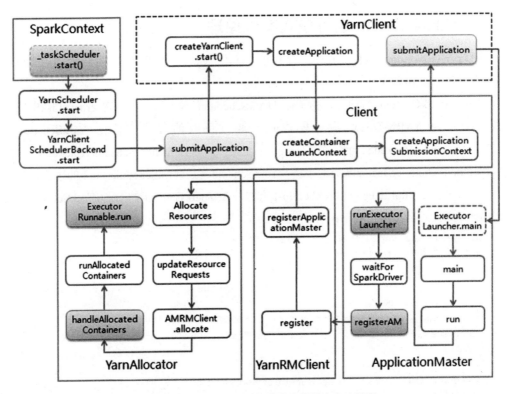

图 6-9　YARN-Client 运行模式应用程序启动类图

其中在 Client.submitApplication 方法中，先向 ResourceManager 确认是否有足够的资源。如果足够的资源，则构造用于启动 ApplicationMaster 环境并提交应用程序到 YARN 集群中。该方法代码如下：

```
def submitApplication(): ApplicationId = {
  var appId: ApplicationId = null
  try {
    launcherBackend.connect()
    setupCredentials()
    //创建 YarnClient，用于和 YARN 集群进行交互
    yarnClient.init(yarnConf)
    yarnClient.start()

    //向 ResourceManager 申请应用程序编号
    val newApp = yarnClient.createApplication()
    val newAppResponse = newApp.getNewApplicationResponse()
    appId = newAppResponse.getApplicationId()
    reportLauncherState(SparkAppHandle.State.SUBMITTED)
```

```
    launcherBackend.setAppId(appId.toString())

    //确认在YARN集群中是否有足够的资源启动ApplicationMaster
    verifyClusterResources(newAppResponse)

    //构造适当的环境用于启动ApplicationMaster
    val containerContext=createContainerLaunchContext(newAppResponse)
    val appContext=createApplicationSubmissionContext(newApp,containerContext)

    //向ResourceManager提交并监控应用程序
    yarnClient.submitApplication(appContext)
    appId
  } catch { ……}
}
```

（2）当 ResourceManager 收到请求后，在集群中选择一个 NodeManager 并启动 ExecutorLauncher，在 ExecutorLauncher 初始化中启动 ApplicationMaster。启动 ExecutorLauncher 是在 Client.createContainerLaunchContext 方法指定的，代码如下：

```
val amClass =
if (isClusterMode) {
  Utils.classForName("org.apache.spark.deploy.yarn.ApplicationMaster").getName
} else {
  Utils.classForName("org.apache.spark.deploy.yarn.ExecutorLauncher").getName
}
```

（3）ApplicationMaster 启动完毕后，在 registerAM 方法中由 ResourceManager 向终端点 DriverEndpoint 发送消息通知 ApplicationMaster 已经启动完毕。通过 YarnAllocator 中的 allocateResources 方法向 ResourceManager 申请资源（Container），其中 ApplicationMaster.registerAM 方法代码如下：

```
private def registerAM(_rpcEnv: RpcEnv, driverRef: RpcEndpointRef, uiAddress:
  String, securityMgr: SecurityManager) = {
  val sc = sparkContextRef.get()

  //获取应用程序和Attempt编号
  val appId = client.getAttemptId().getApplicationId().toString()
  val attemptId = client.getAttemptId().getAttemptId().toString()
  val historyAddress = sparkConf.get(HISTORY_SERVER_ADDRESS).map { text =>
    SparkHadoopUtil.get.substituteHadoopVariables(text,yarnConf) }.map{address
      => s"${address}${HistoryServer.UI_PATH_PREFIX}/${appId}/${attemptId}"}
    .getOrElse("")

  //获取DriverEndpoint终端点引用地址
```

```scala
    val _sparkConf = if (sc != null) sc.getConf else sparkConf
    val driverUrl = RpcEndpointAddress(_sparkConf.get("spark.driver.host"),
      _sparkConf.get("spark.driver.port").toInt,CoarseGrainedSchedulerBackend.
      ENDPOINT_NAME).toString

    //在 ResourceManager 发送消息通知 DriverEndpoint, 通知其 ApplicationMaster 已启动
    allocator = client.register(driverUrl, driverRef, yarnConf, _sparkConf,
      if (sc != null) sc.preferredNodeLocationData else Map(),
      uiAddress, historyAddress, securityMgr)

  //申请运行 Executor 资源
    allocator.allocateResources()
    reporterThread = launchReporterThread()
}
```

（4）在 YarnAllocator.allocateResources 方法中获取可用的 Container，然后调用 YarnAllocator 中 runAllocatedContainers 方法，在该方法中调用 ExecutorRunnable 的 run 方法在 Container 启动 CoarseGrainedExecutorBackend。CoarseGrainedExecutorBackend 启动后会向客户端中的 SparkContext 注册并申请任务集。

```scala
    private def runAllocatedContainers(containersToUse: ArrayBuffer[Container]):
      Unit = {
      for (container <- containersToUse) {
        //更新计数器
        numExecutorsRunning += 1
        val executorHostname = container.getNodeId.getHost
        val containerId = container.getId
        executorIdCounter += 1
        val executorId = executorIdCounter.toString

        executorIdToContainer(executorId) = container
        containerIdToExecutorId(container.getId) = executorId

        //在机器与 container 对应列表中加入当前 container 信息
        val containerSet = allocatedHostToContainersMap.getOrElseUpdate(
          (executorHostname, new HashSet[ContainerId])
        containerSet += containerId
        allocatedContainerToHostMap.put(containerId, executorHostname)

        //在 container 中实例化 ExecutorRunnable, 同时启动 CoarseGrainedExecutorBackend
        val executorRunnable = new ExecutorRunnable( container, conf, sparkConf,
          driverUrl, executorId, executorHostname, executorMemory, executorCores,
          appAttemptId.getApplicationId.toString, securityMgr)
        //加载成功，把 ExecutorRunnable 实例加入线程池中等待使用
```

```
    if (launchContainers) {
      launcherPool.execute(executorRunnable)
    }
  }
}
```

（5）SparkContext 分配任务集给 CoarseGrainedExecutorBackend 执行并跟踪运行状态，该过程和独立运行模式类似，其作业运行调用关系如图 6-10 所示。

（6）应用程序运行完成后，SparkContext 向 ResourceManager 申请注销并关闭。

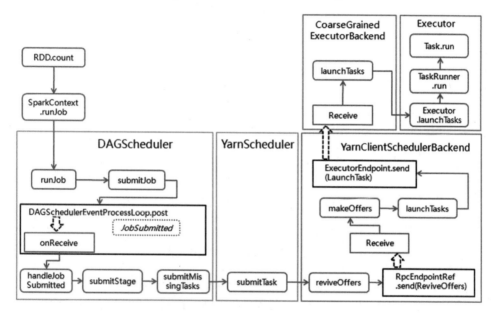

图 6-10　YARN-Client 运行模式作业运行调用关系

6.5.4　YARN-Cluster 运行模式介绍

在 YARN-Cluster 模式中，当用户向 YARN 中提交一个应用程序后，YARN 将分两个阶段运行在该应用程序：第一个阶段是把 Spark 的 Driver 作为一个 Application Master 在 YARN 集群中先启动；第二个阶段是由 Application Master 创建应用程序，然后为它向 Resource Manager 申请资源，并启动 Executor 来运行任务集，同时监控它的整个运行过程，直到运行完成。

YARN-Cluster 的工作流程如图 6-11 所示，分为以下几个步骤：

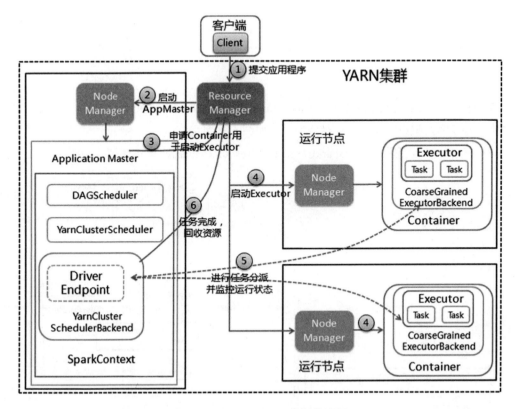

图 6-11　YARN-Cluster 的工作流程

（1）客户端提交应用程序时，启动 Client 向 YARN 中提交应用程序，包括启动 Application Master 的命令、提交给 Application Master 的程序和需要在 Executor 中运行的程序等。

（2）Resource Manager 收到请求后，在集群中选择一个 NodeManager，为该应用程序分配第一个 Container，要求它在这个 Container 中启动应用程序的 Application Master，其中 Application Master 进行 SparkContext 等的初始化。

（3）Application Master 向 Resource Manager 注册，这样用户可以直接通过 Resource Manage 查看应用程序的运行状态，然后它将采用轮询的方式为各个任务申请资源，并监控它们的运行状态直到运行结束。

（4）一旦 Application Master 申请到资源（也就是 Container）后，便与对应的 NodeManager 通信，要求它在获得的 Container 中启动 CoarseGrainedExecutorBackend，CoarseGrainedExecutorBackend 启动后会向 Application Master 中的 SparkContext 注册并申请任务集。这一点和独立运行模式一样，只不过 SparkContext 在 Spark Application 中初始化时，使用

CoarseGrainedSchedulerBackend 配合 YarnClusterScheduler 进行任务的调度，其中 YarnClusterScheduler 只是对 TaskSchedulerImpl 的一个简单包装，增加了对 Executor 的等待逻辑等。

（5）Application Master 中的 SparkContext 分配任务集给 CoarseGrainedExecutorBackend 执行，CoarseGrainedExecutorBackend 运行任务并向 Application Master 汇报运行的状态和进度，以让 Application Master 随时掌握各个任务的运行状态，从而可以在任务失败时重新启动任务；

（6）应用程序运行完成后，Application Master 向 Resource Manager 申请注销并关闭。

6.5.5 YARN-Cluster 运行模式实现原理

前面对 YARN-Cluster 运行模式的运行流程进行了分析，下面将根据运行流程的步骤分析其代码实现原理，具体如下：

（1）在 SparkContext 启动时，初始化 DAGScheduler 调度器，然后在 createTaskScheduler 方法中匹配为 YARN-Cluster 运行模式时，通过反射的方法初始化 YarnClusterScheduler 和 YarnClusterSchedulerBackend 两个对象，其中 YarnClusterSchedulerBackend 类是 CoarseGrainedSchedulerBackend 类的子类，YarnClusterScheduler 是 TaskSchedulerImpl 的子类。在 SparkContext 的 createTaskScheduler 方法中，匹配模式代码及初始化如下：

```
case "yarn" if deployMode == "cluster" =>
  val scheduler = try {
    //任务调度器加载 YarnClusterScheduler 类，使用反射方法创建 TaskSchedulerImpl 实例
    val clazz = Utils.classForName("org.apache.spark.scheduler.cluster.
      YarnClusterScheduler")
    val cons = clazz.getConstructor(classOf[SparkContext])
    cons.newInstance(sc).asInstanceOf[TaskSchedulerImpl]
  } catch {……}
  val backend = try {
    //调度后台进程加载 YarnClusterSchedulerBackend 类，
    //使用反射方法创建 CoarseGrainedSchedulerBackend 实例
    val clazz = Utils.classForName("org.apache.spark.scheduler.cluster.
      YarnClusterSchedulerBackend")
    val cons = clazz.getConstructor(classOf[TaskSchedulerImpl],
      classOf[SparkContext])
    cons.newInstance(scheduler,sc).asInstanceOf[CoarseGrainedSchedulerBackend]
  } catch {……}
  scheduler.initialize(backend)
  (backend, scheduler)
```

（2）参见 YARN-Cluster 启动类图（见图 6-12），通过 spark-submit 脚本提交应用程序时，

会在 SparkSubmit.submit 方法中根据 YARN-Cluster 运行模式匹配启动 Client 类，匹配后在 SparkSubmit.runMain 方法中通过反射初始化 Client 实例，Client 在初始化后向 YARN 集群提交应用程序，提交后会通过 ApplicationMaster.startUserApplication 方法中启动用户提交的应用程序，启动方法是通过反射对传入的类进行初始化并加载 SparkContext。

在 ApplicationMaster.run 方法中，当 YARN-Cluster 运行模式时，会运行 runDriver 方法：

```
if (isClusterMode) {
  //如果是YARN-Cluster运行模式时，启动ApplicationMaster和SparkContext
  runDriver(securityMgr)
} else {
  //如果是YARN-Client运行模式时，启动并注册ApplicationMaster
  runExecutorLauncher(securityMgr)
}
```

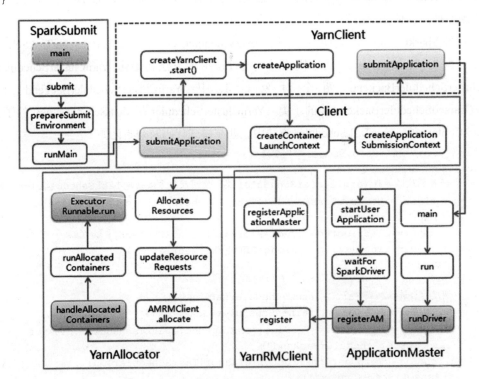

图 6-12　Yarn-Cluster 运行模式应用程序启动类图

在 runDriver 方法会在 ApplicationMaster 中运行 SparkContext，该处是与 YARN-Client 运行模式最大区别。等待 SparkContext 启动完毕后，由 ResourceManager 向终端点 DriverEndpoint 发送消息通知 ApplicationMaster 已经启动完毕。

```
private def runDriver(securityMgr: SecurityManager): Unit = {
  addAmIpFilter()
  //启动用户应用程序,返回该应用程序进程
  userClassThread = startUserApplication()
  //等待 SparkContext 启动,默认情况下等待100s
  val sc = waitForSparkContextInitialized()

  //判断 SparkContext 是否成功启动,如果没有则返回失败
  if (sc == null) {
    finish(FinalApplicationStatus.FAILED, ApplicationMaster.EXIT_SC_NOT_
      INITED, "Timed out waiting for SparkContext.")
  } else {
    rpcEnv = sc.env.rpcEnv
    val driverRef = runAMEndpoint(sc.getConf.get("spark.driver.host"),
      sc.getConf.get("spark.driver.port"),isClusterMode = true)
    //registerAM 方法中,由 ResourceManager 向终端点 DriverEndpoint
    //发送消息通知 ApplicationMaster 已经启动完毕
    registerAM(rpcEnv,driverRef,sc.ui.map(_.appUIAddress).getOrElse(""),
      securityMgr)
    userClassThread.join()
  }
}
```

（3）ApplicationMaster 启动完毕后在 registerAM 方法中，由 ResourceManager 向终端点 DriverEndpoint 发送消息通知 ApplicationMaster 已经启动完毕。通过 YarnAllocator 中的 allocateResources 方法向 ResourceManager 申请资源（Container），该处理与 YARN-Client 步骤 3 相同。

（4）在 YarnAllocator.allocateResources 方法中获取可用的 Container，然后调用 YarnAllocator 中 runAllocatedContainers 方法，在该方法中调用 ExecutorRunnable 的 run 方法在 Container 启动 CoarseGrainedExecutorBackend。CoarseGrainedExecutorBackend 启动后会向客户端中的 SparkContext 注册并申请任务集，该处理与 YARN-Client 步骤 4 相同。

（5）ApplicationMaster 中的 SparkContext 分配任务集给 CoarseGrainedExecutorBackend 执行，CoarseGrainedExecutorBackend 运行任务并向 ApplicationMaster 汇报运行的状态和进度，以让 ApplicationMaster 随时掌握各个任务的运行状态，从而可以在任务失败时重新启动任务，其作业运行调用关系如图 6-13 所示。

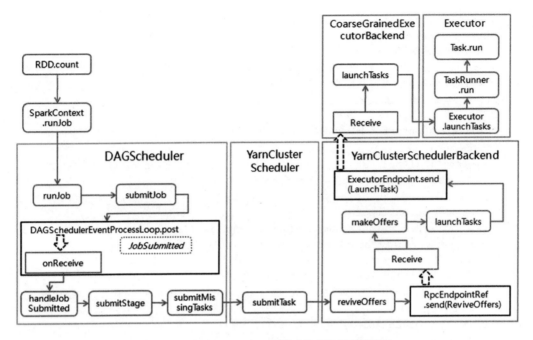

图 6-13　YARN-Cluster 运行模式作业运行调用关系

（6）应用程序运行完成后，ApplicationMaster 向 ResourceManager 申请注销并关闭自己。

6.5.6　YARN-Client 与 YARN-Cluster 对比

理解 YARN-Client 和 YARN-Cluster 区别需要强调一下 Application Master。在 YARN 中，每个 Application 实例都有一个 Application Master 进程，它是 Application 启动的第一个容器。它负责和 Resource Manager 打交道并请求资源，获取资源之后告诉 NodeManager 为其启动 Container。从深层次的含义讲 YARN-Cluster 和 YARN-Client 模式的区别其实就是 Application Master 进程的区别。

- YARN-Client 模式下，Application Master 仅仅向 YARN 请求 Executor，Client 会和请求的 Container 通信来调度它们工作，也就是说 Client 不能离开。
- YARN-Cluster 模式下，Driver 运行在 Application Master 中，它负责向 YARN 申请资源，并监督作业的运行状况。当用户提交了作业之后，就可以关掉 Client，作业会继续在 YARN 上运行，因而 YARN-Cluster 模式不适合运行交互类型的作业。

6.6　Mesos 运行模式

Mesos 可以分为粗粒度和细粒度运行模式，在 Spark 中默认情况下使用 Mesos 细粒度运行模式，可以在配置文件 spark-env.sh 中通过 spark.mesos.coarse 选项设置否使用粗粒度的调度模式。

6.6.1　Mesos 介绍

Mesos 是 Apache 旗下的开源软件，采用了 Master/Save 结构。在 Mesos 使用 ZooKeeper 来解决单点故障，并且将 Master 尽可能轻量化，其元数据可以通过各个 Slave 重新注册而进行重构。在介绍 Mesos 结构之前先介绍其几个重要的概念。

Apache Mesos 由 4 个组件组成，分别是 Mesos Master、Mesos Slave、框架和执行容器，其架构如图 6-14 所示。

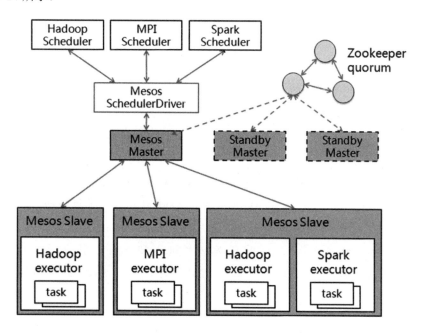

图 6-14　Mesos 高可用运行架构图

Mesos Master 是整个系统的核心，负责管理接入到 Mesos 的各个计算框架和 Slave，并将 Slave 上的资源按照指定的算法分配给计算框架。由于 Master 存在单点故障问题，因此 Mesos 采用了 ZooKeeper 解决该问题。

Mesos Slave 负责接收并执行来自 Mesos Master 的命令，同时也管理节点上的任务，为各个计算框架的任务分配资源。Slave 在启动的时候将资源（CPU 和内存）发送给 Master，当计算框架注册时由 Master 的 Allocator 模块决定将哪些资源分配给该框架。当用户提交作业时，指定每个任务需要的 CPU 个数和内存量，这样 Master 会将任务放到相应的运行容器中运行，以达到资源隔离的效果。

框架指的是用于数据计算的分布式计算框架，如 Hadoop、Spark 等，这些计算框架可通过注册的方式接入 Mesos 中。Mesos 要求可接入的框架必须有一个调度器（Scheduler），该调度器负责框架内部的任务调度。当一个计算框架想要接入 Mesos 时，需要修改其调度器，以便向 Mesos 注册，并获取 Mesos 分配的资源，这样再由其调度器将这些资源分配给框架中的任务。这也是 Mesos 采用所谓的二层调度框架：第一层，由 Mesos 将资源分配给各计算框架；第二层，各计算框架的调度器将资源分配给其内部的任务。Mesos 为了向各种调度器提供统一的接入方式，在 Mesos 内部实现了一个调度器驱动器（MesosSchedulerDriver），计算框架的调度器可调用该驱动器中的接口与 Master 进行交互，完成一系列功能（如注册、资源分配等）。

执行容器主要用于执行计算框架内部的任务。由于不同的框架，启动任务的接口或者方式不同，当一个新的框架要接入 Mesos 时，需要重写该执行容器，告诉 Mesos 如何启动该框架中的任务。为了向各种框架提供统一的执行器编写方式，Mesos 内部实现了一个执行器驱动器（MesosExecutorDriver），计算框架可通过该驱动器的相关接口告诉 Mesos 启动任务的方法。

6.6.2 粗粒度运行模式介绍

在 Mesos 的粗粒度模式中，CoarseMesosSchedulerBackend 实现了 Mesos 的调度器 MScheduler 接口，通过调度器驱动器 MesosSchedulerDriver 将其注册到 Mesos Master 中，用于接收 MesosMaster 分配的资源，在得到资源后通过 Mesos Slave 启动该节点的执行容器 CoarseGrainedExecutorBackend，之后的任务交互过程和 Spark 独立运行模式一样，由 DriverEndpoint 和 CoarseGrainedExecutorBackend 直接交互完成。图 6-15 为 Messos 粗粒度运行模式运行流程。

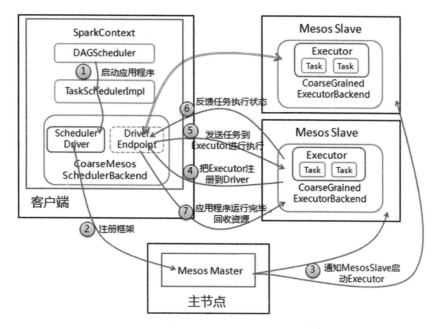

图 6-15 Messos 粗粒度运行模式运行流程

其运行流程如下所述：

（1）启动应用程序，在 SparkContext 启动过程中，初始化 DAGScheduler 和 TaskScheduler 两个调度器，并初始化 Mesos 粗粒度调度后台进程 CoarseMesosSchedulerBackend 实例。CoarseMesosSchedulerBackend 不仅继承 CoarseGrainedSchedulerBackend，通过父类启动 DriverEndpoint 终端点，而且实现了 Mesos 调度器 MScheduler 接口方法，提供 Mesos 调度驱动器回调函数，如注册、资源分配和状态更新等。

（2）在 CoarseMesosSchedulerBackend 内部启动 Mesos 调度驱动器（SchedulerDiver），它是连接计算框架调度器和 Mesos Master 的桥梁。该调度驱动器启动过程中，会把 Spark 作为计算框架注册到 Mesos Master 中。

（3）Mesos Master 对新计算框架注册成功时，会对该框架分配资源，在分配时 Mesos Master 会回调 CoarseMesosSchedulerBackend 中资源分配方法。在该方法中，对分配到的 Mesos Slave 节点中通过./bin/spark-class 启动 CoarseGrainedExecutorBackend 实例。当该进程在各 Slave 节点启动后，SparkContext 和 CoarseGrainedExecutorBackend 之间任务调度处理与独立运行模式处理类似。

（4）CoarseGrainedExecutorBackend 启动时，发送注册 Executor 消息给 SparkContext 的终端点 DriverEndpoint。在 SparkContext 成功注册后，实例化 Executor 对象，等待接收从 Driver 终

端点发送执行任务的消息。

（5）SparkContext 分配任务集给 CoarseGrainedExecutorBackend 执行，任务执行是在 Executor 按照一定调度策略进行的。

（6）CoarseGrainedExecutorBackend 在任务处理过程中，把处理任务的状态发送给 SparkContext 的终端点 DriverEndpoint，SparkContext 根据任务执行不同的结果进行处理。如果任务集处理完毕后，则会继续发送其他的任务集。

（7）应用程序运行完成后，SparkContext 会进行资源回收。先注销在各 Mesos Slave 的 CoarseGrainedExecutorBackend 进程，然后注销 Mesos 调度驱动器（SchedulerDiver）并通知 Mesos Master 回收对该计算框架分配的资源。

6.6.3 粗粒度实现原理

前面对 Mesos 粗粒度运行模式的运行流程进行了分析，下面将根据运行流程的步骤分析其代码实现原理，具体如下：

（1）在 SparkContext 对象启动时，会在内部初始化 TaskScheduler 和 SchedulerBackend，createTaskScheduler 方法中根据 Mesos 粗粒度模式匹配初始化 TaskSchedulerImpl 和 CoarseMesosSchedulerBackend。其匹配代码如下：

```
//Mesos 运行模式匹配字符串
val MESOS_REGEX = """(mesos|zk)://.*""".r
case mesosUrl @ MESOS_REGEX(_) =>
  MesosNativeLibrary.load()
  //启动 TaskSchedulerImpl 调度器
  val scheduler = new TaskSchedulerImpl(sc)
  val coarseGrained = sc.conf.getBoolean("spark.mesos.coarse", false)
  val url = mesosUrl.stripPrefix("mesos://") // strip scheme from raw Mesos URLs
  val backend = if (coarseGrained) {
    //如果是 Mesos 粗粒度运行模式，则实例化粗粒度 Mesos 调度后台进程
    new CoarseMesosSchedulerBackend(scheduler, sc, url, sc.env.securityManager)
  } else {
    //如果是 Mesos 细粒度运行模式，则实例化细粒度 Mesos 调度后台进程
    new MesosSchedulerBackend(scheduler, sc, url)
  }
  scheduler.initialize(backend)
  (backend, scheduler)
```

其中 CoarseMesosSchedulerBackend 不仅继承 CoarseGrainedSchedulerBackend，在该父类中启动 DriverEndpoint 终端点用于与 Executor 进行通信，而且实现了 Mesos 调度器 MScheduler 接

口方法，如 registered、resourceOffers 和 statusUpdate 等，Mesos 会通过调度驱动器回调这些方法。

（2）在 CoarseMesosSchedulerBackend 内部在启动父类后，创建并启动 Mesos 调度驱动器（SchedulerDiver）。在调度驱动器启动过程中，会把 Spark 作为框架注册到 Mesos Master 中，同时把前一步骤实现 CoarseMesosSchedulerBackend 调度器提交到 Mesos Master，提供回调操作。代码如下：

```
override def start() {
  super.start()
  //创建并启动Mesos调度驱动器，并把粗粒度Mesos调度后台进程提交到Mesos Master
  val driver = createSchedulerDriver(
    master,CoarseMesosSchedulerBackend.this,sc.sparkUser,sc.appName,sc.conf)
  startScheduler(driver)
}
```

（3）Mesos Master 对新计算框架注册成功时，会对该框架分配资源，在分配时 Mesos Master 会回调 CoarseMesosSchedulerBackend 中资源分配方法。在该方法中，对分配到的 Mesos Slave 节点中通过$SPARK_HOME/bin/spark-class 启动 CoarseGrainedExecutorBackend 实例。当该进程在各 Slave 节点启动后，CoarseGrainedExecutorBackend 根据 SparkContext 引用地址发送注册消息，注册成功后启动执行容器 Executor，然后 SparkContext 和 CoarseGrainedExecutorBackend 建立联系，它们之间任务调度处理与独立运行模式处理类似。启动过程调用类关系图如图 6-16 所示。

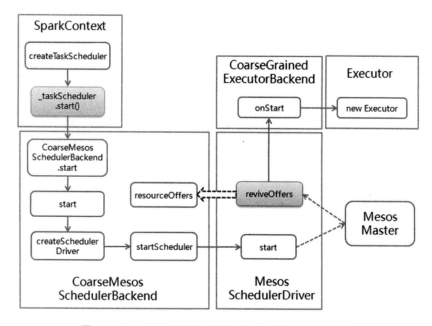

图 6-16　Messos 粗粒度运行模式启动过程类调用关系图

资源调度代码如下：

```scala
override def resourceOffers(d: SchedulerDriver, offers: JList[Offer]) {
  stateLock.synchronized {
    val filters = Filters.newBuilder().setRefuseSeconds(5).build()
    //遍历 Mesos 分配的资源构造任务执行脚本
    for (offer <- offers) {
      //获取 Mesos Slave 的编号、内存和 CPU 资源
      val slaveId = offer.getSlaveId.getValue
      val mem = getResource(offer.getResourcesList, "mem")
      val cpus = getResource(offer.getResourcesList, "cpus").toInt
      val id = offer.getId.getValue

      //如果 Mesos Slave 资源满足运行运行条件时，则分配任务
      if (taskIdToSlaveId.size < executorLimit && totalCoresAcquired < maxCores
        &&
        meetsConstraints && mem >= calculateTotalMemory(sc) && cpus >= 1 &&
        failuresBySlaveId.getOrElse(slaveId, 0) < MAX_SLAVE_FAILURES &&
        !slaveIdsWithExecutors.contains(slaveId)) {

        //获取该 Mesos Slave 能够使用的 CPU 和内存资源
        val (remainingResources, cpuResourcesToUse) = partitionResources(
          offer.getResourcesList, "cpus", cpusToUse)
        val (_, memResourcesToUse) = partitionResources(remainingResources,
          "mem", calculateTotalMemory(sc))

        //构造在 Mesos 执行任务，在该任务中将在 Slave 创建 CoarseGrainedExecutorBackend
        //进程，该进程将执行 SparkContext 发送过来的任务集
        val taskBuilder = MesosTaskInfo.newBuilder()
          .setTaskId(TaskID.newBuilder().setValue(taskId.toString).build())
          .setSlaveId(offer.getSlaveId)
          .setCommand(createCommand(offer, cpusToUse+extraCoresPerSlave, taskId))
          .setName("Task " + taskId)
          .addAllResources(cpuResourcesToUse)
          .addAllResources(memResourcesToUse)
        }

        //接受该 Mesos Slave 资源并提交任务
        slaveIdToHost(offer.getSlaveId.getValue) = offer.getHostname
        d.launchTasks(Collections.singleton(offer.getId),
          Collections.singleton(taskBuilder.build()), filters)
      } else {
        d.declineOffer(offer.getId)
      }
```

......
}

步骤4~7与独立运行模式相似,参考其描述,这里不赘述。

6.6.4 细粒度运行模式介绍

对于 Mesos 的细粒度模式,MesosSchedulerBackend 直接继承自 SchedulerBackend,它实现了 Mesos 的调度器 MScheduler 接口,并注册到 Mesos 资源调度的框架中用于接收 Mesos Master 的资源分配。不同的是在接受资源后,MesosSchedulerBackend 启动的是基于单个任务的执行容器,并在 MesosExecutorbackend 的执行容器中直接启动和执行对应的任务。图 6-17 为 Messos 细粒度运行模式运行流程。

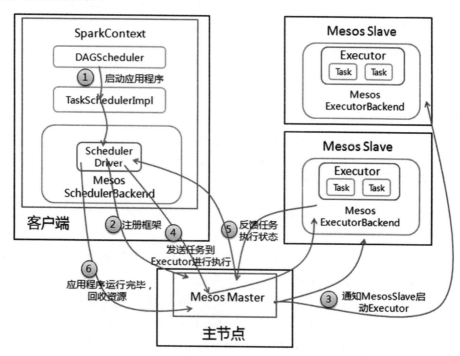

图 6-17 Messos 细粒度运行模式运行流程

其运行流程如下:

(1)启动应用程序,在 SparkContext 启动过程中,初始化 DAGScheduler 和 TaskScheduler 两个调度器,并初始化 Mesos 细粒度调度后台进程 MesosSchedulerBackend 实例。其中 MesosSchedulerBackend 不仅实现了 Spark 中 SchedulerBackend 接口方法,而且实现了 Mesos 调

度器 MScheduler 接口方法，提供如注册、资源分配和状态更新等回调函数。

（2）在 MesosSchedulerBackend 内部启动 Mesos 调度驱动器（SchedulerDiver），它是连接计算框架调度器和 Mesos Master 的桥梁。在该调度驱动器启动过程中，会把 Spark 作为计算框架注册到 Mesos Master 中。

（3）Mesos Master 对新计算框架注册成功时，会对该框架分配资源，在分配时 Mesos Master 会回调 MesosSchedulerBackend 中资源分配方法。在该方法中，对分配到的 Mesos Slave 节点中通过/bin/spark-class 启动 MesosExecutorBackend 实例。MesosExecutorBackend 实现了 Spark 的 ExecutorBackend 和 MesosExecutor 接口。当其实例化过程中将向 Mesos Master 注册，并在其内部启动执行容器 Executor，等待执行发送过来的任务。

（4）SparkContext 通过调度驱动器（SchedulerDiver）把任务集发送给 Mesos Master，由 Mesos Master 把任务集重新封装后，以单个任务分发到 Slave 的执行容器 Executor 中进行执行。

（5）MesosExecutorBackend 在任务处理过程中，把处理任务的状态发送给 Mesos Master，然后由再 Mesos 调度驱动器转发给 SparkContext。SparkContext 根据任务执行不同的结果进行处理。如果任务集处理完毕后，则会继续发送其他的任务集。

（6）应用程序运行完成后，SparkContext 内部会进行资源回收，然后注销 Mesos 调度驱动器（SchedulerDiver），并通知 Mesos Master 回收对该计算框架分配的资源。

6.6.5 细粒度实现原理

前面对 Mesos 细粒度运行模式的运行流程进行了分析，下面将根据运行流程的步骤分析其代码实现原理，具体如下：

（1）在 SparkContext 对象启动时，会在内部初始化 TaskScheduler 和 SchedulerBackend，createTaskScheduler 方法中根据 Mesos 粗粒度模式匹配初始化 TaskSchedulerImpl 和 MesosSchedulerBackend。其运行模式匹配代码参见 6.6.3 粗粒度实现原理的步骤 1。

（2）在 MesosSchedulerBackend 内部启动 Mesos 调度驱动器（SchedulerDiver），它是连接计算框架调度器和 Mesos Master 的桥梁。该调度驱动器启动过程中，会把 Spark 作为框架注册到 Mesos Master 中，同时把前一步骤实现 CoarseMesosSchedulerBackend 调度器提交到 Mesos，提供回调操作。代码如下：

```
override def start() {
  classLoader = Thread.currentThread.getContextClassLoader
  //创建并启动 Mesos 调度驱动器，并把细粒度 Mesos 调度后台进程提交到 Mesos Master
  val driver = createSchedulerDriver(master, MesosSchedulerBackend.this,
    sc.sparkUser, sc.appName, sc.conf)
```

```
    startScheduler(driver)
}
```

（3）Mesos Master 对新计算框架注册成功时，会对该框架分配资源，在分配时 Mesos Master 会回调 MesosSchedulerBackend 中的 reviveOffers 方法进行资源分配。在该方法中，对分配到的 Mesos Slave 节点中通过 $SPARK_HOME/bin/spark-class 启动 MesosExecutorBackend 实例。MesosExecutorBackend 实现了 Spark 的 ExecutorBackend 和 MesosExecutor 接口，当其实例化过程中将向 Mesos Master 注册，并在其内部启动执行容器 Executor，等待执行发送过来的任务。代码和启动类调用关系图如图 6-18 所示。

```
def createExecutorInfo( availableResources: JList[Resource],
    execId: String): (MesosExecutorInfo, JList[Resource]) = {
  ......
  //创建命令行实例，用于构造 Mesos Slave 执行环境
  val command = CommandInfo.newBuilder().setEnvironment(environment)
  val uri = sc.conf.getOption("spark.executor.uri").orElse(Option
    (System.getenv("SPARK_EXECUTOR_URI")))
  //设置在 Mesos Slave 启动的执行器后台进程为 MesosExecutorBackend
  val executorBackendName = classOf[MesosExecutorBackend].getName
  if (uri.isEmpty) {
    //如果未指定 Spark 安装包指定位置，则设置启动路径为 Spark 在 Mesos Slave 安装路径
    val executorPath = new File(executorSparkHome,
      "/bin/spark-class").getCanonicalPath
    command.setValue(s"$prefixEnv $executorPath $executorBackendName")
  } else {
    //如果指定 Spark 安装包指定位置，则设置启动路径为该 Spark 安装包路径
    val basename = uri.get.split('/').last.split('.').head
    command.setValue(s"cd ${basename}*; $prefixEnv ./bin/spark-class
      $executorBackendName")
    command.addUris(CommandInfo.URI.newBuilder().setValue(uri.get))
  }
  //实例化 Mesos 执行容器创建器
  val builder = MesosExecutorInfo.newBuilder()

  //获取该 Mesos Slave 能够使用的 CPU 和内存资源
  val (resourcesAfterCpu, usedCpuResources) = partitionResources(availableResources,
    "cpus", scheduler.CPUS_PER_TASK)
  val (resourcesAfterMem, usedMemResources) = partitionResources(resourcesAfterCpu,
    "mem", calculateTotalMemory(sc))
  builder.addAllResources(usedCpuResources)
  builder.addAllResources(usedMemResources)
  sc.conf.getOption("spark.mesos.uris").map { uris => setupUris(uris, command) }
```

```
        //获取并返回Mesos执行容器创建的完整信息
        val executorInfo = builder
          .setExecutorId(ExecutorID.newBuilder().setValue(execId).build())
          .setCommand(command) .setData(ByteString.copyFrom(createExecArg())))
      }
      (executorInfo.build(), resourcesAfterMem)
    }
```

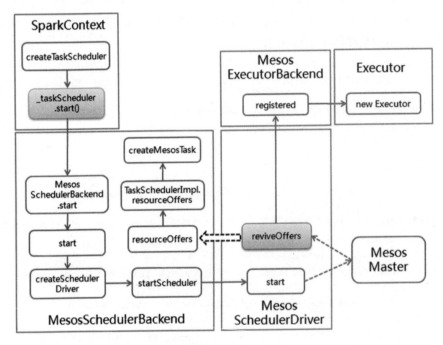

图 6-18 Messos 粗粒度运行模式启动类调用关系图

（4）SparkContext 通过调度驱动器（SchedulerDiver）把任务集发送给 Mesos Master，由 Mesos Master 把任务集重新封装后，以单个任务分发到 Slave 的执行容器 Executor 中进行执行。

（5）MesosExecutorBackend 在任务处理过程中，把处理任务的状态发送给 Mesos Master，然后由再 Mesos 调度驱动器转发给 SparkContext 。SparkContext 根据任务执行不同的结果进行处理，如果任务集处理完毕后会继续发送其他的任务集。

（6）应用程序运行完成后，SparkContext 内部会进行资源回收，然后注销 Mesos 调度驱动器（SchedulerDiver）并通知 Mesos Master 回收对该计算框架分配的资源。

6.6.6 Mesos 粗粒度和 Mesos 细粒度对比

Mesos 粗粒度运行模式中，Spark 应用程序在注册到 Mesos 时会分配对应系统资源，在执行过程中由 SparkContext 和 Executor 直接进行交互。该模式优点是由于资源长期持有减少了资源调度的时间开销，缺点是该模式中 Mesos 无法感知资源使用的变化，易造成系统资源的闲置，无法被 Mesos 其他框架所使用，从而造成了资源浪费。

而在细粒度的运行模式中，Spark 应用程序是以单个任务的粒度发送到 Mesos 中执行，在执行过程中 SparkContext 并不能 Executor 直接进行交互，而是由 Mesos Master 进行统一的调度管理，这样能够根据整个 Mesos 集群资源使用的情况动态调整。该模式的优点是系统资源能够得到充分利用，缺点是该模式中每个任务都需要从 Mesos 获取资源，调度延迟较大，对于 Mesos Master 开销较大。

6.7 实例演示

在下面几个实例中，我们使用独立运行、YARN-Client 和 YARN-Cluster3 种运行模式计算搜狗的日志数据，并观察集群中数据存储位置和处理节点的关系。一般情况下，Spark 集群的节点中 40%的内存用于计算，60%的内存用于缓存结果。为了能够直观地感受数据使用缓存和没有使用缓存的区别，在该演示中将使用大小为 1GB 的 Sogou3.txt 数据文件。

6.7.1 独立运行模式实例

在该实例中将使用独立运行模式处理，在运行模式中分别由客户端、Master 节点和 Worker 节点组成。在 Spark Shell 提交计算搜狗日志行数代码时，所在机器作为客户端启动应用程序，然后向 Master 注册应用程序，由 Master 通知 Worker 在节点启动 Executor，Executor 启动后向客户端的 Driver 注册，最后由 Driver 发送执行任务给 Executor 并监控任务执行情况。该程序代码中，在触发计算行数动作之前，需设置缓存代码，这样在执行计算行数行为时进行缓存数据，缓存后再次运行计算行数。

1. 使用 HDFS 命令观察 Sogou3.txt 数据存放节点的位置

```
$cd /app/spark/hadoop-2.7.2/bin
$hdfs fsck /sogou/SogouQ3.txt -files -blocks -locations
```

通过图 6-19 可以看到该文件被分隔为 9 个块放在集群中。

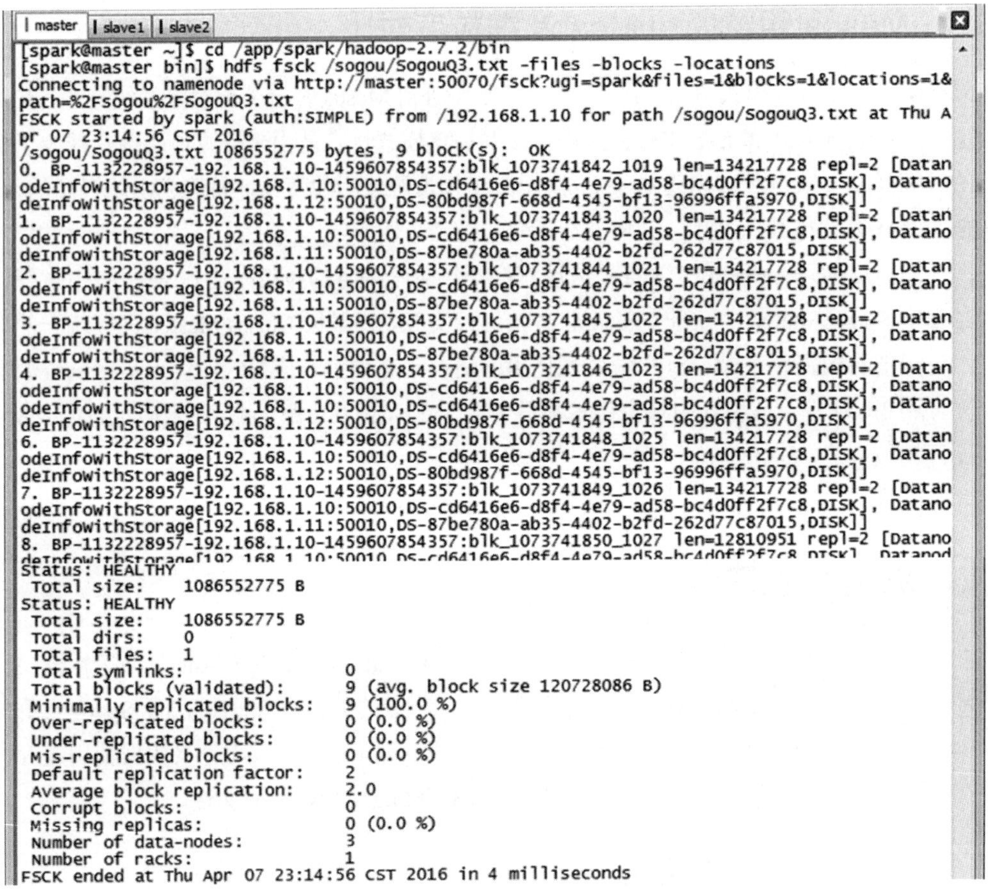

```
| master | slave1 | slave2 |
[spark@master ~]$ cd /app/spark/hadoop-2.7.2/bin
[spark@master bin]$ hdfs fsck /sogou/SogouQ3.txt -files -blocks -locations
Connecting to namenode via http://master:50070/fsck?ugi=spark&files=1&blocks=1&locations=1&
path=%2Fsogou%2FSogouQ3.txt
FSCK started by spark (auth:SIMPLE) from /192.168.1.10 for path /sogou/SogouQ3.txt at Thu A
pr 07 23:14:56 CST 2016
/sogou/SogouQ3.txt 1086552775 bytes, 9 block(s):  OK
0. BP-1132228957-192.168.1.10-1459607854357:blk_1073741842_1019 len=134217728 repl=2 [Datan
odeInfoWithStorage[192.168.1.10:50010,DS-cd6416e6-d8f4-4e79-ad58-bc4d0ff2f7c8,DISK], Datano
deInfoWithStorage[192.168.1.12:50010,DS-80bd987f-668d-4545-bf13-96996ffa5970,DISK]]
1. BP-1132228957-192.168.1.10-1459607854357:blk_1073741843_1020 len=134217728 repl=2 [Datan
odeInfoWithStorage[192.168.1.10:50010,DS-cd6416e6-d8f4-4e79-ad58-bc4d0ff2f7c8,DISK], Datano
deInfoWithStorage[192.168.1.11:50010,DS-87be780a-ab35-4402-b2fd-262d77c87015,DISK]]
2. BP-1132228957-192.168.1.10-1459607854357:blk_1073741844_1021 len=134217728 repl=2 [Datan
odeInfoWithStorage[192.168.1.10:50010,DS-cd6416e6-d8f4-4e79-ad58-bc4d0ff2f7c8,DISK], Datano
deInfoWithStorage[192.168.1.11:50010,DS-87be780a-ab35-4402-b2fd-262d77c87015,DISK]]
3. BP-1132228957-192.168.1.10-1459607854357:blk_1073741845_1022 len=134217728 repl=2 [Datan
odeInfoWithStorage[192.168.1.10:50010,DS-cd6416e6-d8f4-4e79-ad58-bc4d0ff2f7c8,DISK], Datano
deInfoWithStorage[192.168.1.11:50010,DS-87be780a-ab35-4402-b2fd-262d77c87015,DISK]]
4. BP-1132228957-192.168.1.10-1459607854357:blk_1073741846_1023 len=134217728 repl=2 [Datan
odeInfoWithStorage[192.168.1.10:50010,DS-cd6416e6-d8f4-4e79-ad58-bc4d0ff2f7c8,DISK], Datano
deInfoWithStorage[192.168.1.12:50010,DS-80bd987f-668d-4545-bf13-96996ffa5970,DISK]]
6. BP-1132228957-192.168.1.10-1459607854357:blk_1073741848_1025 len=134217728 repl=2 [Datan
odeInfoWithStorage[192.168.1.10:50010,DS-cd6416e6-d8f4-4e79-ad58-bc4d0ff2f7c8,DISK], Datano
deInfoWithStorage[192.168.1.12:50010,DS-80bd987f-668d-4545-bf13-96996ffa5970,DISK]]
7. BP-1132228957-192.168.1.10-1459607854357:blk_1073741849_1026 len=134217728 repl=2 [Datan
odeInfoWithStorage[192.168.1.10:50010,DS-cd6416e6-d8f4-4e79-ad58-bc4d0ff2f7c8,DISK], Datano
deInfoWithStorage[192.168.1.11:50010,DS-87be780a-ab35-4402-b2fd-262d77c87015,DISK]]
8. BP-1132228957-192.168.1.10-1459607854357:blk_1073741850_1027 len=12810951 repl=2 [Datano
deInfoWithStorage[192.168.1.10:50010,DS-cd6416e6-d8f4-4e79-ad58-bc4d0ff2f7c8,DISK], Datanod
Status: HEALTHY
 Total size:    1086552775 B
Status: HEALTHY
 Total size:    1086552775 B
 Total dirs:    0
 Total files:   1
 Total symlinks:                0
 Total blocks (validated):      9 (avg. block size 120728086 B)
 Minimally replicated blocks:   9 (100.0 %)
 Over-replicated blocks:        0 (0.0 %)
 Under-replicated blocks:       0 (0.0 %)
 Mis-replicated blocks:         0 (0.0 %)
 Default replication factor:    2
 Average block replication:     2.0
 Corrupt blocks:                0
 Missing replicas:              0 (0.0 %)
 Number of data-nodes:          3
 Number of racks:               1
FSCK ended at Thu Apr 07 23:14:56 CST 2016 in 4 milliseconds
```

图 6-19　搜狗数据存储在集群中分布情况

2．启动 Spark-Shell

通过如下命令启动 Spark Shell，在演示当中每个 Executor 分配 1GB 内存。

```
$cd /app/hadoop/spark-1.1.0/bin
$./spark-shell --master spark://master:7077 --executor-memory 1g
```

通过 Spark 的监控界面查看 Executors 的情况，可以观察到有 1 个 Driver 和 3 个 Executor，其中 slave1 和 slave2 各启动一个 Executor，而 master 启动一个 Executor 和一个 Driver。在独立运行模式下，Driver 中运行 SparkContect，也就是 DAGSheduler 和 TaskSheduler 等进程是运行在 Master 节点上的，它们负责进行调度节点和任务的分配及管理。

3. 读取文件后计算数据集条数，并计算过程中使用 cache() 方法对数据集进行缓存

```
scala> val sogou =sc.textFile("hdfs://master:9000/sogou/SogouQ3.txt")
sogou: org.apache.spark.rdd.RDD[String] = hdfs://master:9000/sogou/SogouQ3.txt
  MapPartitionsRDD[1] at textFile at <console>:25

scala> sogou.cache()
res0: sogou.type = hdfs://master:9000/sogou/SogouQ3.txt MapPartitionsRDD[1] at
  textFile at <console>:25

scala> sogou.count()
```

通过页面监控可以看到该作业分为 8 个任务，其中一个任务的数据来源于两个数据分片，其他的任务各对应一个数据分片，即显示 7 个任务获取数据的类型为（NODE_LOCAL），1 个任务获取数据的类型为任何位置（ANY），如图 6-20 所示。通过结果可以看到任务运行尽可能靠近数据存储的节点，任务的首选位置一般是其数据存储的位置。

- 任务 0：运行在 master，对应数据块 0，数据块位于 master 和 slave2。
- 任务 1：运行在 slave1，对应数据块 1，数据块位于 master 和 slave1。
- 任务 2：运行在 master，对应数据块 2，数据块位于 master 和 slave1。
- 任务 3：运行在 master，对应数据块 3，数据块位于 master 和 slave1。
- 任务 4：运行在 slave2，对应数据块 4，数据块位于 master 和 slave2。
- 任务 5：运行在 master，对应数据块 5，数据块位于 master 和 slave2。
- 任务 6：运行在 master，对应数据块 6，数据块位于 master 和 slave2。
- 任务 7：运行在 master，对应数据块 7 和数据块 8，数据块 7 位于 master 和 slave1，数据块 8 位于 master 和 slave2。

Tasks

Index ▲	ID	Attempt	Status	Locality Level	Executor ID / Host	Launch Time	Duration	GC Time	Input Size / Records	Errors
0	0	0	SUCCESS	NODE_LOCAL	0 / master	2016/04/07 23:28:04	26 s	1 s	0.0 B (hadoop) / 1234177	
1	1	0	SUCCESS	NODE_LOCAL	2 / slave1	2016/04/07 23:28:04	1.6 min	49 s	0.0 B (hadoop) / 1236118	
2	3	0	SUCCESS	NODE_LOCAL	0 / master	2016/04/07 23:28:33	23 s	0.9 s	128.1 MB (hadoop) / 1234730	
3	4	0	SUCCESS	NODE_LOCAL	0 / master	2016/04/07 23:28:57	26 s	2 s	128.1 MB (hadoop) / 1236238	
4	2	0	SUCCESS	NODE_LOCAL	1 / slave2	2016/04/07 23:28:04	2.8 min	1.1 min	0.0 B (hadoop) / 1235232	
5	5	0	SUCCESS	NODE_LOCAL	0 / master	2016/04/07 23:29:23	26 s	0.1 s	128.1 MB (hadoop) / 1234753	
6	6	0	SUCCESS	NODE_LOCAL	0 / master	2016/04/07 23:29:49	16 s	0.3 s	128.1 MB (hadoop) / 1235626	
7	7	0	SUCCESS	ANY	0 / master	2016/04/07 23:30:05	47 s	2 s	140.2 MB (hadoop) / 1353196	

图 6-20 计算数据集条数任务分布情况

在存储监控界面中，我们可以看到缓存份数为 3，分别为数据块 0、1、4，大小为 906.3M，缓存率为 38%，如图 6-21 所示。

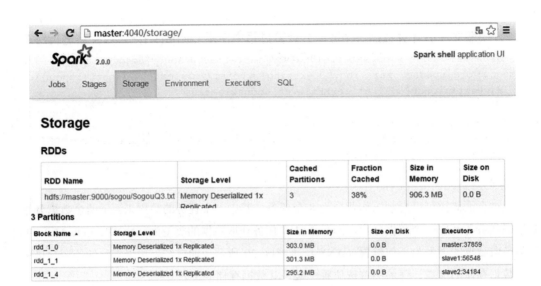

图 6-21　计算数据集条数数据缓存情况

5. 再次读取文件后计算数据集条数，此次计算使用缓存的数据，对比前后

```
scala> sogou.count()
```

通过如图 6-22 所示的页面监控可以看到该作业还是分为 8 个任务，其中 3 个任务数据来自内存（PROCESS_LOCAL），也就是缓存在内存中的 3 份数据（数据块 0、1、4），4 个任务数据来自本机（NODE_LOCAL），其他 1 个任务来自任何位置（ANY）。任务所耗费的时间多少排序为：ANY> NODE_LOCAL> PROCESS_LOCAL，对比看出使用内存的数据比使用本机或任何位置的速度至少会快 2 个数量级。

图 6-22　使用缓存中数据计算数据集条数任务列表

运行结果得到数据集的数量为 1000 万笔数据，第一次共花费了 3.8min，第二次整个作业的花费时间为 39s，比没有缓存将近提高了一个数量级，如图 6-23 所示。

Job Id	Description	Submitted	Duration	Stages: Succeeded/Total	Tasks (for all stages): Succeeded/Total
1	count at <console>:30	2016/04/07 23:37:49	39 s	1/1	8/8
0	count at <console>:30	2016/04/07 23:28:04	3.8 min	1/1	8/8

图 6-23　使用缓存中数据计算数据集条数执行结果

由于刚才例子中数据只是部分缓存（缓存率为 38%），如果完全缓存速度能够得到进一步提升，从这体验到 Spark 非常耗内存，不过也够快、够锋利！

6.7.2　YARN-Client 实例

在该实例中将使用 YARN-Client 运行模式进行处理，在该运行模式中分别由客户端和 YARN 集群组成。在 Spark Shell 提交计算搜狗日志行数代码时，所在机器作为客户端启动应用程序，然后由客户端向 YARN 集群的 Resource Manager 申请启动 Application Master，启动完毕后与 SparkContext 取得通信，由 SparkContext 并向 ResourceManager 申请资源。若申请到资源后，SparkContext 便与对应的 NodeManager 通信，告知所在节点启动 Executor，启动后由 SparkContext 分配任务，Executor 在运行过程中随时汇报状态和进度。

1．启动 Spark 集群、Spark Shell 和 YARN

在执行前需要启动 Spark 集群和 YARN，然后通过如下命令启动 Spark Shell，其中分配 3 个 Executor，每个 Executor 拥有 1GB 内存。

```
$cd /app/spark/spark-2.0.0/bin
$./spark-shell --master yarn --deploy-mode client--num-executors 3
  --executor-memory 1024m
```

在 Spark Shell 中输入执行命令后，其执行过程如下：

（1）上传执行 JAR 包到 HDFS 中。

使用 hadoop 命令查看/user/spark/.sparkStaging/application_1465981853155_0001 目录 JAR 的情况（需根据实际应用程序编号进行确定）。

```
$hadoop fs -ls /user/spark/.sparkStaging/application_1465981853155_0001
Found 203 items
……
-rw-… 2 spark supergroup 6247462 2016-06-15 17:23 /…/spark-catalyst_2.11-2.0.0.jar
-rw-… 2 spark supergroup 11377899 2016-06-15 17:24 /…/spark-core_2.11-2.0.0.jar
-rw-… 2 spark supergroup677257 2016-06-15 17:24 /…/spark- graphx_2.11-2.0.0.jar
-rw-… 2 spark supergroup 1295219 2016-06-15 17:24 /…/spark-hive_2.11-2.0.0.jar
```

```
-rw-… 2 spark supergroup 5512139 2016-06-15 17:23 /…/spark-mllib_2.11-2.0.0.jar
-rw-… 2 spark supergroup   57442 2016-06-15 17:23 /…/spark-repl_2.11-2.0.0.jar
-rw-… 2 spark supergroup 5331329 2016-06-15 17:23 /…/spark-sql_2.11-2.0.0.jar
……
```

（2）启动 Application Master，注册 Executor。

在 Application Master 启动后，与 SparkContext 取得联系，由 Driver 申请资源并与相关的 NodeManager 通信，在获得的 Container 上启动 Executor。从图 6-24 可以看到在 master 上运行 Driver，并在 master、slave1 和 slave2 各启动了 1 个 Executor。

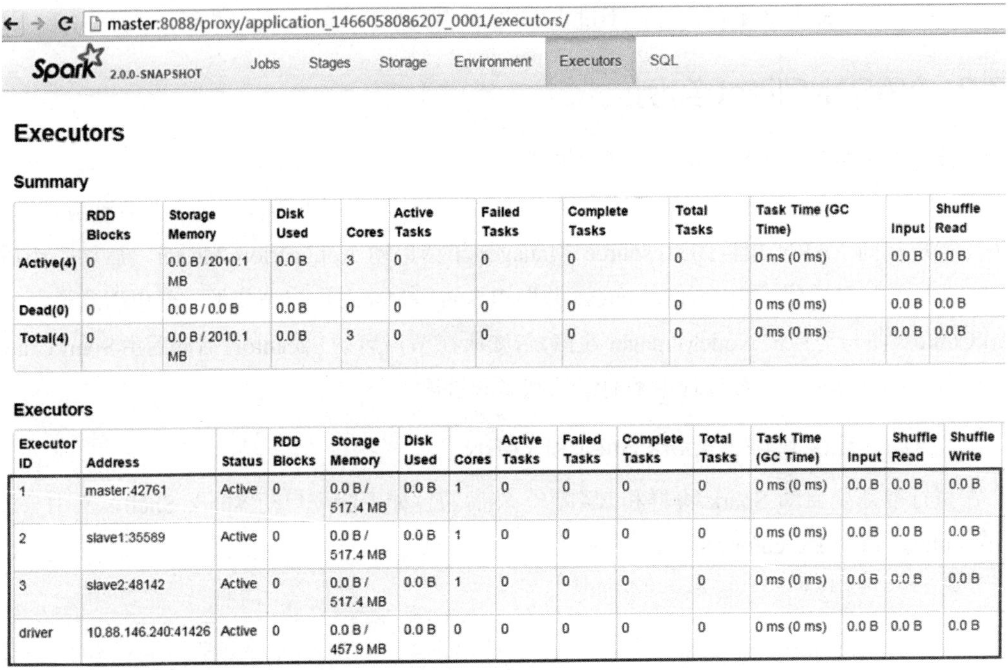

Executors

Summary

	RDD Blocks	Storage Memory	Disk Used	Cores	Active Tasks	Failed Tasks	Complete Tasks	Total Tasks	Task Time (GC Time)	Input	Shuffle Read
Active(4)	0	0.0 B / 2010.1 MB	0.0 B	3	0	0	0	0	0 ms (0 ms)	0.0 B	0.0 B
Dead(0)	0	0.0 B / 0.0 B	0.0 B	0	0	0	0	0	0 ms (0 ms)	0.0 B	0.0 B
Total(4)	0	0.0 B / 2010.1 MB	0.0 B	3	0	0	0	0	0 ms (0 ms)	0.0 B	0.0 B

Executors

Executor ID	Address	Status	RDD Blocks	Storage Memory	Disk Used	Cores	Active Tasks	Failed Tasks	Complete Tasks	Total Tasks	Task Time (GC Time)	Input	Shuffle Read	Shuffle Write
1	master:42761	Active	0	0.0 B / 517.4 MB	0.0 B	1	0	0	0	0	0 ms (0 ms)	0.0 B	0.0 B	0.0 B
2	slave1:35589	Active	0	0.0 B / 517.4 MB	0.0 B	1	0	0	0	0	0 ms (0 ms)	0.0 B	0.0 B	0.0 B
3	slave2:48142	Active	0	0.0 B / 517.4 MB	0.0 B	1	0	0	0	0	0 ms (0 ms)	0.0 B	0.0 B	0.0 B
driver	10.88.146.240:41326	Active	0	0.0 B / 457.9 MB	0.0 B	0	0	0	0	0	0 ms (0 ms)	0.0 B	0.0 B	0.0 B

图 6-24　Yarn-Client 运行模式中资源分配情况

（3）查看启动结果。

YARN-Client 模式中，Driver 在客户端本地运行，这种模式可以使得 Spark Application 和客户端进行交互，因为 Driver 在客户端所以可以通过 webUI 访问 Driver 的状态，在 Spark 中默认是 http://master:4040，而 YARN 通过 http://master:8088 访问。其中，YARN 监控页面如图 6-25 所示。在该界面上的应用程序列表中单击对应的 ApplicationMaster 连接，可以查看该应用程序详细情况。由于该应用程序是 Spark Shell，所以该页面与 Spark 监控页面类似。

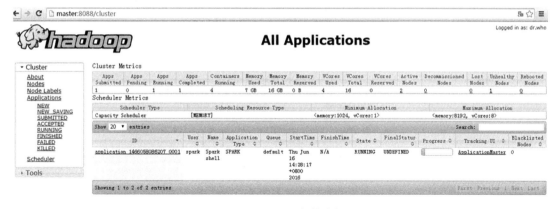

图 6-25　Yarn-Client 运行模式监控页面

（4）运行过程及结果分析。

1）读取文件后计算数据集条数，计算过程中使用 cache()方法对数据集进行缓存。
```
scala> val sogou=sc.textFile("hdfs://master:9000/sogou/SogouQ3.txt")
scala> sogou.cache()
scala> sogou.count()
```
通过页面监控可以看到该作业分为 8 个任务，其中一个任务的数据来源于两个数据分片，其他的任务各对应一个数据分片，即显示 7 个任务获取数据的位置为本地节点（NODE_LOCAL），1 个任务获取数据的位置为同机架节点（RACK_LOCAL），整个作业耗费 108.6s，如图 6-26 所示。

Tasks

Index	ID	Attempt	Status	Locality Level	Executor ID / Host	Launch Time	Duration	GC Time	Input Size / Records
0	2	0	SUCCESS	NODE_LOCAL	2 / slave1	2016/06/16 16:13:29	18 s	2 s	128.1 MB (hadoop) / 1234177
1	0	0	SUCCESS	NODE_LOCAL	1 /master	2016/06/16 16:13:29	1.7 min	30 s	128.1 MB (hadoop) / 1236118
2	1	0	SUCCESS	NODE_LOCAL	3 / slave2	2016/06/16 16:13:29	16 s	2 s	128.1 MB (hadoop) / 1234730
3	3	0	SUCCESS	NODE_LOCAL	2 / slave1	2016/06/16 16:13:42	17 s	2 s	128.1 MB (hadoop) / 1236238
4	4	0	SUCCESS	NODE_LOCAL	2 / slave1	2016/06/16 16:13:53	13 s	1 s	128.1 MB (hadoop) / 1235232
5	5	0	SUCCESS	NODE_LOCAL	1 / master	2016/06/16 16:14:07	18 s	3 s	128.1 MB (hadoop) / 1234753
6	6	0	SUCCESS	NODE_LOCAL	3 / slave2	2016/06/16 16:14:07	15 s	3 s	128.1 MB (hadoop) / 1235626
7	7	0	SUCCESS	RACK_LOCAL	2 / slave1	2016/06/16 16:14:14	21 s	2 s	140.2 MB (hadoop) / 1353126

图 6-26　计算数据集条数任务详细情况

2）查看数据缓存情况。

通过如图 6-27 所示的监控界面可以看到，和独立运行模式一样，38%的数据已经缓存在内存中。

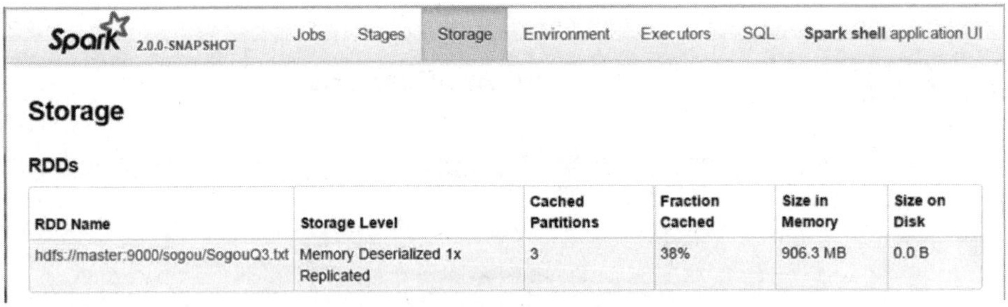

图 6-27　计算数据集条数数据缓存情况

3）再次读取文件后计算数据集条数，此次计算使用缓存的数据，对比前后

```
scala> sogou.count()
```

通过如图 6-28 所示的页面监控可以看到该作业还是分为 8 个任务，其中 3 个任务数据来自内存（PROCESS_LOCAL），4 个任务数据来自本机（NODE_LOCAL），1 个任务数据来自机架（RACK_LOCAL）。对比在内存中的运行速度最快，速度比在本机至少快 1 个数量级。

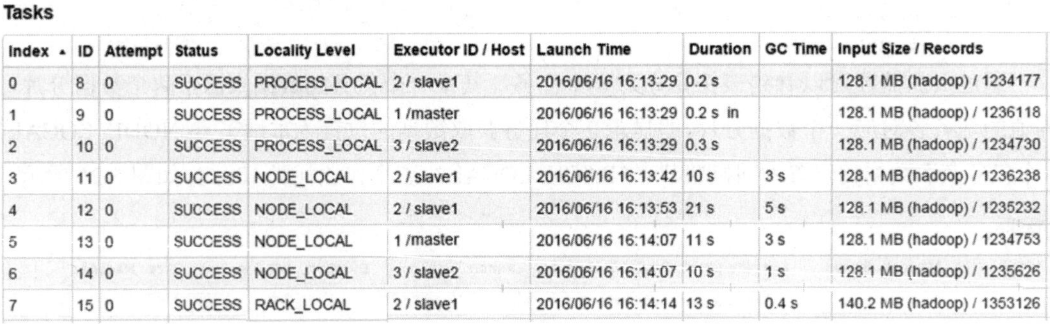

图 6-28　使用缓存计算数据集条数任务详细情况

YARNClientClusterScheduler 替代了独立运行模式下的 TaskScheduler 进行任务管理，在任务结束后通知 YARN 集群进行资源的回收，最后关闭 SparkContect。部分缓存数据运行过程耗费了 29.77s，比没有缓存速度提升不少。

6.7.3　YARN-Cluster 实例

在该实例中将使用 YARN-Cluster 运行模式进行处理，在该运行模式中分别由客户端（仅进行提交）和 YARN 集群组成。

1. 对程序进行打包

由于 YARN-Cluster 运行模式中，需要把程序代码打包提交到集群运行，在提交之前需要打包运行代码,可参见 2.4.2 节的内容,把运行程序打包为 sparklearning.ja,并把该文件移动到 Spark 根目录下。其程序代码如下:

```scala
import org.apache.spark.{SparkConf, SparkContext}

object QueryResult{
  def main(args: Array[String]) {
    if (args.length == 0) {
      System.err.println("Usage: QueryResult <inputPath><outPath>")
      System.exit(1)
    }

    val conf = new SparkConf().setAppName("QueryResult")
    val sc = new SparkContext(conf)

    //Session查询次数排行榜
    val inputDataRdd = sc.textFile(args(0)).map(_.split("\t")).filter
      (_.length==6)
    val sortedResult=inputDataRdd.map(x=>(x(1),1)).reduceByKey(_+_).map(x=>
      (x._2,x._1)).sortByKey(false).map(x=>(x._2,x._1))
    sortedResult.saveAsTextFile(args(1))
    sc.stop()
  }
}
```

2. 运行程序

通过如下命令执行程序,在演示当中分配 3 个 Executor、每个 Executor 为 1024MB 的内存。

```
$cd /app/spark/spark-2.0.0
$./bin/spark-submit --class chpater6.QueryResult \
--master yarn\
--deploy-mode cluster \
--executor-memory 1024m \
sparklearning.jar \
hdfs://master:9000/sogou/SogouQ2.txthdfs://master:9000/chapter6/output
```

(1) 把相关的资源上传到 HDFS 中,相对于 YARN-Client 多了 sparklearning.jar 文件。

命令提交应用程序后,在提交界面后显示把运行 JAR 包提交 HDFS 中,也可使用 hadoop 命令查看/user/spark/.sparkStaging/application_1465981853155_0002 目录 JAR 的情况(需根据实际应用程序编号进行确定)。

```
$hadoop fs -ls /user/spark/.sparkStaging/application_1465981853155_0002
Found 204 items
……
```

```
-rw-… 2 spark supergroup 62474622 2016-06-17 14:48 /…/spark-catalyst_2.11-2.0.0.jar
-rw-… 2 spark supergroup 11377899 2016-06-17 14:48 /…/spark- core_2.11-2.0.0.jar
-rw-… 2 spark supergroup   677257 2016-06-17 14:48 /…/spark- graphx_2.11-2.0.0.jar
-rw-… 2 spark supergroup  1295219 2016-06-17 14:48 /…/spark- hive_2.11-2.0.0.jar
-rw-… 2 spark supergroup  5512139 2016-06-17 14:48 /…/spark- mllib_2.11-2.0.0.jar
……
-rw-… 2 spark supergroup  5331329 2016-06-17 14:49 /…/spark-yarn_2.11-2.0.0.jar
-rw-… 2 spark supergroup    13871 2016-06-17 14:49 /…/sparklearning_2.11-2.0.0.jar
……
```

（2）YARN 集群接管运行。

首先 YARN 集群中由 Resource Manager 分配 Container 启动 Application Master，进而启动 SparkContext，并申请运行资源，在获取运行节点上启动 Executor，然后由 SparkContext 的 YarnClusterScheduler 进行任务的分发和监控，最终在任务执行完毕时由 YarnClusterScheduler 通知 ResourceManager 进行资源的回收。提交应用程序到 Resource Manager 并启动 Application Master 相关启动日志如下：

```
16/06/17 15:35:19 INFO yarn.Client: Submitting application 24 to ResourceManager
16/06/17 15:35:19 INFO impl.YarnClientImpl: Submitted application
    application_1465981853155_0002
16/06/17 15:35:20 INFO yarn.Client: Application report for
    application_1465981853155_0002 (state: ACCEPTED)
16/06/17 15:35:20 INFO yarn.Client:
    client token: N/A
    diagnostics: N/A
    ApplicationMaster host: N/A
    ApplicationMaster RPC port: -1
    queue: default
    start time: 1466148919637
    final status: UNDEFINED
    tracking URL: http://master:8088/proxy/application_1465981853155_0002/
    user: spark
```

3. 运行结果

在 YARN-Cluster 模式中命令界面只负责应用的提交，SparkContext 和作业运行均在 YARN 集群中，可以从 http://master:8088 中查看到具体运行过程。在该界面上的应用程序列表中单击对应 ApplicationMaster 连接，可以查看该应用程序详细情况。如果应用程序正在处理中，该页面与 Spark 监控页面类似。

通过如图 6-29 所示的页面监控可以看到该作业分为 5 个调度阶段，具体可以查看输出文件调度阶段。该调度阶段包含 8 个任务，其中 5 个任务获取数据的位置为本地节点

（NODE_LOCAL），3 个任务获取数据的位置为同机架节点（RACK_LOCAL），整个作业耗费 334s。

Tasks

Index ▲	ID	Status	Locality Level	Executor	Launch Time	Dura	GC	Input Size / Records	Write Time	Shuffle Write Size
0	1	SUCCESS	NODE_LOCAL	3 / slave1	2016/06/17 15:36:38	48 s	3 s	128.1 MB (hadoop) / 1234177	1 s	10.8 MB / 343928
1	0	SUCCESS	NODE_LOCAL	2 / slave2	2016/06/17 15:36:33	22 s	2 s	128.1 MB (hadoop) / 1236118	0.6 s	10.9 MB / 345480
2	3	SUCCESS	NODE_LOCAL	2 / slave2	2016/06/17 15:36:56	21 s	1 s	128.1 MB (hadoop) / 1234730	93 ms	10.9 MB / 346942
3	2	SUCCESS	NODE_LOCAL	1 / slave1	2016/06/17 15:36:38	48 s	3 s	128.1 MB (hadoop) / 1236238	1 s	11.2 MB / 354687
4	4	SUCCESS	NODE_LOCAL	2 / slave2	2016/06/17 15:37:18	17 s	1 s	128.1 MB (hadoop) / 1235232	0.3 s	10.8 MB / 344468
5	5	SUCCESS	RACK_LOCAL	3 / slave1	2016/06/17 15:37:29	34 s	2 s	128.1 MB (hadoop) / 1234753	0.1 s	11.2 MB / 357008
6	6	SUCCESS	RACK_LOCAL	1 / slave1	2016/06/17 15:37:29	34 s	2 s	128.1 MB (hadoop) / 1235626	0.1 s	11.0 MB / 349554
7	7	SUCCESS	RACK_LOCAL	2 / slave2	2016/06/17 15:37:36	25 s	2 s	140.2 MB (hadoop) / 1353126	99 ms	12.2 MB / 386718

图 6-29　计算 Session 查询次数排行榜

对保存到 HDFS 的结果进行合并，合并完查看结果，命令如下：

```
$cd /home/spark/work
$hdfs dfs -getmerge /chapter6/output result.txt
$ll result.txt
-rw-r--r-- 1 spark spark 99786369 Jun 17 15:54 result.txt
$cat result.txt | less
(ac65768b987c20b3b25cd35612f61892,20385)
(9faa09e57c277063e6eb70d178df8529,11653)
(02a8557754445a9b1b22a37b40d6db38,11528)
(cc7063efc64510c20bcdd604e12a3b26,2571)
(b64b0ec03efd0ca9cef7642c4921658b,2355)
(7a28a70fe4aaff6c35f8517613fb5c67,1292)
(b1e371de5729cdda9270b7ad09484c4f,1277)
(f656e28e7c3e10c2b733e6b68385d5a2,1241)
……
```

6.8　小结

在本章中详细介绍了 Spark 所支持的运行架构，用户能够根据不同需求，部署对应的运行架构。首先概况地介绍的 Spark 运行架构，Spark 应用程序一般由三部分，包括 SparkContext、Cluster Manager 和 Executor。SparkContext 用于负责和 ClusterManager 通信，进行资源的申请、任务的分配和监控等，负责作业执行的全生命周期管理，而 ClusterManager 提供了资源的分配和管理，在不同的运行模式下担任的角色有所不同。同时也介绍了两个重要类 TaskScheduler 和

SchedulerBackend，TaskScheduler 负责具体任务的调度执行，而 SchedulerBackend 则负责应用程序运行期间与底层资源调度系统交互。

然后介绍了本地运行模式、伪分布运行模式和独立运行模式。在本地运行模式中，Spark 所有进程都运行在一台机器的 JVM 中，Spark 默认设置为 Local 模式，也可以通过 local[N]开启 N 个线程进行多线程运行。伪分布运行模式是在一台机器中模拟集群运行，该模式中 Master、Worker 和 SpartContext 在一台机器上。而独立运行模式是 Spark 自身实现的资源调度框架，由客户端、Master 节点和 Worker 节点组成，其中 SparkContext 既可以运行在 Master 节点上中，也可以运行在本地客户端。

紧接着介绍了 YARN 运行模式和 Mesos 运行模式，YARN 运行模式中根据 Driver 在集群中的位置分为两种模式：YARN-Client 模式和 YARN-Cluster 模式，而 Mesos 运行模式则分为粗粒度和细粒度运行模式，在 Spark 中默认情况下使用 Mesos 细粒度运行模式。

最后演示了独立运行模式、YARN-Client 和 YARN-Cluster 三个实例。在独立运行模式演示了在 Spark Shell 计算了搜狗日志数据的行数，而在 YARN-Client 和 YARN-Cluster 实例中，对独立运行模式中计数的代码进行打包，通过 UI 界面监控运行过程。

第三篇 组件篇

第 7 章

Spark SQL

实际上"Spark SQL"这个名字并不恰当。根据 Spark 官方文档的定义：Spark SQL 是一个用于处理结构化数据的 Spark 组件——该定义强调的是"结构化数据"，而非"SQL"。Spark 1.3 更加完整地表达了 Spark SQL 的愿景：让开发者用更精简的代码处理尽量少的数据，同时让 Spark SQL 自动优化执行过程，以达到降低开发成本，提升数据分析执行效率的目的。

7.1 Spark SQL 简介

7.1.1 Spark SQL 发展历史

Spark SQL 是 Spark 1.0 版本中加入的组件，是 Spark 生态系统中最活跃的组件之一。它能够利用 Spark 进行结构化数据的存储和操作，结构化数据既可以来自外部结构化数据源（当前支持 Hive、JSON 和 Parquet 等操作，Spark 1.2 版本开始对 JDBC/ODBC 等的支持），也可以通过向已有 RDD 增加 Schema 的方式得到。

Spark SQL 提供了方便的调用接口，用户可以同时使用 Scala、Java 和 Python 语言开发基于 Spark SQL API 的数据处理程序，并通过 SQL 语句来与 Spark 代码交互。当前 Spark SQL 使用 Catalyst 优化器来对 SQL 语句进行优化，从而得到更有效的执行方案，并且可以将结果存储到 Parquet 格式等兼容的外部存储系统中。更重要的是，基于 Spark 的 DataFrame，Spark SQL 可以和 Spark Streaming、GraphX 和 MLlib 等子框架无缝集成，这样就可以在一个技术堆栈中对数据进行批处理、实时流处理和交互式查询等多种业务处理。

介绍 Spark SQL，不得不提 Hive 和 Shark，Hive 是 Shark 的前身，Shark 是 Spark SQL 的前

身。根据伯克利实验室提供的测试数据，Shark 在基于内存计算的性能比 Hive 高出 100 倍，即使是基于磁盘计算，它的性能也比 Hive 高出 10 倍，而 Spark SQL 的性能比 Shark 又有较大的提升。

Hive 是建立在 Hadoop 上的数据仓库基础构架，也是最早运行在 Hadoop 上 SQL on Hadoop 工具之一。它提供了一系列的工具，可以用来进行数据提取转化加载（ETL），这是一种可以存储、查询和分析存储在 Hadoop 中的大规模数据的机制。Hive 还定义了简单的类 SQL 查询语言，称为 HQL。它允许熟悉 SQL 的用户查询数据，同时这个语言也允许熟悉 MapReduce 开发者的开发自定义的 Mapper 和 Reducer 来处理复杂的分析工作。正由于 Hive 大大简化了对大规模数据集的分析门槛，所以它很快就流行起来，成为 Hadoop 生态系统重要的一份子。但是由于 Hive 基于 MapReduce 进行处理，在该过程中大量的中间磁盘落地过程消耗了大量的 I/O，这大大降低了运行效率，基于此，大量优化的 SQL on Hadoop 工具出现了，其中表现最为突出的工具之一就是 Shark。

Shark 是由伯克利实验室技术团队开发的 Spark 生态环境组件之一，它扩展了 Hive 并修改了 Hive 架构中的内存管理、物理计划和执行 3 个模块，使之可以运行在 Spark 引擎上，大大加快了在内存和磁盘上的查询。

图 7-1 为 EC2 中 100 个节点 Shark 内存/磁盘、Hive 处理时间的对比情况。

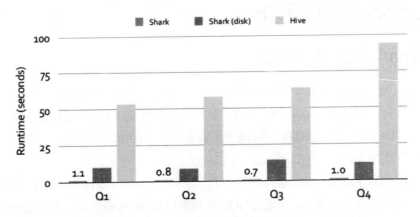

图 7-1　EC2 中 100 节点 Shark 内存/磁盘、Hive 处理时间对比

Shark 直接建立在 Apache Hive 代码库上，所以它支持几乎所有 Hive 特点。它支持现有的 Hive SQL 语言、Hive 数据格式（SerDes）和用户自定义函数（UDF），调用外部脚本查询，并

采用 Hive 解析器、查询优化器等。但正是由于 Shark 的整体设计架构对 Hive 的依赖性太强，难以支持其长远发展，比如不能和 Spark 的其他组件进行很好的集成，无法满足 Spark 的一栈式解决大数据处理的需求，Databricks 公司在 Spark Summit 2014 上宣布 Shark 已经完成了其学术使命全面转向 Spark SQL。

图 7-2 为 Hive 和 Shark 的架构比较。

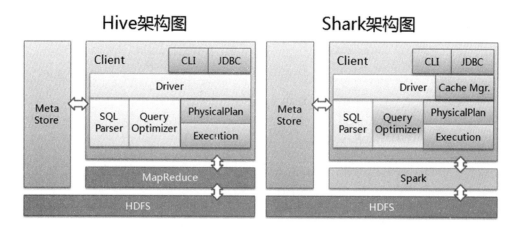

图 7-2　Hive 和 Shark 的架构比较

相比于 Shark 对 Hive 的过度依赖，Spark SQL 在 Hive 兼容层面仅依赖 HQL Parser、Hive Metastore 和 Hive SerDes。也就是说，从 HQL 被解析成抽象语法树（AST）起，就全部由 Spark SQL 接管了，执行计划生成和优化都由 Catalyst 负责。借助 Scala 的模式匹配等函数式语言特性，利用 Catalyst 开发执行计划优化策略比 Hive 要简洁得多。此外，除了兼容 HQL、加速现有 Hive 数据的查询分析以外，Spark SQL 还支持直接对原生 RDD 对象进行关系查询。同时，除了 HQL 以外，Spark SQL 还内建了一个精简的 SQL Parser 以及一套 Scala DSL。也就是说，如果只是使用 Spark SQL 内建的 SQL 方言或 Scala DSL 对原生 RDD 对象进行关系查询，用户在开发 Spark 应用时完全不需要依赖 Hive 的任何东西。当然，在重新 Spark SQL 代码的时候，Spark SQL 技术人员还吸取了 Shark 的一些优点，如内存列存储（In-Memory Columnar Storage）等。这些优点使得 Spark SQL 无论在数据兼容性、性能优化、组件扩展等方面都得到了极大的方便。

值得一提的是，Hive 社区也推出了一个 Hive on Spark 的项目——将 Hive 的执行引擎换成 Spark，如图 7-3 所示。不过从目标上看，Hive on Spark 更注重针对 Hive 彻底地向下兼容性，而 Spark SQL 更注重 Spark 与其他组件的互操作和多元化数据处理。

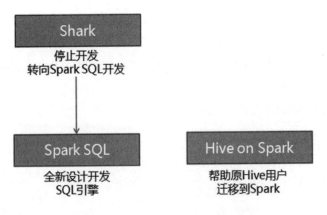

图 7-3　Spark SQL 和 Hive on Spark

7.1.2　DataFrame/Dataset 介绍

Spark 的 RDD API 比传统的 MapReduce API 在易用性有了巨大的提升，但是对于没有 MapReduce 和函数式编程经验的新手，RDD API 还是存在一定的门槛。另一方面，数据分析人员所使用的 R 和 Pandas 等传统数据分析工具虽然提供直观的 API，却由于这些工具只能处理单机的数据，无法胜任大数据处理任务。为了解决这两个问题，从 Spark SQL 1.3 版本开始在原有 SchemaRDD 的基础上提供了与 R 和 Pandas 风格类似的 DataFrame API，新的 DataFrame API 不仅大大降低了新手学习门槛高，而且还支持 Scala、Java 和 Python 3 种语言，最重要的是脱胎于 SchemaRDD 的 DataFrame 支持分布式大数据处理。

在 Spark 中，DataFrame 是一种以 RDD 为基础的分布式数据集，类似于传统数据库中的二维表格。与 RDD 的主要区别在于：前者带有 Schema 元数据，即 DataFrame 所表示的二维表数据集的每一列都带有名称和类型，其结构对比如图 7-4 所示。由于无法知道 RDD 数据集内部的结构，Spark 作业执行职能在调度阶段层面进行简单通用的优化，而对于 DataFrame 带有数据集内部的结构，可以根据这些信息进行针对性的优化，最终实现优化运行效率。

图 7-4 直观地展示了 RDD 和 DataFrame 的区别：左边的 RDD[Person]虽然以 Person 为类型参数，但 Spark 框架本身不了解 Person 类的内部结构，而右边的 DataFrame 却提供了详细的结构信息，使得 Spark SQL 可以清楚地知道该数据集中包含哪些列，每列的名称和类型分别是什么，这和关系数据库中的物理表类似。有了这些元数据，Spark SQL 的查询优化器就可以进行针对性的优化。在 Spark SQL 中引入 DataFrame 能带来的如下好处。

Name	Age	Height
String	Int	Double
String	Int	Double
String	Int	Double
String	Int	Double
String	Int	Double
String	Int	Double

RDD[Person]　　　　　　　　　　DataFrame

图 7-4　RDD 与 DataFrame 的结构对比

1．精简代码

DataFrame 最大的优点之一就是能够使用更精简的代码。下面分别使用 Hadoop MapReduce、Python RDD API 和 Python DataFrame API 对 people 表中人员按照名字求平均年龄代码片段。很明显，Hadoop MapReduce 的代码量最多，也不容易理解代码逻辑；Python RDD API 实现的代码精简了许多，但还是不容易理解代码逻辑；而 Python DataFrame API 实现的代码显然比 Python RDD API 的代码更加精炼，更重要的是，对于有 SQL 基础的，都可以非常容易看懂代码实现的逻辑，大大增加了代码的可读性。另外，从 Spark1.3 版本起，Spark 提供了 Python、Scala 和 Java 3 种语言的 DataFrame API，可以根据用户熟悉程度进行选择。

2．提升执行效率

使用 DataFrame API，不仅仅能够精简代码，还可以提升执行效率。图 7-5 对比了用 Scala、Python RDD API 和 DataFrame API 实现的累加 1 千万整数对的 4 段程序的性能对比。从图 7-5 中可以看到，Python DataFrame API 相对于 Python RDD API 的执行效率有了近 5 倍的提升，这是因为 DataFrame API 实际上仅仅组装了一段体积小巧的逻辑查询计划，Python 端只需将查询计划发送到 JVM 端即可，计算任务的大头都由 JVM 端负责。而在使用 Python RDD API 时，Python VM 和 JVM 之间需要进行大量的跨进程数据交换，从而拖慢了 Python RDD API 的速度。

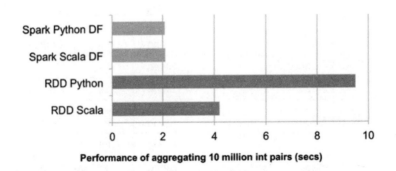

图 7-5 不同语言之间使用 RDD 与 DF 执行效率对比

值得注意的是，不仅 Python RDD API 有了显著的性能提升，即便是使用 Scala DataFrame API 的版本也要比 PythonRDD API 快一倍。这个例子的操作极为简单，查询优化器起的作用不大，但为什么有加速效果呢？这是因为 PythonRDD API 是函数式的，强调不变形，在大部分场景下倾向于创建新对象而不是修改老对象，这点虽然可以实现整洁的 API，但使得 Spark 在应用程序运行过程中倾向创建大量临时对象，对 GC 造成较大压力。而在 Spark SQL 中尽可能重用对象，这样虽然在内部打破了不变性，但在数据返回给用户时，还会重新转化为不可变数据，这样用户利用 DataFrame API 能够使用到优化效果。

3．减少数据读取

分析大数据最有效的方法就是忽略无关的数据，根据查询条件进行恰当的裁剪。对于一些比较"智能"数据格式，Spark SQL 可以根据数据文件中附带的统计信息来进行剪枝。在这类数据格式中，数据是分段保存的，每段数据都带有最大值、最小值和 NULL 值数量等统计信息，当统计信息表名某一数据段肯定不包括符合查询条件的目标数据时，该数据段就可以直接跳过（例如某整数列 a 某段的最大值为 100，而查询条件要求 a > 200）。此外，Spark SQL 也可以充分利用 RCFile、ORC 和 Parquet 等列式存储格式的优势，仅扫描查询真正涉及的列，而忽略其余列。

DataFrame 在提供结构化数据抽象的同时，还能通过其接口实现多种外部数据源读取，然后对这些获取的数据进行混合处理，处理完毕后以特定的格式写回数据源或直接予以某种形式的展现。Spark 1.2 版本中引入的外部数据源 API 正是为了解决这一问题而产生的，Spark SQL 外部数据源 API 的优势在于：可以将查询中的各种信息下推至数据源处，从而充分利用数据源自身的优化能力来完成列剪枝、过滤条件下推等优化，实现减少 I/O、提高执行效率的目的。图 7-6 是 Spark 1.3 支持的各种数据源的一个概览（其中左侧是 Spark SQL 内置支持的数据源，右侧为社区开发者贡献的数据源）。在外部数据源 API 的帮助下，DataFrame 实际上成为了各种数据格式和存储系统进行数据交换的中间媒介：在 Spark SQL 内，来自各处的数据都被加载为 DataFrame 混合、统一成单一形态，再以之基础进行数据分析和价值提取。

图 7-6　DataFrame 支持外部数据源

7.2　Spark SQL 运行原理

7.2.1　通用 SQL 执行原理

在传统关系型数据库中，最基本的 SQL 查询语句如 SELECT fieldA, fieldB, fieldC FROM tableA WHERE fieldA > 10，由 Projection（fieldA、fieldB、fieldC）、Data Source（tableA）和 Filter（fieldA > 10）三部分组成，分别对应 SQL 查询过程中的 Result、Data Source 和 Operation，也就是说 SQL 语句按 Result→Data Source→Operation 的次序来描述的，如图 7-7 所示。

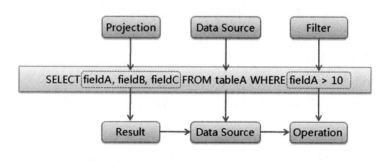

图 7-7　通过 SQL 执行顺序

但在实际执行 SQL 语句的过程中是按照 Operation→Data Source→Result 的顺序来执行，与 SQL 语法顺序刚好相反，其具体执行过程如下。

（1）词法和语法解析（Parse）：对读入的 SQL 语句进行词法和语法解析（Parse），分辨出 SQL 语句中哪些词是关键词（如 SELECT、FROM 和 WHERE）、哪些是表达式、哪些是 Projection、哪些是 Data Source 等，判断 SQL 语句是否规范，并形成逻辑计划。

（2）绑定（Bind）：将 SQL 语句和数据库的数据字典（列、表和视图等）进行绑定（Bind），如果相关的 Projection 和 Data Source 等都存在的话，则表示这个 SQL 语句是可以执行的。

（3）优化（Optimize）：一般的数据库会提供几个执行计划，这些计划一般都有运行统计数据，数据库会在这些计划中选择一个最优计划。

（4）执行（Execute）：执行前面的步骤获取的最优执行计划，返回从数据库中查询的数据集。

一般来说，关系数据库在运行过程中，会在缓冲池缓存解析过的 SQL 语句，在后续的过程中如果能够命中缓存 SQL 就可以直接返回可执行的计划，比如重新运行刚运行过的 SQL 语句，可能直接从数据库的缓冲池中获取返回结果。

7.2.2　SparkSQL 运行架构

Spark SQL 对 SQL 语句的处理和关系型数据库采用了类似的方法，Spark SQL 先会将 SQL 语句进行解析（Parse）形成一个 Tree，然后使用 Rule 对 Tree 进行绑定、优化等处理过程，通过模式匹配对不同类型的节点采用不同的操作。而 Spark SQ 的查询优化器是 Catalyst，它负责处理查询语句的解析、绑定、优化和生成物理计划等过程，Catalyst 是 Spark SQL 最核心的部分，其性能优劣将决定整体的性能。

Spark SQL 由 Core、Catalyst、Hive 和 Hive-Thriftserver4 个部分组成。

- Core：负责处理数据的输入/输出，从不同的数据源获取数据（如 RDD、Parquet 文件和 JSON 文件等），然后将查询结果输出成 Data Frame。
- Catalyst：负责处理查询语句的整个处理过程，包括解析、绑定、优化、物理计划等。
- Hive：负责对 Hive 数据的处理。
- Hive-thriftserver：提供 CLI 和 JDBC/ODBC 接口等。

在介绍 Spark SQL 代码分析之前先介绍 Tree 和 Rules 两个重要概念。

1．Tree

Tree 的具体操作是通过 TreeNode 来实现的，Tree 是 Catalyst 执行计划表示的数据结构。

Logical Plans、Expressions 和 Physical Operators 都可以使用 Tree 来表示，Tree 具备一些 Scala Collection 的操作能力和树遍历能力。在 Spark SQL 中会根据语法生成一棵树，该树一直在内存里维护，不会保存到磁盘以某种格式的文件存在，且无论是 Analyzer 分析过的逻辑计划还是 Optimizer 优化过的逻辑计划，树的修改是以替换已有节点的方式进行的。

Tree 内部带一个 Children: Seq[BaseType]方法，可以返回一系列孩子节点。Tree 提供 UnaryNode、BinaryNode 和 LeafNode 三种特质（trait）。

- UnaryNode：表示一元节点，即只有一个子节点；
- BinaryNode：表示二元节点，即有左右子节点的二叉节点；
- LeafNode：叶子节点，没有子节点的节点。

针对不同的节点，Tree 提供了不同的操作方法，对 UnaryNode 可以进行 Limit 和 Filter 等操作；对 BinaryNode 可以进行 Join 和 Union 等操作；对于 LeafNode 主要用户命令类操作，如 Set Command 等。

而对 Tree 的遍历操作，主要是借助各个 Tree 之间的关系，使用 transformDown、transformUp 将 Rule 应用到给定的树段，并对匹配节点实施变化的方法（使用的是 Tree 节点中的 Transform 方法），其中 transformDown 是默认的前序遍历；也可以使用 transformChildrenDown、transformChildrenUp 对一个给定的节点进行操作，通过迭代将 Rule 应用到该节点以及子节点。

Tree 有两个子类继承体系，即 QueryPlan 和 Expression。

- QueryPlan 下面的两个子类分别是 LogicalPlan（逻辑执行计划）和 SparkPlan（物理执行计划）。QueryPlan 内部带有 output: Seq[Attribute]、transformExpressionDown 和 transformExpressionUp 等方法，它的主要子类体系是 LogicalPlan，即逻辑执行计划表示，它在 Catalyst 优化器里有详细实现。LogicalPlan 内部带一个 reference:Set[Attribute]，主要方法为 resolve(name:String): Option[NamedExpression]，用于分析生成对应的 NamedExpression。对于 SparkPlan，即物理执行计划表示，需要用户在系统中自己实现。LogicalPlan 本身也有许多具体子类，也分为 UnaryNode、BinaryNode 和 LeafNode 三类。
- Expression 是表达式体系，是指不需要执行引擎计算，而可以直接计算或处理的节点，包括 Cast 操作、Projection 操作、四则运算和逻辑操作符运算等。

2. Rule

Rule[TreeType <: TreeNode[_]]是一个抽象类，子类需要复写 apply(plan: TreeType)方法来制定处理逻辑。对于 Rule 的具体实现是通过 RuleExecutor 完成的，凡是需要处理执行计划树进行

实施规则匹配和节点处理的，都需要继承 RuleExecutor[TreeType]抽象类。

在 RuleExecutor 的实现子类（如 Analyzer 和 Optimizer）中会定义 Batch、Once 和 FixedPoint。其中每个 Batch 代表着一套规则，这样可以简便地、模块化地对 Tree 进行 Transform 操作。Once 和 FixedPoint 是配备的策略，相对应的是对 Tree 进行一次操作或多次的迭代操作（如对某些 Tree 进行多次迭代操作的时候，达到 FixedPoint 次数迭代或达到前后两次的树结构没变化才停止操作）。RuleExecutor 内部提供了一个 Seq[Batch]属性，里面定义的是该 RuleExecutor 的处理逻辑，具体的处理逻辑由具体 Rule 子类实现。最后，RuleExecutor 的 apply(plan: TreeType): TreeType 方法会按照 Batch 顺序和 Batch 内的 Rules 顺序，对传入的计划节点进行迭代处理。

对于 Rule 的使用在下面例子中进行说明，在 Analyzer 过程中处理由解析器（SqlParser）生成的未绑定逻辑计划 Tree 时，就定义了多种 Rules 应用到该 Unresolved 逻辑计划 Tree 上，如图 7-8 所示。

Analyzer 过程中使用了自身定义的多个 Batch，如 MultiInstanceRelations、Resolution、Check Analysis 和 AnalysisOperators；每个 Batch 又有不同的 Rule 构成，如 Check Analysis 由 CheckResolution 和 CheckAggregation 构成；每个 Rule 又有自己相对应的处理函数，同时要注意的是，不同的 Rule 使用的次数是不同的，如 MultiInstanceRelations 这个 Batch 中 Rule 只应用了一次（Once），而 AnalysisOperators 这个 Batch 中的 Rule 应用了多次（fixedPoint = FixedPoint(100)，也就是说最多应用 100 次，除非前后迭代结果一致而退出）。

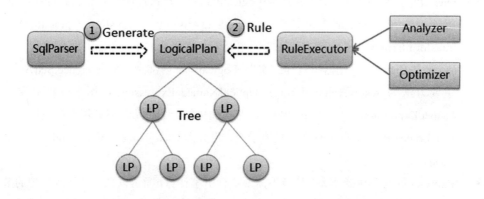

图 7-8　Spark SQL 中生成未绑定逻辑计划 Tree

介绍完两个概念后，我们再来看 Spark SQL 运行架构如图 7-9 所示。通过该图分析 Spark SQL 主要组件的功能以及 SQL 语句执行的运行流程。

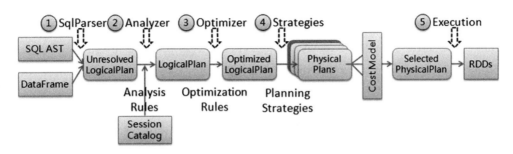

图 7-9　Spark SQL 运行架构图

（1）将 SQL 语句通过词法和语法解析生成未绑定的逻辑计划（包含 Unresolved Relation、Unresolved Function 和 Unresolved Attribute），然后在后续步骤中使用不同的 Rule 应用到该逻辑计划上。

（2）Analyzer 使用 Analysis Rules，配合数据元数据（如 SessionCatalog 或 Hive Metastore 等），完善未绑定的逻辑计划的属性而转换成绑定的逻辑计划。具体流程是先实例化一个 Simple Analyzer，然后遍历预先定义好的 Batch，通过父类 Rule Exector 的执行方法运行 Batch 里面的 Rules，每个 Rule 会对未绑定的逻辑计划进行处理，有些可以通过一次解析处理，有些需要多次迭代，迭代直到达到 FixedPoint 次数迭代或达到前后两次的树结构没变化才停止操作。

（3）Optimizer 使用 Optimization Rules，将绑定的逻辑计划进行合并、列裁剪和过滤器下推等优化工作后生成优化的逻辑计划。

（4）Planner 使用 Planning Strategies，对优化的逻辑计划进行转换（Transform）生成可以执行的物理计划。根据过去的性能统计数据，选择最佳的物理执行计划 CostModel，最后可以执行的物理执行计划树，即得到 SparkPlan。

（5）在最终真正执行物理执行计划前，还要进行 preparations 规则处理，最后调用 SparkPlan 的 execute 执行计算 RDD。

7.2.3　SQLContext 运行原理分析

在前面我们对 Spark SQL 运行架构进行分析，知道从输入 SQL 语句到生成 DataSet 分为 5 个步骤，但实际运行过程中在输入 SQL 语句之前，Spark 还有加载 SessionCatalog 步骤。下面我们将以例子结合源码来分析 SQL 语句的执行整个流程，运行流程图参见图 7-11。

1．使用 SessionCatalog 保存元数据

在解析 SQL 语句之前先需要初始化 SQLContext，它定义了 Spark SQL 执行的上下文，并把元数据保存在 SessionCatalog 中，这些元数据包括表名称、表字段名称和字段类型等。在

SessionCatalog 中保存的是表名和逻辑计划对应的哈希列表，这些数据将在解析未绑定的逻辑计划上使用。下面将使用例子讲解 SessionCatalog 注册执行过程。在该例子中，先定义 SQLContext 和隐含类，然后使用 SparkContext.parallelize 方法创建员工 RDD，再通过 toDF 方法把该 RDD 转换为员工 DataSet，最后把该 DataSet 注册到 SQLContext 的 SessionCatalog 中。

```
//引入隐含类用于把 RDD 转换成 DataSet
scala> val sqlContext = new org.apache.spark.sql.SQLContext(sc)
scala> import sqlContext.implicits._

//定义员工类，添加员工数据形成 DataSet，并注册到 SessionCatalog 中
scala> case class Employee(name: String, age: Int, provinceId: Int)
scala> val employee = sc.parallelize(Employee("zhangsan",31,"beijing")::
  Employee("wangwu",28,"hubei")::Employee("lisi",44,"tianjin")::Employee("li
  ping",23,"guangdong")::Nil).toDF()
scala> employee.registerTempTable("employee")
```

最后一句调用 DataSet.registerTempTable 方法，该方法继续调用 SQLContext.registerDataFrameAsTable 方法，该方法最终调用的是 SessionCatalog.createTempTable 方法进行实现。在 createTempTable 方法中，先验证表名是否符合规范或包含保留关键字，如果验证通过把表名和对应的逻辑计划放入到哈希列表中，代码如下：

```
protected[this] val tempTables = new mutable.HashMap[String, LogicalPlan]
def createTempTable(name: String,tableDefinition: LogicalPlan,overrideIfExists:
  Boolean): Unit = {
  val table = formatTableName(name)
  if (tempTables.contains(table) && !overrideIfExists) {
    throw new AnalysisException(s"Temporary table '$name' already exists.")
  }
  tempTables.put(table, tableDefinition)
}
```

例子中 SessionCatalog 哈希列表保存了员工表数据，其中 employee 表对应的 LogicalRDD 包括 3 个字段，分别为 name#0、age#1 和 province#2。

```
"employee" -> LogicalRDD [name#0,age#1,province#2], MapPartitionsRDD[1] at
  rddToDataFrameHolder at <console>:28
```

2. 使用 Antlr 生成未绑定的逻辑计划

从 Spark2.0 版本起使用 Antlr（Another Tool for Language Recognition）进行词法和语法解析，它为 Java、C++和 C#等语言提供了一个通过语法描述来自动构造语言的识别器（Recognizer）、编译器（Parser）和解释器（Translator）的框架。使用 Antlr 生成未绑定的逻辑计划分为两个阶段：第一阶段的过程为词法分析（Lexical Analysis），负责将符号（Token）分组成符号类（Token class or token type）；而第二阶段就是真正的 Parser，默认 Antlr 会构建出一棵分析树（Parse Tree）或者叫做语法树（Syntax Tree）。图 7-10 为赋值表达式的解析过程。

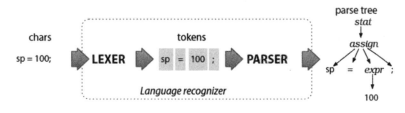

图 7-10　Antlr 赋值表达式的解析过程

在 SQLContext 类中定义了 SQL 解析方法 parseSql，该方法调用特质 ParserInterface 定义的 parsePlan 方法，其中定义解析代码如下：

```
protected[sql] def parseSql(sql: String): LogicalPlan = sessionState.sqlParser.
    parsePlan(sql)
```

具体 SQL 解析在 AbstractSqlParser 抽象类中的 parse 方法进行实现，解析完毕后生成语法树。该语法树会根据系统初始化的 AstBuilder 解析生成表达式、逻辑计划或表标识等对象。相关类图如图 7-11 所示。

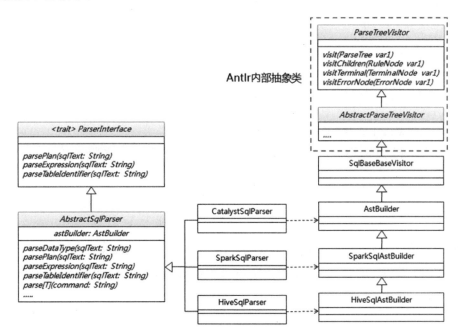

图 7-11　Spark SQL 解析调用类图

在 AbstractSqlParse 的 parse 方法中，先实例化词法解析器 SqlBaseLexer 和语法解析器 SqlBaseParser，然后尝试使用 Antlr 较快的解析模式 SLL。如果解析失败，则会再尝试使用普通解析模式 LL，解析完毕后返回解析结果，具体代码如下：

```
protected def parse[T](command: String)(toResult: SqlBaseParser => T): T = {
  //实例化词法解析器
  val lexer = new SqlBaseLexer(new ANTLRNoCaseStringStream(command))
  lexer.removeErrorListeners()
  lexer.addErrorListener(ParseErrorListener)

  //实例化语法解析器
  val tokenStream = new CommonTokenStream(lexer)
  val parser = new SqlBaseParser(tokenStream)
  parser.addParseListener(PostProcessor)
  parser.removeErrorListeners()
  parser.addErrorListener(ParseErrorListener)

  try {
    try {
      //先使用较快的 SLL 模式进行解析，成功返回结果
      parser.getInterpreter.setPredictionMode(PredictionMode.SLL)
      toResult(parser)
    }
    catch {
      case e: ParseCancellationException =>
        //如果解析失败，复位
        tokenStream.reset() // rewind input stream
        parser.reset()

        //然后使用 LL 模式进行解析，成功返回结果
        parser.getInterpreter.setPredictionMode(PredictionMode.LL)
        toResult(parser)
    }
  }
  catch { …… }
}
```

值得一提的，词法解析器 SqlBaseLexer 和语法解析器 SqlBaseParser 是通过 Antlr 定义自动生成而来的，这些文件位于 sql/catalyst/target/generated-sources/antlr4/org/apache/spark/sql/catalyst/parser 目录下，生成的代码列表如图 7-12 所示。

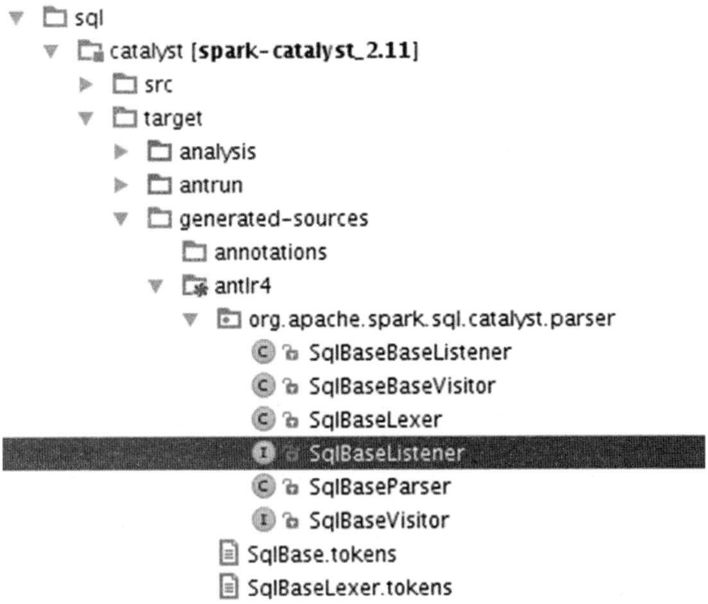

```
    ▼ 🗀 sql
        ▼ 🗀 catalyst [spark-catalyst_2.11]
            ▶ 🗀 src
            ▼ 🗀 target
                ▶ 🗀 analysis
                ▶ 🗀 antrun
                ▼ 🗀 generated-sources
                    🗀 annotations
                    ▼ 🗀 antlr4
                        ▼ 🗀 org.apache.spark.sql.catalyst.parser
                            © SqlBaseBaseListener
                            © SqlBaseBaseVisitor
                            © SqlBaseLexer
                            ⓘ SqlBaseListener
                            © SqlBaseParser
                            ⓘ SqlBaseVisitor
                    🗋 SqlBase.tokens
                    🗋 SqlBaseLexer.tokens
```

图 7-12 Antlr 生成的代码列表

在该步骤将对前面例子的员工 DataSet 进行查询，语句如下：

```
scala> val query = sqlContext.sql("select name,age from (select * from employee
  where province='beijing') a where a.age >=20 and a.age <40")
```

通过执行 query.queryExecution.logical 命令获取该步骤生成的未绑定逻辑计划，可以看到通过 SQL 解析生成一棵按照关键字生成的树。

```
res2: org.apache.spark.sql.catalyst.plans.logical.LogicalPlan =
'Project ['name,'age]
'Filter (('a.age >= 20) && ('a.age < 40))
'Subquery a
'Project [*]
'Filter ('province = beijing)
'UnresolvedRelation [employee], None
```

3．使用 Analyzer 绑定逻辑计划

在该阶段 Analyzer 使用 Analysis Rules，结合 SessionCatalog 元数据，对未绑定的逻辑计划进行解析，生成了已绑定的逻辑计划。在该处理过程中，先实例化一个 Analyzer，在 Analyzer 中定义了 FixedPoint 和 Seq[Batch] 两个变量，其中 FixedPoint 为迭代次数的上限，而 Seq[Batch] 为所定义需要执行批处理的序列，每个批处理由一系列 Rule 和策略所组成。策略一般分为 Once 和 FixedPoint（可以理解为迭代次数），其代码如下：

```
class Analyzer(catalog: SessionCatalog, conf: CatalystConf, maxIterations: Int
  = 100) extends RuleExecutor[LogicalPlan] with CheckAnalysis {
  val fixedPoint = FixedPoint(maxIterations)
  lazy val batches: Seq[Batch] = Seq(
    Batch("Substitution", fixedPoint,
      CTESubstitution,
      WindowsSubstitution,
      EliminateUnions),
    Batch("Resolution", fixedPoint,
      ResolveRelations ::
      ResolveReferences ::
      ResolveDeserializer ::
      ResolveNewInstance ::
      ResolveUpCast ::
      ResolveGroupingAnalytics ::
      ResolvePivot ::
      ResolveOrdinalInOrderByAndGroupBy ::
      ResolveMissingReferences ::
      ResolveGenerate ::
      ResolveFunctions ::
      ResolveAliases ::
      ResolveSubquery ::
      ResolveWindowOrder ::
      ResolveWindowFrame ::
      ResolveNaturalAndUsingJoin ::
      ExtractWindowExpressions ::
      GlobalAggregates ::
      ResolveAggregateFunctions ::
      TimeWindowing ::
      HiveTypeCoercion.typeCoercionRules ++
      extendedResolutionRules : _*),
    Batch("Nondeterministic", Once,
      PullOutNondeterministic),
    Batch("UDF", Once,
      HandleNullInputsForUDF),
    Batch("Cleanup", fixedPoint,
      CleanupAliases)
  )
  ……
}
```

Analyzer 解析主要是根据这些批处理定义的 Rule 来对未绑定的逻辑计划进行解析的，而 Analyzer 类本身并没有定义执行 Rule 的方法，需要调用父类 RuleExecutor.execute 来执行这些

Rule，如图 7-13 所示。

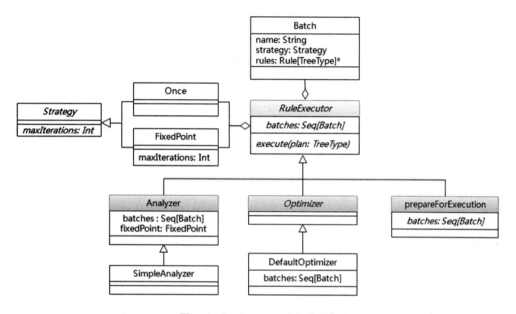

图 7-13　RuleExecutor 相关类图

在 execute 方法中，执行这些 Rule 为串行操作，迭代这些 Rule 处理未绑定逻辑计划直到达到 FixedPoint 次数或迭代前后两次的已绑定逻辑计划相同才停止操作，其中 RuleExecutor.execute 代码如下：

```
def execute(plan: TreeType): TreeType = {
  var curPlan = plan

  batches.foreach { batch =>
    val batchStartPlan = curPlan
    var iteration = 1
    var lastPlan = curPlan
    var continue = true

    //用 Rule 迭代处理直到达到 FixedPoint 次数或迭代前后两次结果相同
    while (continue) {
      curPlan = batch.rules.foldLeft(curPlan) {
        case (plan, rule) =>
          val startTime = System.nanoTime()
          //使用 Rule 对前一次处理的逻辑计划进行绑定
          val result = rule(plan)
          val runTime = System.nanoTime() - startTime
```

```
          RuleExecutor.timeMap.addAndGet(rule.ruleName, runTime)
          …….
          result
        }
        iteration += 1

        //当迭代次数大于策略设置的最大次数将退出
        if (iteration > batch.strategy.maxIterations) {
          if (iteration != 2) {
            logInfo(s"Max iterations (${iteration - 1}) reached for batch
              ${batch.name}")
          }
          continue = false
        }
        lastPlan = curPlan
    }
    …….
  }
  curPlan
}
```

在该步骤中通过执行 query.queryExecution.analyzed 命令生成已绑定逻辑计划如下结果，我们可以清楚地看到逻辑计划中的表、字段均替换为 Catalog 中表的元数据。

```
res3: org.apache.spark.sql.catalyst.plans.logical.LogicalPlan =
Project [name#0,age#1]
 Filter ((age#1 >= 20) && (age#1 < 40))
  Subquery a
   Project [name#0,age#1,province#2]
    Filter (province#2 = beijing)
     Subquery employee
      LogicalRDD    [name#0,age#1,province#2],    MapPartitionsRDD[1]    at
        rddToDataFrameHolder at <console>:28
```

4．使用 Optimizer 优化逻辑计划

Optimizer（优化器）的实现和处理方式同 Analyzer 类似，在该类中定义一系列 Rule 并同样继承于 RuleExecutor。利用这些 Rule 对逻辑计划和 Expression 进行迭代处理，从而达到树的节点进行合并和优化。其中主要的优化策略总结起来是合并、列裁剪和过滤器下推等几大类，其代码如下：

```
abstract class Optimizer extends RuleExecutor[LogicalPlan] {
  def batches: Seq[Batch] = {
    Batch("Finish Analysis", Once,
      EliminateSubqueryAliases,
```

```
    ComputeCurrentTime,
    DistinctAggregationRewriter) ::
Batch("Union", Once,
    CombineUnions) ::
Batch("Replace Operators", FixedPoint(100),
    ReplaceIntersectWithSemiJoin,
    ReplaceDistinctWithAggregate) ::
Batch("Aggregate", FixedPoint(100),
    RemoveLiteralFromGroupExpressions) ::
Batch("Operator Optimizations", FixedPoint(100),
    //操作下推
    SetOperationPushDown,
    SamplePushDown,
    ReorderJoin,
    OuterJoinElimination,
    PushPredicateThroughJoin,
    PushPredicateThroughProject,
    PushPredicateThroughGenerate,
    PushPredicateThroughAggregate,
    LimitPushDown,
    ColumnPruning,
    InferFiltersFromConstraints,
    //操作合并
    CollapseRepartition,
    CollapseProject,
    CombineFilters,
    CombineLimits,
    CombineUnions,
    //常量合并和维数降低
    NullPropagation,
    OptimizeIn,
    ConstantFolding,
    LikeSimplification,
    BooleanSimplification,
    SimplifyConditionals,
    RemoveDispensableExpressions,
    PruneFilters,
    EliminateSorts,
    SimplifyCasts,
    SimplifyCaseConversionExpressions,
    EliminateSerialization) ::
Batch("Decimal Optimizations", FixedPoint(100),
    DecimalAggregates) ::
Batch("Typed Filter Optimization", FixedPoint(100),
```

```
      EmbedSerializerInFilter) ::
    Batch("LocalRelation", FixedPoint(100),
      ConvertToLocalRelation) ::
    Batch("Subquery", Once,
      OptimizeSubqueries) :: Nil
  }
  ......
}
```

Optimizer 的优化策略不仅有对已绑定的逻辑计划进行优化，而且对逻辑计划中的 Expression 也进行了优化。其原理就是遍历树，然后应用优化 Rule，但是注意一点，对逻辑计划处理是先序遍历(pre-order)，而对 Expression 处理的时候是后序遍历(post-order)。

在该步骤中通过执行 query.queryExecution.optimizedPlan 命令生成优化的逻辑计划如下：

```
res4: org.apache.spark.sql.catalyst.plans.logical.LogicalPlan =
Project [name#0,age#1]
 Filter ((province#2 = beijing) && ((age#1 >= 20) && (age#1 < 40)))
  LogicalRDD [name#0,age#1,province#2], MapPartitionsRDD[1] at
    rddToDataFrameHolder at <console>:28
```

5. 使用 SparkPlanner 生成可执行的物理计划

SparkPlanner 使用 Planning Strategies 对优化的逻辑计划进行转换（Transform），生成可以执行的物理计划。在 QueryExecution 类代码中，调用 SparkPlanner.plan 方法对优化的逻辑计划进行处理，而 SparkPlanner 并未定义 plan 方法，实际是调用 SparkPlanner 祖父类 QueryPlanner 的 plan 方法，然后会返回一个 Iterator[PhysicalPlan]，其代码如下：

```
lazy val optimizedPlan: LogicalPlan = sqlContext.sessionState.optimizer.
    execute(withCachedData)
lazy val sparkPlan: SparkPlan = {
  SQLContext.setActive(sqlContext)
  planner.plan(ReturnAnswer(optimizedPlan)).next()
}
```

SparkPlanner 继承于 SparkStrategies，而 SparkStrategies 继承了 QueryPlanner。其中 SparkStrategies 包含了一系列特定的 Strategies，这些 Strategies 是继承自 QueryPlanner 中定义的 GenericStrategy。在 SparkPlanner 通过改写祖父类 QueryPlanner 中的 strategies 策略变量，在该变量中定义了转变成物理计划所执行的策略。

```
class SparkPlanner(val sparkContext: SparkContext,val conf: SQLConf,
  val extraStrategies: Seq[Strategy])extends SparkStrategies {
  def numPartitions: Int = conf.numShufflePartitions

  def strategies: Seq[Strategy] =
```

```
      extraStrategies ++ (
      FileSourceStrategy ::
      DataSourceStrategy ::
      DDLStrategy ::
      SpecialLimits ::
      Aggregation ::
      ExistenceJoin ::
      EquiJoinSelection ::
      InMemoryScans ::
      BasicOperators ::
      BroadcastNestedLoop ::
      CartesianProduct ::
      DefaultJoin :: Nil)
    ……
}
```

QueryPlanner 是 SparkPlanner 的祖父类，该类中定义了策略的抽象类 GenericStrategy，并实现了 plan 方法，其代码如下：

```
abstract class GenericStrategy[PhysicalPlan <: TreeNode[PhysicalPlan]] extends
    Logging {
  //传入一个逻辑计划，生成一个物理计划序列
  def apply(plan: LogicalPlan): Seq[PhysicalPlan]
}

abstract class QueryPlanner[PhysicalPlan <: TreeNode[PhysicalPlan]] {
  def strategies: Seq[GenericStrategy[PhysicalPlan]]
  //返回一个占位符
  protected def planLater(plan: LogicalPlan) = this.plan(plan).next()

  def plan(plan: LogicalPlan): Iterator[PhysicalPlan] = {
    //遍历所有策略，把所有策略作用于逻辑计划，生成物理计划集合
    val iter = strategies.view.flatMap(_(plan)).toIterator
    assert(iter.hasNext, s"No plan for $plan")
    iter
  }
}
```

在该步骤中通过执行 query.queryExecution.sparkPlan 命令生成可执行的物理计划：

```
res5: org.apache.spark.sql.execution.SparkPlan =
Project [name#0,age#1]
 Filter ((province#2 = beijing) && ((age#1 >= 20) && (age#1 < 40)))
  PhysicalRDD [name#0,age#1,province#2], MapPartitionsRDD[1] at
     rddToDataFrameHolder at <console>:28
```

6. 使用 QueryExecution 执行物理计划

在该步骤中先调用 SparkPlanner.preparations 对物理计划进行准备工作，规范返回数据行的格式等，然后调用 SparkPlan.execute 执行物理计划，从数据库中查询数据并生成 RDD。

SparkPlanner 的 preparations 其实是一个 RuleExecutor[SparkPlan]，它会调用 RuleExecutor 的 execute 方法对前面生成的物理计划应用 Rule 进行匹配，最终生成一个 SparkPlan。

```
protected def preparations: Seq[Rule[SparkPlan]] = Seq(
  python.ExtractPythonUDFs,
  PlanSubqueries(sqlContext),
  EnsureRequirements(sqlContext.conf),
  CollapseCodegenStages(sqlContext.conf),
  ReuseExchange(sqlContext.conf))
```

SparkPlan 继承于 QueryPlan，SparkPlan 中定义了 SQL 语句执行的 execute 方法，执行完 execute 返回的是一个 RDD，之后可以运行 Spark 作业对该 RDD 进行操作。其中在 QueryExecution 类中，调用 execute 方法代码如下：

```
lazy val toRdd: RDD[InternalRow] = executedPlan.execute()
```

该步骤中通过下面语句，打印出所有满足条件的数据：

```
scala> query.map(_.getValuesMap[Any](List("name", "age"))).collect().
  foreach(println)
```

7.2.4 HiveContext 介绍

由于历史原因，Spark 技术团队根据该情况开发了 Hive on Spark 项目来兼容 Hive。在这个框架中，HiveContext 是 Spark 提供的用户接口，Spark SQL 使用 HiveContext 很容易实现对 Hive 中的数据进行访问查询，而且 HiveContext 继承自 SQLContext，所以在 HiveContext 的的运行过程中，不仅可以使用改写的方法和变量，而且也可以使用 SQLContext 中的方法和变量。我们可以根据图 7-14 来简单比较一下 SQLContext 和 HiveContext 运行过程的异同。

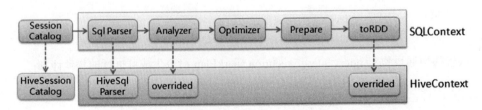

图 7-14　SQL 和 HiveQL 运行过程的比较

通过观察图 7-14，我们可以知道 HiveContext 的运行过程和 SQLContext 的运行过程基本一致，只是改写了 SessionCatalog、Analyzer、Optimizer 和 toRDD，比如 HiveContext 的 Catalog 指向的是 HiveSessionCatalog 等。

1. 使用 HqlParser 生成未绑定的逻辑计划

在该步骤中，由 HiveContext.parseSql 方法生成未绑定的逻辑计划，而该方法实际调用的是其父类 SQLContext.parseSql 解析方法，它的解析过程和 SQL 查询的解析过程类似，均使用 Antlr 进行词法和语法解析。其中在 HiveContext 定义 parseSql 方法代码如下：

```
protected[sql] override def parseSql(sql: String): LogicalPlan = {
  executionHive.withHiveState {
    super.parseSql(substitutor.substitute(hiveconf, sql))
  }
}
```

需要注意的是，在这个解析过程中，HiveSqlAstBuilder 对 SparkSqlAstBuilder 的方法进行重写，具体参见 SQLContext 解析类图。

2. 使用 Analyzer 绑定逻辑计划

Analyzer 解析器结合 Hive 的元数据 Metastore 进行绑定，生成已绑定的逻辑计划，这里的 Catalog 的实例就是 HiveSessionCatalog，其代码如下：

```
def analyze(tableName: String) {
  //通过表名获取表标识符
  val tableIdent = sessionState.sqlParser.parseTableIdentifier(tableName)
  //在 HiveSessionCatalog 中，根据表标识符获取表关系信息
  val relation = EliminateSubqueryAliases(sessionState.catalog.lookupRelation
    (tableIdent))

  relation match {
    case relation: MetastoreRelation =>
      val stagingDir = metadataHive.getConf(HiveConf.ConfVars.STAGINGDIR.
        varname, HiveConf.ConfVars.STAGINGDIR.defaultStrVal)
      ……
      val tableParameters = relation.hiveQlTable.getParameters
      val oldTotalSize = Option(tableParameters.get(StatsSetupConst.TOTAL_SIZE
        )).map (_.toLong).getOrElse(0L)
      val newTotalSize = getFileSizeForTable(hiveconf, relation.hiveQlTable)
      //如果获取的表数量和记录的表数量不一致，则更新 Hive 的元数据表
      if (newTotalSize > 0 && newTotalSize != oldTotalSize) {
        sessionState.catalog.alterTable(
          relation.table.copy(properties = relation.table.properties +
            (StatsSetupConst.TOTAL_SIZE -> newTotalSize.toString)))
```

```
        }
      case otherRelation =>
        throw new UnsupportedOperationException(s"Analyze only works for Hive
          tables, but $tableName is a ${otherRelation.nodeName}")
    }
 }
```

3. 使用 Optimizer 优化逻辑计划

Optimizer 对已绑定的逻辑计划进行优化，生成优化的逻辑计划，除了在查询中使用的 SQLContext 优化器的操作，其他的优化使用的是 Hive 自身的执行引擎。

4. 使用 SparkPlanner 生成可执行的物理计划

该步骤和 SQLContext 处理类似，SparkPlanner 使用 Planning Strategies 对优化的逻辑计划进行转换（Transform），生成可以执行的物理计划。

5. 使用 QueryExecution 执行物理计划

该步骤和 SQLContext 处理类似，先调用父类 QueryExecution 中的 preparations 方法将 PhysicalPlan 转换成可执行物理计划（SparkPlan），再调用 SparkPlan.execute 方法执行可执行物理计划，查询数据库返回 RDD。

7.3 使用 Hive-Console

从 Spark SQL1.0 版本开始提供了一个调试工具 hive/console，通过该工具可以详细查看 Spark SQL 的运行计划，该工具是提供给开发人员使用，在下载的安装包并没有提供，需要使用 SBT 对进行 Spark 源代码编译。由于 hive/console 使用到 Hive，在编译 Spark 源代码前需要编译 Hive 源代码，其中编译 Hive 参见附录 B。

7.3.1 编译 Hive-Console

1. 获取 Spark 源代码

由于首次运行 Hive-Console 需要在 Spark 源代码进行编译，建议使用 2.2.2 节中编译过的 Spark 源代码，可以减少下载依赖包或编译的时间。获取源代码后把源代码移动到 /app/compile 目录，并命名为 spark-2.0.0-hive。

2. 配置环境变量

（1）使用如下命令打开/etc/profile 文件：

```
$sudo vi /etc/profile
```

（2）设置如下参数：

```
export HADOOP_HOME=/app/spark/hadoop-2.7.2
export HIVE_HOME=/app/compile/hive-1.2.1-src
export HIVE_DEV_HOME=/app/compile/hive-1.2.1-src
```

（3）生效配置并验证：

```
$source /etc/profile
$echo $HIVE_DEV_HOME
```

3. 运行 sbt 进行编译

运行 hive/console 不是启动 Spark，而是进入到 Spark 根目录下使用 sbt/sbt hive/console 进行首次运行编译，编译以后下次可以直接启动。编译 Spark 源代码的时候，需要从网上下载依赖包，所以整个编译过程机器必须保证在联网状态。编译命令如下：

```
$cp -r /app/compile/spark-2.0.0-src /app/compile/spark-2.0.0-hive
$cd /app/compile/spark-2.0.0-hive
$build/sbt hive/console
```

编译时间较长，在编译过程中可能出现速度慢或者中断，可以再次启动编译，编译程序会在上次的编译中断处继续进行编译，整个编译过程耗时与网速紧密相关。

图 7-15 为 hive/console 编译成功的界面。

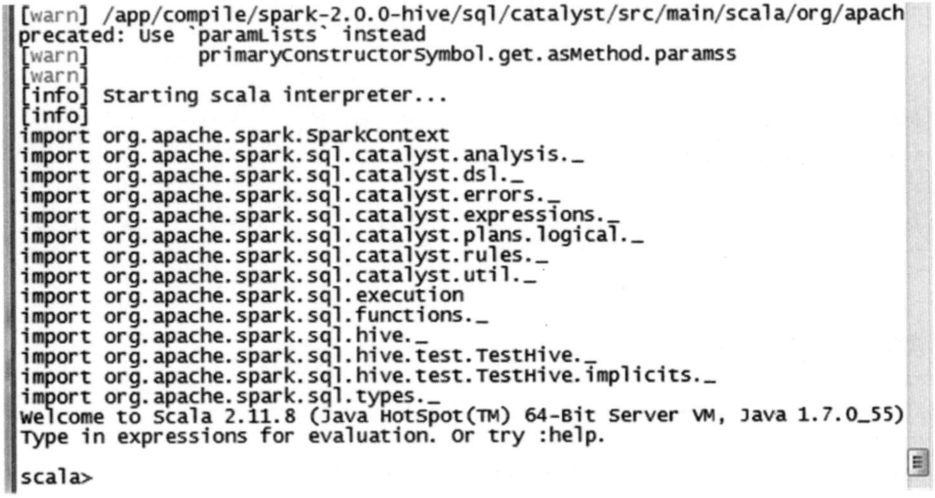

图 7-15 hive/console 编译成功的界面

通过如下命令查看最终编译完成整个目录大小，可以看到目录大小为 295MB 左右，如图 7-16 所示。

```
$du -s /app/compile/spark-2.0.0-hive
```

```
[spark@master ~]$ cd /app/compile/spark-2.0.0-hive
[spark@master spark-2.0.0-hive]$ du -s /app/compile/spark-2.0.0-hive
295540   /app/compile/spark-2.0.0-hive
[spark@master spark-2.0.0-hive]$ _
```

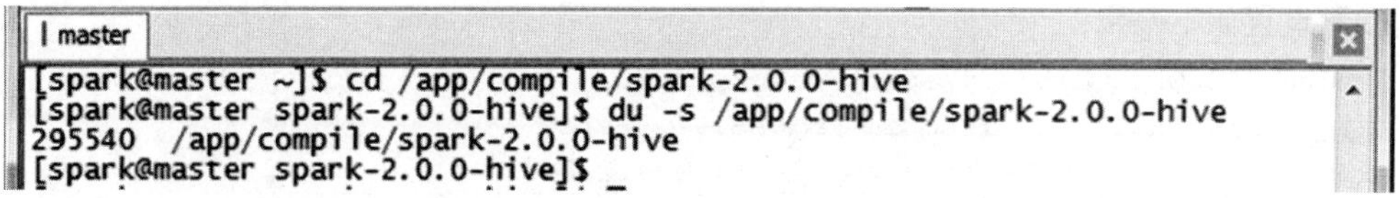

图 7-16　hive/console 编译成功目录大小

7.3.2　查看执行计划

使用 Hive-Console 查看执行计划前，进入到 Spark 根目录下，使用如下命令启动 Hive-Console：

```
$cd /app/compile/spark-2.0.0-hive
$build/sbt hive/console
```

首先定义 Person 类，在该类中定义 name、age 和 state 三个列，然后把该类注册为 people 表并装载数据，最后通过查询到数据存放到 query 中。

```
//定义 Person 类，往该类加载数据注册到表中
scala> case class Person(name:String, age:Int, state:String)
scala> sparkContext.parallelize(Person("Michael",29,"CA")::Person("Andy",30,
    "NY")::Person("Justin",19,"CA")::Person("Justin",25,"CA")::Nil).toDF().reg
    isterTempTable("people")

scala> val query= sql("select * from people")
16/04/14 16:13:13 INFO HiveSqlParser: Parsing command: select * from people
query: org.apache.spark.sql.DataFrame = [name: string, age: int …… 1 more field]

//查看 schema
scala> query.printSchema
root
 |-- name: string (nullable = true)
 |-- age: integer (nullable = false)
 |-- state: string (nullable = true)

//查看 query 具体内容
scala> query.collect()
16/04/14 16:13:18 INFO SparkContext: Starting job: collect at <console>:51
…………
res2: Array[org.apache.spark.sql.Row] = Array([Michael,29,CA], [Andy,30,NY],
    [Justin,19,CA], [Justin,25,CA])
```

```
//查看 Unresolved LogicalPlan
scala> query.queryExecution.logical
res4: org.apache.spark.sql.catalyst.plans.logical.LogicalPlan =
'Project [*]
+- 'UnresolvedRelation `people`, None

//查看 Analyzed LogicalPlan
scala> query.queryExecution.analyzed
res5: org.apache.spark.sql.catalyst.plans.logical.LogicalPlan =
Project [name#3,age#4,state#5]
+- SubqueryAlias people
   +- LogicalRDD [name#3,age#4,state#5], MapPartitionsRDD[1] at
      rddToDatasetHolder at <console>:52

//查看优化后的 LogicalPlan
scala> query.queryExecution.optimizedPlan
res6: org.apache.spark.sql.catalyst.plans.logical.LogicalPlan =
LogicalRDD [name#3,age#4,state#5], MapPartitionsRDD[1] at rddToDatasetHolder at
    <console>:52

//查看物理计划
scala> query.queryExecution.sparkPlan
res7: org.apache.spark.sql.execution.SparkPlan =
Scan ExistingRDD[name#3,age#4,state#5]

//显示查询结果
scala> query.show
16/04/14 16:16:28 INFO SparkContext: Starting job: show at <console>:51
......................
+-------+---+-----+
|   name|age|state|
+-------+---+-----+
|Michael| 29|   CA|
|  Andy| 30|   NY|
| Justin| 19|   CA|
| Justin| 25|   CA|
+-------+---+-----+
```

7.3.3 应用 Hive-Console

在应用 Hive-Console 前需要参照附录 A "安装 MySql 数据库"和附录 B "安装 Hive"。在执行过程中，可以对比 Hive 进行数据查询的操作，可以观察 Hive 使用的是 Hadoop 的 MapReduce

作为计算引擎,而 Hive-Console 使用的 Spark 作为计算引擎,可以明显感到它们之间处理速度的差异!

1. 启动 HDFS 和 hive

在 master 节点中,先启动 HDFS,然后启动 Hive:

```
$cd /app/spark/hadoop-2.7.2/sbin
$./start-dfs.sh
$nohup hive --service metastore > metastore.log 2>&1 &
```

2. 创建配置文件并启动 Spark Shell

在$SPARK_HOME/conf 目录下创建 hive-site.xml 文件,如图 7-17 所示。

```
<configuration>
  <property>
    <name>hive.metastore.uris</name>
    <value>thrift://master:9083</value>
     <description>Thrift URI for the remote metastore. Used by metastore client
       to connect to remote metastore.</description>
  </property>
</configuration>
```

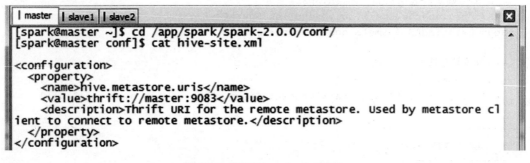

图 7-17 配置 hive-site.xml 文件

修改配置后需要重新启动 Spark Shell:

```
$cd /app/spark/spark-2.0.0/sbin
$./start-all.sh
$cd /app/spark/spark-2.0.0/bin
$./spark-shell --master spark://master:7077 --executor-memory 1024m
 --driver-memory 1024m
```

3. 查看数据库表

要使用 hiveContext,需要先构建 hiveContext:

```
scala> val hiveContext = new org.apache.spark.sql.hive.HiveContext(sc)
hiveContext: org.apache.spark.sql.hive.HiveContext = org.apache.spark.sql.hive.
  HiveContext@3097209
```

然后就可以对 Hive 数据进行操作了。下面我们将使用 Hive 中的销售数据（参见 15.4 节），首先切换数据库到 Hive 并查看有几张表：

```
//销售数据演示
scala> hiveContext.sql("use hive")
res0: org.apache.spark.sql.DataFrame = []

scala> hiveContext.sql("show tables").collect().foreach(println)
[tbdate,false]
[tbstock,false]
[tbstockdetail,false]
```

4．计算所有订单中每年的销售单数、销售总额

```
//所有订单中每年的销售单数、销售总额
//三张表连接后以 count(distinct a.ordernumber)计销售单数，sum(b.amount)计销售总额
scala> hiveContext.sql("select c.theyear,count(distinct a.ordernumber),
  sum(b.amount) from tbStock a join tbStockDetail b on a.ordernumber =
  b.ordernumber join tbDate c on a.dateid=c.dateid group by c.theyear order by
  c.theyear").collect().foreach(println)
[2004,1094,3265696]
[2005,3828,13247234]
[2006,3772,13670416]
[2007,4885,16711974]
[2008,4861,14670698]
[2009,2619,6322137]
[2010,94,210924]
```

通过监控页面，查看任务运行情况，整个查询分为 4 个作业，花费大约 1min，如图 7-18 所示。

Completed Jobs (4)

Job Id	Description	Submitted	Duration	Stages: Succeeded/Total	Tasks (for all stages): Succeeded/Total
3	collect at <console>:28	2016/04/14 16:43:34	5 s	2/2 (2 skipped)	208/208 (202 skipped)
2	collect at <console>:28	2016/04/14 16:43:01	33 s	3/3	402/402
1	run at ThreadPoolExecutor.java:1145	2016/04/14 16:42:33	26 s	1/1	2/2
0	run at ThreadPoolExecutor.java:1145	2016/04/14 16:42:32	24 s	1/1	2/2

图 7-18　计算所有订单中每年的销售单数、销售总额作业情况

其中，作业 3 的调度阶段 8 详细情况如图 7-19 所示。

Summary Metrics for 8 Completed Tasks

Metric	Min	25th percentile	Median	75th percentile	Max
Duration	10 ms	14 ms	32 ms	0.3 s	0.3 s
GC Time	0 ms	0 ms	0 ms	0 ms	0 ms
Shuffle Read Size / Records	0.0 B / 0	78.0 B / 1	78.0 B / 1	78.0 B / 1	80.0 B / 1

Aggregated Metrics by Executor

Executor ID ▲	Address	Task Time	Total Tasks	Failed Tasks	Succeeded Tasks	Shuffle Read Size / Records
0	10.88.146.240:39626	0.5 s	3	0	3	233.0 B / 3
1	10.88.146.241:49318	0.5 s	4	0	4	314.0 B / 4
2	10.88.146.242:46318	0.5 s	1	0	1	0.0 B / 0

Tasks

Index ▲	ID	Attempt	Status	Locality Level	Executor ID / Host	Launch Time	Duration	GC Time	Shuffle Read Size / Records	Errors
0	607	0	SUCCESS	ANY	1 / slave1	2016/04/14 16:43:39	0.3 s		80.0 B / 1	
1	608	0	SUCCESS	ANY	0 / master	2016/04/14 16:43:39	0.3 s		78.0 B / 1	
2	609	0	SUCCESS	ANY	1 / slave1	2016/04/14 16:43:39	15 ms		78.0 B / 1	
3	610	0	SUCCESS	ANY	0 / master	2016/04/14 16:43:39	32 ms		78.0 B / 1	
4	611	0	SUCCESS	ANY	1 / slave1	2016/04/14 16:43:39	10 ms		78.0 B / 1	

图 7-19　作业 3 的调度阶段 8 详细情况

7.4　使用 SQLConsole

7.4.1　启动 HDFS 和 Spark Shell

在集群中，启动 HDFS 和 Spark 集群，并使用如下命令启动 Spark Shell：
```
$cd /app/spark/spark-2.0.0/bin
$./spark-shell --master spark://master:7077 --executor-memory 1024m
 --driver-memory 1024m
```

7.4.2　与 RDD 交互操作

Spark 提供了两种方式将 RDD 转换成 DataFrames：
- 通过定义 Case Class，使用反射推断 Schema（Case Class 方式）。
- 通过可编程接口，定义 Schema，并应用到 RDD 上（createDataFrame 方式）。

前者使用简单、代码简洁，适用于已知 Schema 的源数据上；后者使用较为复杂，但可以在程序运行过程中实行，适用于未知 Schema 的 RDD 上。

1. 使用 Case Class 定义 RDD 演示

对于 Case Class 方式，首先要定义 Case Class，在 RDD 的 Transform 过程中使用 Case Class 可以隐式转化成 SchemaRDD，然后再使用 registerTempTable 注册成表。注册成表后就可以在 sqlContext 对表进行操作，如 select、insert、join 等。注意，case class 既可以是嵌套的，也可以使用类似 Sequences 或 Arrays 之类复杂的数据类型。

下面的例子是定义一个符合数据文件 people.txt 类型的 Case Class（Person），然后将数据文件读入后隐式转换成 SchemaRDD：people，并将 people 在 sqlContext 中注册成 people 表，最后对表进行查询，找出年龄在 13～19 岁的人名。

（1）上传测试数据。

在 HDFS 中创建 /chapter7 目录，把配套资源 /chapter7/people.txt 上传到该目录上：

```
$hadoop fs -mkdir /chapter7
$hadoop fs -copyFromLocal /home/spark/work/chapter7/people.txt /chapter7
$hadoop fs -ls /chapter7
Found 1 items
-rw-r--r--   2 spark supergroup 32 2016-04-15 10:05 /chapter7/people.txt
```

（2）定义 SQLContext 并引入包。

```
//创建 sqlContext 对象
scala> val sqlContext=new org.apache.spark.sql.SQLContext(sc)
sqlContext:org.apache.spark.sql.SQLContext=org.apache.spark.sql.SQLContext@4286

//引入这些包用于把 RDD 转换为 DataFrame
scala> import sqlContext.implicits._
import sqlContext.implicits._
```

（3）定义 Person 类，读入数据并注册为临时表。

```
//使用 Case Clase 创建 Person 类
scala> case class Person(name: String, age: Int)
defined class Person

//创建 Persion 的 RDD，并注册为 people 表
scala> val people = sc.textFile("hdfs://master:9000/chapter7/people.txt").
  map(_.split(",")).map(p => Person(p(0), p(1).trim.toInt)).toDF()
people: org.apache.spark.sql.DataFrame = [name: string, age: int]
scala> people.registerTempTable("people")
```

（4）在查询年龄在 13～19 岁的人员

```
scala> val teenagers = sqlContext.sql("SELECT name, age FROM people WHERE age >=
  13 AND age <= 19")
```

```
//按照字段索引号，打印结果
scala> teenagers.map(t => "Name: " + t(0)).collect().foreach(println)
Name: Justin

//按照字段名称，打印结果
scala> teenagers.map(t => "Name: " + t.getAs[String]("name")).collect().
  foreach(println)
Name: Justin
```

2. 使用 createDataFrame 定义 RDD 演示

createDataFrame 方式比较复杂，通常有 3 步过程：

- 从源 RDD 创建 rowRDD。
- 创建与 rowRDD 匹配的 Schema。
- 将 Schema 通过 createDataFrame 应用到 rowRDD。

（1）导入包创建 Schema。

```
//导入 SparkSQL 的数据类型和 Row
scala> val sqlContext = new org.apache.spark.sql.SQLContext(sc)
scala> import org.apache.spark.sql.Row
scala> import org.apache.spark.sql.types.{StructType,StructField,StringType}

//从 HDFS 读取 people 数据
scala> val people = sc.textFile("hdfs://master:9000/chapter7/people.txt")
people: org.apache.spark.rdd.RDD[String] = hdfs://master:9000/chapter7/
  people.txt MapPartitionsRDD[12] at textFile at <console>:29

//创建于数据结构匹配的 Schema
scala> val schemaString = "name age"
scala> val schema =
     |   StructType(schemaString.
     |     split(" ").map(fieldName => StructField(fieldName, StringType, true)))
schema: org.apache.spark.sql.types.StructType = StructType(StructField(name,
  StringType,true), StructField(age,StringType,true))
```

（2）创建 rowRDD 并读入数据。

```
//把 people 的数据转换为 RowsRDD
scala> val rowRDD = people.map(_.split(",")).map(p => Row(p(0), p(1).trim))
rowRDD: org.apache.spark.rdd.RDD[org.apache.spark.sql.Row] =
  MapPartitionsRDD[14] at map at <console>:31

//使用 createDataFrame 方法将 rowRDD 转换为 DataFrame
scala> val peopleDataFrame = sqlContext.createDataFrame(rowRDD, schema)
peopleDataFrame: org.apache.spark.sql.DataFrame = [name: string, age: string]
```

```
//把 DataFrame 注册为 people 表
scala> peopleDataFrame.registerTempTable("people")
```
（3）查询获取数据。
```
//使用 SQL 语句查询 people 表中数据
scala> val results = sqlContext.sql("SELECT name FROM people")
results: org.apache.spark.sql.DataFrame = [name: string]

//打印查询结果
scala> results.map(t => "Name: " + t(0)).collect().foreach(println)
Name: Michael
Name: Andy
Name: Justin
```

7.4.3 读取 JSON 格式数据

SparkSQL1.1.0 开始提供对 JSON 文件格式的支持，这意味着开发者可以使用更多的数据源，如鼎鼎大名的 NOSQL 数据库 MongDB 等。sqlContext 可以从 jsonFile 或 jsonRDD 获取 schema 信息，来构建 SchemaRDD，注册成表后就可以使用。

（1）上传测试数据。

在 HDFS 中创建 /chapter7 目录，把配套资源/chapter7/people.json 上传到该目录上。
```
$hadoop fs -mkdir /chapter7
$hadoop fs -copyFromLocal /home/spark/work/chapter7/people.json /chapter7
$hadoop fs -ls /chapter7
Found 2 items
-rw-r--r-- 2 spark supergroup 73 2016-04-15 10:05 /chapter7/people.json
-rw-r--r-- 2 spark supergroup 32 2016-04-15 10:05 /chapter7/people.txt
```
（2）读取数据并注册 jsonTable 表。
```
scala> val sqlContext=new org.apache.spark.sql.SQLContext(sc)
sqlContext:org.apache.spark.sql.SQLContext=org.apache.spark.sql.SQLContext@186

//通过指定路径读取 json 数据，该路径可以为单一文件或保存文本文件的目录
scala> val people=sqlContext.read.json("hdfs://master:9000/chapter7/people.json")
people: org.apache.spark.sql.DataFrame = [age: bigint, name: string]

//通过 printSchema 方法查看 people 结构信息
scala> people.printSchema()
root
 |-- age: long (nullable = true)
 |-- name: string (nullable = true)
```

```
//把 people 注册为临时表
scala> people.registerTempTable("people")
```
（3）查询年龄大于等于 25 的人名。
```
scala> val teenagers = sqlContext.sql("SELECT name FROM people WHERE age >= 25")
teenagers: org.apache.spark.sql.DataFrame = [name: string]

scala>teenagers.show
+----+
|name|
+----+
|Andy|
+----+
```

7.4.4 读取 Parquet 格式数据

由于 parquet 文件中保留了 schema 的信息，所以不需要使用 Case Class 来隐式转换。sqlContext 读入 parquet 文件后直接转换成 SchemaRDD，也可以将 SchemaRDD 保存成 parquet 文件格式。

（1）读取 parquet 文件，注册成临时表。

```
scala> val sqlContext = new org.apache.spark.sql.SQLContext(sc)
scala> import sqlContext.implicits._

scala> val parquetFile = sqlContext.read.parquet("/home/spark/work/chapter7/
    wiki_parquet")
parquetFile: org.apache.spark.sql.DataFrame = [id: int, title: binary ...... 3 more
    fields]
scala> parquetFile.registerTempTable("parquetWiki")
```

（2）读取 parquet 文件，注册成临时表。

```
scala> parquetFile.show(10)
```

```
scala> parquetFile.show(10)
+--------+--------------+----------+--------------+--------------+
|      id|         title|  modified|          text|      username|
+--------+--------------+----------+--------------+--------------+
|12380864|[4A 61 6D 65 73 2...|1391974325|[7B 7B 6D 75 6C 7...|            []|
|12393262|[4C 69 73 74 20 6...|1398823696|[54 68 69 73 20 2...|            []|
|12401596|[4A 6F 73 65 70 6...|1381672714|[27 27 27 4A 6F 7...|[57 61 61 63 73 7...|
|12403348|[4E 79 63 74 61 6...|1365669177|[7B 7B 54 61 78 6...|[41 64 64 62 6F 74]|
|12424813|[4D 61 67 69 63 2...|1366901154|[7B 7B 63 6F 6F 7...|[53 64 66 65 69 6...|
|12438539|[50 75 62 6C 69 6...|1394444182|[5B 5B 46 69 6C 6...|[4A 69 6D 31 31 3...|
|12457321|[47 61 72 72 69 7...|1379656141|[7B 7B 47 65 6F 6...|[48 6D 61 69 6E 73]|
|12471532|[50 61 6E 61 65 6...|1361692514|[7B 7B 44 69 73 5...|[4D 65 72 6C 49 7...|
|12490755|[55 6E 69 76 65 7...|1396407464|[27 27 27 55 6E 6...|[4F 64 20 4D 69 7...|
|12505173|[44 61 6E 69 65 6...|1398220440|[7B 7B 49 6E 66 6...|[4D 61 63 6F 66 65]|
+--------+--------------+----------+--------------+--------------+
only showing top 10 rows
```

7.4.5 缓存演示

sparkSQL 的 cache 可以使用两种方法来实现：
- CacheTable()方法。
- CACHE TABLE 命令。

千万不要先使用 cache SchemaRDD，然后 registerAsTable；使用 RDD 的 cache()将使用原生态的 cache，而不是针对 SQL 优化后的内存列存储。

（1）对 rddTable 表进行缓存。

```
scala> val sqlContext = new org.apache.spark.sql.SQLContext(sc)
scala> import sqlContext.implicits._
scala> case class Person(name: String, age: Int)
scala> val people = sc.textFile("hdfs://master:9000/chapter7/people.txt").
  map(_.split(",")).map(p => Person(p(0), p(1).trim.toInt)).toDF()
scala> people.registerTempTable("people")

scala> sqlContext.cacheTable("people")
scala> sqlContext.sql("SELECT name FROM people WHERE age >= 13 AND age <=
  19").map(t => "Name: " + t(0)).collect().foreach(println)
Name: Justin
```

在监控界面上看到该表数据已经缓存，如图 7-20 所示。

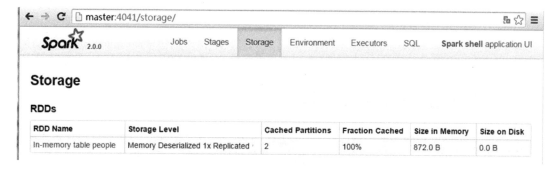

图 7-20 数据已经缓存查看界面

（2）解除缓存。

```
scala> sqlContext.uncacheTable("people")
```

或者使用

```
scala> sqlContext.sql("UNCACHE TABLE people")
res8: org.apache.spark.sql.DataFrame = []
```

7.4.6 DSL 演示

SparkSQL 除了支持 HiveQL 和 SQL-92 语法外，还支持 DSL（Domain Specific Language）。在 DSL 中，使用 Scala 符号 ' +标示符表示基础表中的列，Spark 的 execution engine 会将这些标示符隐式转换成表达式。另外可以在 API 中找到很多 DSL 相关的方法，如 where()、select()、limit() 等，详细资料可以查看 Catalyst 模块中的 DSL 子模块，下面为其中定义几种常用的方法。

```
scala> val sqlContext = new org.apache.spark.sql.SQLContext(sc)
scala> import sqlContext.implicits._
scala> case class Person(name: String, age: Int)

//读取people数据，注册成临时表
scala> val people = sc.textFile("hdfs://master:9000/chapter7/people.txt").
  map(_.split(",")).map(p => Person(p(0), p(1).trim.toInt)).toDF()
people: org.apache.spark.sql.DataFrame = [name: string, age: int]
scala> people.registerTempTable("people")

//使用DSL语句进行查询
scala> val teenagers_dsl = people.where('age >= 10).where('age <= 19).select('
  name)
teenagers_dsl: org.apache.spark.sql.DataFrame = [name: string]

//打印结果
scala> teenagers_dsl.map(t => "Name: " + t(0)).collect().foreach(println)
Name: Justin
```

7.5 使用 Spark SQL CLI

CLI（Command-Line Interface，命令行界面）是指可在用户提示符下输入可执行指令的界面，它通常不支持鼠标，用户通过键盘输入指令，计算机接收到指令后予以执行。Spark SQL CLI 指的是使用命令界面直接输入 SQL 命令，然后发送到 Spark 集群进行执行，在界面中显示运行过程和最终的结果。

Spark SQL CLI 和 ThriftServer，使得 Hive 用户，还有用惯了命令行的 RDBMS 数据库管理员较容易地上手，真正意义上进入了 SQL 时代。

7.5.1 配置并启动 Spark SQL CLI

1. 创建并配置 hive-site.xml

在运行 Spark SQL CLI 中需要使用到 Hive Metastore，故需要在 Spark 中添加其 uris。具体方法是在 SPARK_HOME/conf 目录下创建 hive-site.xml 文件，然后在该配置文件中，添加 hive.metastore.uris 属性，具体如下：

```xml
<configuration>
  <property>
    <name>hive.metastore.uris</name>
    <value>thrift://master:9083</value>
     <description>Thrift URI for the remote metastore. Used by metastore client
       to connect to remote metastore.</description>
  </property>
</configuration>
```

2. 启动 HDFS 和 Hive

在使用 Spark SQL CLI 之前需要启动 Hive Metastore（如果数据存放在 HDFS 文件系统，还需要启动 Hadoop 的 HDFS），使用如下命令可以使 Hive Metastore 启动后运行在后台，可以通过 jobs 查询：

```
$cd /app/spark/hadoop-2.7.2/sbin
$./start-dfs.sh
$nohup hive --service metastore > metastore.log 2>&1 &
```

3. 启动 Spark 集群和 Spark SQL CLI

通过如下命令启动 Spark 集群和 Spark SQL CLI：

```
$cd /app/spark/spark-2.0.0
$sbin/start-all.sh
$bin/spark-sql --master spark://master:7077 --executor-memory 1g
 --driver-memory 1g
```

这时可以使用 HQL 语句对 Hive 数据进行查询，另外可以使用 COMMAND，如使用 set 进行设置参数：默认情况下，SparkSQL Shuffle 的时候是 200 个 partition，可以使用如下命令修改该参数：

```
SET spark.sql.shuffle.partitions=20;
```

运行同一个查询语句，参数改变后，Task（partition）的数量就由 200 变成了 20。

7.5.2 实战 Spark SQL CLI

1. 获取订单每年的销售单数、销售总额

```
//设置任务个数，在这里修改为 20 个
spark-sql>SET spark.sql.shuffle.partitions=20;
spark-sql>use hive;

spark-sql>select c.theyear,count(distinct a.ordernumber),sum(b.amount) from
  tbStock a join tbStockDetail b on a.ordernumber=b.ordernumber join tbDate c on
  a.dateid=c.dateid group by c.theyear order by c.theyear;
```

图 7-21 是获取订单每年的销售单数、销售总额的结果显示页面。

图 7-21 获取订单每年的销售单数、销售总额结果

2. 计算所有订单每年的总金额

```
spark-sql>select   c.theyear,count(distinct   a.ordernumber),sum(b.amount)from
  tbStock a join tbStockDetail  b on a.ordernumber=b.ordernumber join tbDate c
  on a.dateid=c.dateid group by c.theyear order by c.theyear;
```

图 7-22 是计算所有订单每年的总金额的结果显示页面。

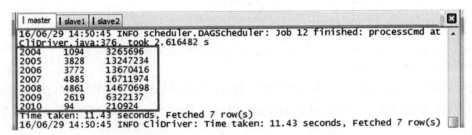

图 7-22 计算所有订单每年的总金额结果

3. 计算所有订单每年最大金额订单的销售额

```
spark-sql>select   c.theyear,max(d.sumofamount) from tbDate c join (select
  a.dateid,a.ordernumber,sum(b.amount) as sumofamount from tbStock a join
  tbStockDetail   b on  a.ordernumber=b.ordernumber  group  by  a.dateid,
```

```
           a.ordernumber ) d  on c.dateid=d.dateid group by c.theyear order by c.theyear;
```
图 7-23 是计算所有订单每年最大金额订单的销售额的结果显示页面。

```
| master | slave1 | slave2 |
16/06/29 14:54:00 INFO scheduler.TaskSchedulerImpl: Removed TaskSet 41.0, whos
e tasks have all completed, from pool
2004    23612
2005    38180
2006    36124
2007    159126
2008    55828
2009    25810
2010    13063
Time taken: 10.419 seconds, Fetched 7 row(s)
16/06/29 14:54:00 INFO CliDriver: Time taken: 10.419 seconds, Fetched 7 row(s)
```

图 7-23 计算所有订单每年最大金额订单的销售额

7.6 使用 Thrift Server

ThriftServer 是一个 JDBC/ODBC 接口，用户可以通过 JDBC/ODBC 连接 ThriftServer 来访问 Spark SQL 的数据。ThriftServer 在启动的时候，会启动了一个 Spark SQL 的应用程序，而通过 JDBC/ODBC 连接进来的客户端共同分享这个 Spark SQL 应用程序的资源，也就是说，不同的用户之间可以共享数据。ThriftServer 启动时还开启一个侦听器，等待 JDBC 客户端的连接和提交查询。所以，在配置 ThriftServer 的时候，至少要配置 ThriftServer 的主机名和端口。如果要使用 Hive 数据的话，还要提供 Hive Metastore 的 uris。

7.6.1 配置并启动 Thrift Server

1. 创建并配置 hive-site.xml

在$SPARK_HOME/conf 目录下修改 hive-site.xml 配置文件（如果在 Spark SQL CLI 中已经添加，可省略）。由于 slave1 作为 Thrift Server 服务器，该配置文件在 master 节点配置完毕后，需要复制到 slave1 对应的配置目录下，命令如下：

```
$cd /app/spark/spark-2.0.0/conf
$sudo vi hive-site.xml
$scp hive-site.xml spark@slave1:/app/spark/spark-2.0.0/conf/
```

设置 master 为 Metastore 服务器，slave1 为 Thrift Server 服务器，配置内容如下：

```
<configuration>
  <property>
    <name>hive.metastore.uris</name>
    <value>thrift://master:9083</value>
```

```xml
    <description>Thrift URI for the remote metastore. Used by metastore client
      to connect to remote metastore.</description>
  </property>
  <property>
    <name>hive.server2.thrift.min.worker.threads</name>
    <value>5</value>
    <description>Minimum number of Thrift worker threads</description>
  </property>
  <property>
    <name>hive.server2.thrift.max.worker.threads</name>
    <value>500</value>
    <description>Maximum number of Thrift worker threads</description>
  </property>
  <property>
    <name>hive.server2.thrift.port</name>
    <value>10000</value>
    <description>Port number of HiveServer2 Thrift interface. Can be overridden
      by setting $HIVE_SERVER2_THRIFT_PORT</description>
  </property>
  <property>
    <name>hive.server2.thrift.bind.host</name>
    <value>slave1</value>
    <description>Bind host on which to run the HiveServer2 Thrift interface.Can
      be overridden by setting$HIVE_SERVER2_THRIFT_BIND_HOST</description>
  </property>
</configuration>
```

2. 启动 Hive

在 master 节点中，在后台启动 Hive Metastore（如果数据存放在 HDFS 文件系统，还需要启动 Hadoop 的 HDFS）：

```
$nohup hive --service metastore > metastore.log 2>&1 &
```

3. 启动 Spark 集群和 Thrift Server

在 master 节点启动 Spark 集群：

```
$cd /app/spark/spark-2.0.0/sbin
$./start-all.sh
```

在 slave1 节点上进入 SPARK_HOME/sbin 目录，使用如下命令启动 Thrift Server：

```
$cd /app/spark/spark-2.0.0/sbin
$./start-thriftserver.sh --master spark://master:7077 --executor-memory 1g
```

注意：Thrift Server 需要按照配置在 slave1 启动！

7.6.2　基本操作

1．远程客户端连接

用户可以在任意节点启动 bin/beeline，用!connect jdbc:hive2://slave1:10000 连接 ThriftServer，因为没有采用权限管理，所以用户名用运行 bin/beeline 的用户 spark，密码为 spark：

```
$cd /app/spark/spark-2.0.0/bin
$./beeline
beeline>!connect jdbc:hive2://slave1:10000
```

图 7-24 为 Thrift Server 的启动过程。

```
[spark@master ~]$ cd /app/spark/spark-2.0.0/bin
[spark@master bin]$ ./beeline
Beeline version 1.2.1.spark2 by Apache Hive
beeline> !connect jdbc:hive2://slave1:10000
Connecting to jdbc:hive2://slave1:10000
Enter username for jdbc:hive2://slave1:10000: spark
Enter password for jdbc:hive2://slave1:10000: *****
16/06/29 15:05:15 INFO jdbc.Utils: Supplied authorities: slave1:10000
16/06/29 15:05:15 INFO jdbc.Utils: Resolved authority: slave1:10000
16/06/29 15:05:15 INFO jdbc.HiveConnection: Will try to open client transport
Connected to: Spark SQL (version 2.0.0)
Driver: Hive JDBC (version 1.2.1.spark2)
Transaction isolation: TRANSACTION_REPEATABLE_READ
0: jdbc:hive2://slave1:10000>
```

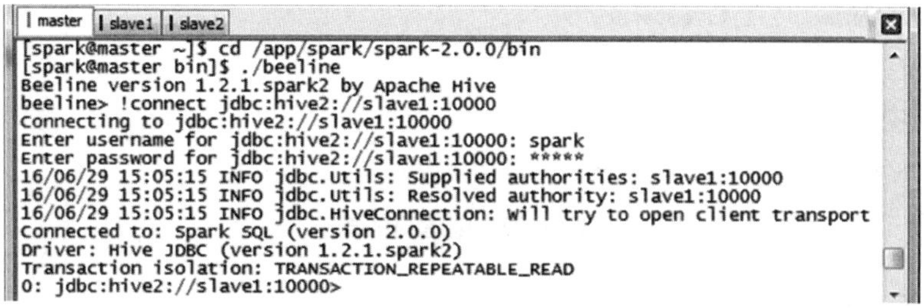

图 7-24　Thrift Server 的启动过程

2．基本操作

（1）使用 hive 数据库，代码如下：

```
0: jdbc:hive2://slave1:10000> show databases;
0: jdbc:hive2://slave1:10000> use hive;
0: jdbc:hive2://slave1:10000> show databases;
```

使用 hive 数据库并查看表如图 7-25 所示。

```
0: jdbc:hive2://slave1:10000> show databases;
+---------------+--+
| databaseName  |
+---------------+--+
| default       |
| hive          |
+---------------+--+
2 rows selected (4.818 seconds)
0: jdbc:hive2://slave1:10000> use hive;
+---------+--+
| Result  |
+---------+--+
+---------+--+
No rows selected (0.141 seconds)
0: jdbc:hive2://slave1:10000> show tables;
+----------------+--------------+--+
|   tableName    | isTemporary  |
+----------------+--------------+--+
| tbdate         | false        |
| tbstock        | false        |
| tbstockdetail  | false        |
+----------------+--------------+--+
3 rows selected (0.229 seconds)
```

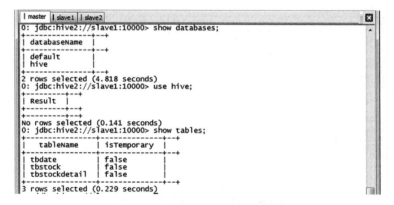

图 7-25　使用 hive 数据库并查看表

（2）创建表 testThrift，代码如下：

```
0:jdbc:hive2://slave1:10000>create table testThrift(field1 String , field2 Int);
0:jdbc:hive2://slave1:10000>show tables;
```

创建测试表如图 7-26 所示。

```
0: jdbc:hive2://slave1:10000> create table testThrift(field1 String , field2 I
nt);
+---------+--+
| Result  |
+---------+--+
+---------+--+
No rows selected (2.892 seconds)
0: jdbc:hive2://slave1:10000> show tables;
+----------------+--------------+--+
|   tableName    | isTemporary  |
+----------------+--------------+--+
| tbdate         | false        |
| tbstock        | false        |
| tbstockdetail  | false        |
| testthrift     | false        |
+----------------+--------------+--+
4 rows selected (0.154 seconds)
```

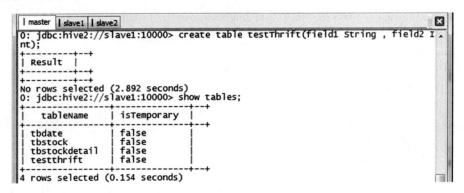

图 7-26　创建测试表

7.6.3　交易数据实例

1．把 tbStockDetail 表中金额大于 3000 插入到 testThrift 表中

其代码如下：

```
0: jdbc:hive2://slave1:10000>insert into table testThrift select
  ordernumber,amount from tbStockDetailwhere amount>3000;
0: jdbc:hive2://slave1:10000> select * from testThrift;
```

图 7-27 为使用 Thrift Server 查询数据的结果显示页面。

```
0: jdbc:hive2://slave1:10000> insert into table testThrift select ordernumber,
amount from tbStockDetail where amount > 3000;
+---------+--+
| Result  |
+---------+--+
+---------+--+
No rows selected (18.74 seconds)
0: jdbc:hive2://slave1:10000> select * from testThrift;
+----------------+---------+--+
|     field1     | field2  |
+----------------+---------+--+
| GHSL00000743   | 5025    |
| GHSL00001503   | 10989   |
| HMJSL00000706  | 4490    |
| HMJSL00000706  | 10776   |
| HMJSL00000826  | 5988    |
| HMJSL00001769  | 6578    |
| HMJSL00002224  | 3440    |
```

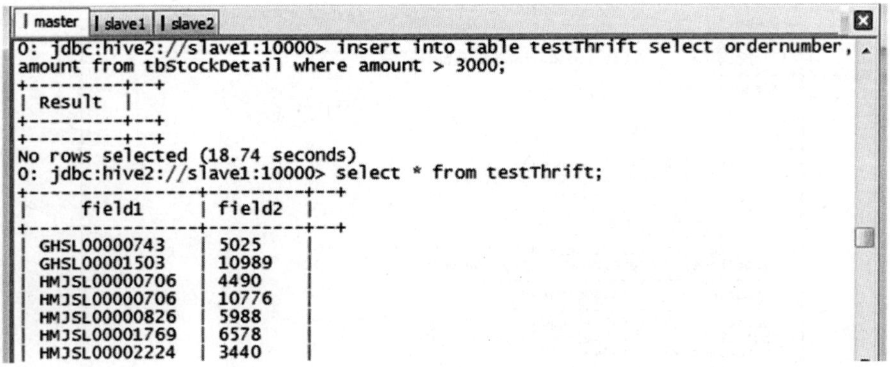

图 7-27　使用 Thrift Server 查询数据

2. 重新创建 testThrift 表中，把年度最大订单插入到该表中

其代码如下：

```
0:jdbc:hive2://slave1:10000>drop table testThrift;
0:jdbc:hive2://slave1:10000>create table testThrift(field1 String , field2 Int);
0:jdbc:hive2://slave1:10000>insert into table testThrift select
    c.theyear,max(d.sumofamount) from tbDate c join (select
    a.dateid,a.ordernumber,sum(b.amount) as sumofamount from tbStock a join
    tbStockDetail  b on a.ordernumber=b.ordernumber group by
    a.dateid,a.ordernumber ) d  on c.dateid=d.dateid group by c.theyear sort by
    c.theyear;
0:jdbc:hive2://slave1:10000>select * from testThrift;
```

图 7-28 为计算年度最大订单插入测试表中的结果显示页面。

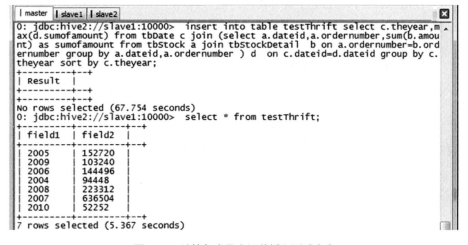

图 7-28　计算年度最大订单插入测试表中

3. 计算所有订单每年的订单数

其代码如下：

```
0: jdbc:hive2://slave1:10000>select c.theyear,count(distinct a.ordernumber)
    from tbStock a join tbStockDetail  b on a.ordernumber=b.ordernumber join
    tbDate c on a.dateid=c.dateid group by c.theyear order by c.theyear;
```

图 7-29 为计算所有订单每年的订单数的结果显示页面。

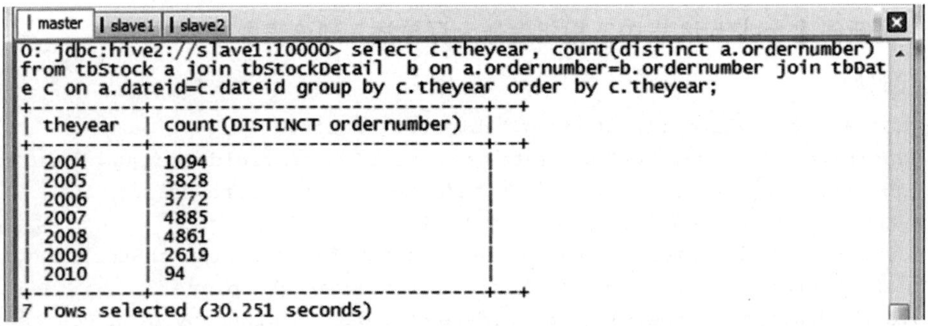

```
| master  | slave1 | slave2 |
0: jdbc:hive2://slave1:10000> select c.theyear, count(distinct a.ordernumber)
from tbStock a join tbStockDetail   b on a.ordernumber=b.ordernumber join tbDat
e c on a.dateid=c.dateid group by c.theyear order by c.theyear;
+----------+------------------------+
| theyear  | count(DISTINCT ordernumber) |
+----------+------------------------+
| 2004     | 1094                   |
| 2005     | 3828                   |
| 2006     | 3772                   |
| 2007     | 4885                   |
| 2008     | 4861                   |
| 2009     | 2619                   |
| 2010     | 94                     |
+----------+------------------------+
7 rows selected (30.251 seconds)
```

图 7-29　计算所有订单每年的订单数

从上可以看出，ThriftServer 可以连接多个 JDBC/ODBC 客户端，并相互之间可以共享数据。顺便提一句，ThriftServer 启动后处于监听状态，用户可以使用 Ctrl+C 组合键退出 ThriftServer；而 beeline 的退出使用!q 命令。

7.6.4　使用 IDEA 开发实例

有了 ThriftServer，开发人员可以非常方便地使用 JDBC/ODBC 来访问 SparkSQL。下面是一个 Scala 代码，查询表 tbStockDetail，返回 amount>3000 的单据号和交易金额。

1. 在 IDEA 创建 chapter7 包和类 SparkSQL4JDBC

在该类中查询 tbStockDetail 金额大于 3000 的订单，代码如下：

```scala
import java.sql.DriverManager

object SparkSQL4JDBC {
  def main(args: Array[String]) {
    Class.forName("org.apache.hive.jdbc.HiveDriver")
    val conn = DriverManager.getConnection("jdbc:hive2://slave1:10000/",
      "spark", " spark")
    try {
      val statement = conn.createStatement
      statement.executeQuery("use hive")
      val rs = statement.executeQuery("select ordernumber,amount from
        tbStockDetailwhere amount > 3000")
      while (rs.next) {
        val ordernumber = rs.getString("ordernumber")
        val amount = rs.getString("amount")
        println("ordernumber = %s, amount = %s".format(ordernumber, amount))
      }
    } catch {
```

```
        case e: Exception => e.printStackTrace
      }
      conn.close
    }
}
```

2. 查看运行结果

在 IDEA 中可以观察到,在运行日志窗口中没有运行过程的日志,只显示查询结果,如图 7-30 所示。

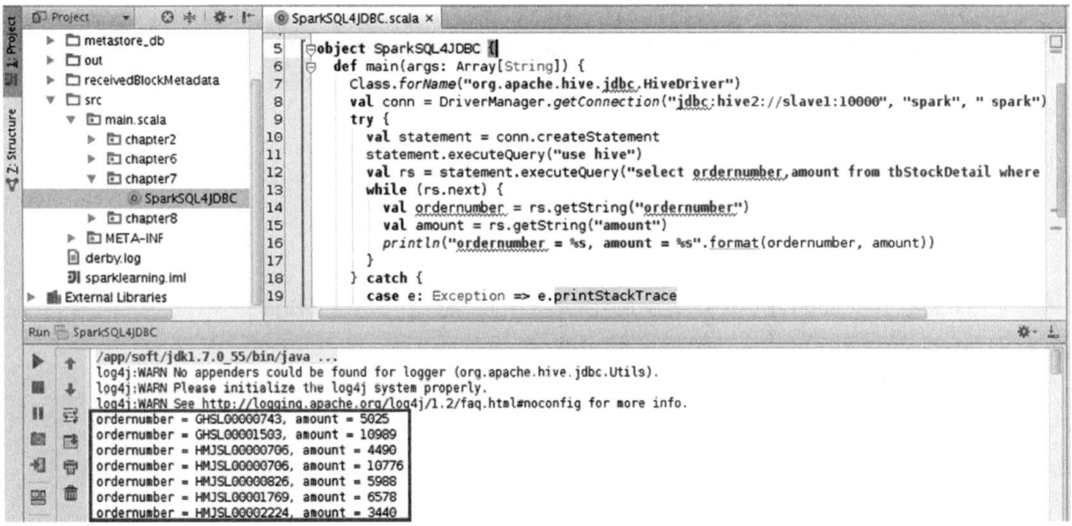

图 7-30 在 IDEA 中使用 Thrift Server 运行结果

7.7 实例演示

7.7.1 销售数据分类实例

1. 例子介绍

在该实例中,对某种商品各城市销情况进行统计并进行分类,把分类情况使用图形展示出来。实现该需求,首先把某种商品各城市销售额数据导入到 Hive 中,然后使用 Spark SQL 把数据从 Hive 中读取出来,接着利用 MLlib 中的 KMeans 聚类算法对读取的数据集进行处理,最终使用 SparkR 对处理的数据进行展示。

2. 程序代码

在 Spark SQL 中，读取销售数据并利用 MLlib 中的 KMeans 聚类算法，先生成模型，然后对进行分类计算。

```scala
import org.apache.log4j.{Level, Logger}
import org.apache.spark.mllib.clustering.KMeans
import org.apache.spark.mllib.linalg.Vectors
import org.apache.spark.sql.SparkSession

object SaleDataDeal {
  def main(args: Array[String]): Unit = {
    if (args.length != 3) {
      println("Usage: /path/to/spark/bin/spark-submit --master spark://master:9000 " +
        "--driver-memory 1g --class chapter11.SaleDataDeal " +
        "sparklearning.jar numClusters numIterations saveAsTextFilePath")
      sys.exit(1)
    }
    //屏蔽不必要的日志显示在终端上
    Logger.getLogger("org.apache.hadoop").setLevel(Level.ERROR)
    Logger.getLogger("org.apache.spark").setLevel(Level.ERROR)
    Logger.getLogger("org.eclipse.jetty.server").setLevel(Level.OFF)

    //设置运行环境
    val spark = SparkSession.builder.enableHiveSupport
      .appName(s"${this.getClass.getSimpleName}")
      .getOrCreate()

    //使用sparksql查出每个店的销售数量和金额
    spark.sql("SET spark.sql.shuffle.partitions=20")
    spark.sql("use hive")
    val sqldata = spark.sql("select a.locationid, sum(b.qty) totalqty,sum(b.amount) totalamount " +
      "from tbStock a join tbStockDetail b on a.ordernumber=b.ordernumber group by a.locationid").cache()

    //将查询数据转换成向量
    val vectors = sqldata.rdd.map(r => Vectors.dense(r.getLong(1).toDouble,
      r.getLong(2).toDouble ))

    //对数据集聚类，进行迭代形成数据模型
    val numClusters = args(0).toInt
    val numIterations = args(1).toInt
```

```
    val model = KMeans.train(vectors, numClusters, numIterations)

    //打印聚族中心点
    println("Cluster centers:")
    model.clusterCenters.foreach(println)

    //交叉评估,返回数据集和结果
    val result = sqldata.rdd.map {
      line =>
        val linevectore = Vectors.dense(line.getLong(1).toDouble, line.
            getLong(2).toDouble)
        val prediction = model.predict(linevectore)
        line(0) + " " +line.getLong(1) + " " + line.getLong(2) + " " + prediction
    }
    println("Prediction detail:")
    result.collect().foreach(println)
    result.saveAsTextFile(args(2))
  }
}
```

3. 运行过程

(1) 打包运行

对运行代码进行打包,生成打包文件 sparklearning.jar 并移动到 Spark 根目录下。在运行前需要启动 HDFS、Hive 和 Spark 集群,然后使用 spark-submit 脚本提交程序到集群中,其运行命令如下所示:

```
$cd /app/spark/spark-2.0.0
$./bin/spark-submit --class chapter7.SaleDataDeal \
  --master spark://master:7077 \
  --executor-memory 1024m \
  sparklearning.jar \
  4 20 /chapter7/SaleDataDealResult
```

提交 jar 包运行需要三个参数,分别为聚族的格式、迭代次数和计算结果保存文件的路径。

(2) 执行并观察输出

通过执行程序结果如下:

```
Cluster centers:
[6485.428571428571,826045.7142857142]
[45571.0,9172191.666666666]
[25130.5,4049019.5]
[30305.666666666664,6201936.333333333]
Prediction detail:
```

```
ZY 27049 6307626 3
ZHAO 24996 4587718 2
RM 21627 3010800 2
BYYZ 4401 533953 0
ZM 30513 4267355 2
TM 4363 708261 0
HUAXIN 4186 487099 0
DY 355 55195 0
TAIHUA 41002 9180132 1
LJ 47375 9865094 1
TS 17961 2236106 0
HL 13426 1637889 0
GUIHE 30446 6571570 3
DOGNGUAN 33422 5726613 3
TY 23386 4330205 2
LZ 706 123817 0
YINZUO 48336 8471349 1
```

4．使用 R 进行展示

使用如下命令对保存在 HDFS 的结果进行合并：

```
$cd /home/spark/work
$hdfs dfs -getmerge /chapter7/SaleDataDealResult SaleDataDealResult.txt
```

合并完毕后，使用 RServer 对这些数据进行显示，具体脚本如下：

（1）UI.R

```
shinyUI(pageWithSidebar(
  headerPanel('销售数据聚类模型图'),
  sidebarPanel(),
  mainPanel(
    plotOutput('plot1')
  )
))
```

（2）Server.R

```
palette(c("#E41A1C", "#377EB8", "#4DAF4A", "#984EA3",
          "#FF7F00", "#FFFF33", "#A65628", "#F781BF", "#999999"))
library(plyr)
shinyServer(function(input, output, session) {

  # Combine the selected variables into a new data frame
  output$plot1 <- renderPlot({
    data<-read.table("/home/spark/work/chapter11/SaleDataDealResult.txt
      ",header=F)    names(data)<-c("地区","销量","总金额","cluster")
```

```
        centers<-ddply(data[2:3],.(data$cluster),.fun=function(x)
            apply(x,2,mean))
        par(mar = c(5.1, 4.1, 0, 1))
        plot(data[,2:3],
             col = data$cluster+1,
             pch = 20, cex = 3)
        points(centers[,-1], pch = 4, cex = 4, lwd = 4)
    })
})
```

显示图形如图 7-31 所示：

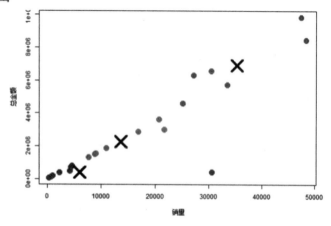

图 7-31　销售数据聚类模型图

7.7.2　网店销售数据统计

在该实例中对用户信息和在线销售明细数据进行关联，将按照年份、指定时间范围和用户、按照省份销售额排名等不同的统计方式，通过对网购数据统计，能够更加清楚地了解到不同地区用户的消费能力、商品的受欢迎程度，进而制订针对性的营销策略，提高网站的销售额。在下面演示中，首先把用户信息和在线销售明细数据上传到 HDFS（测试数据生成代码在/chapter7 目录中的 OnLineTradeUserGenerator.scala 和 OnLineTradeDetailGenernator.scala 两个文件），然后把 OnLineTradeStatistics.scala 打包成 sparklearning.jar，最后通过 spark-submit 命令执行统计任务。

1. 实现代码

在 chapter7 包中添加 **OnLineTradeStatistics** 对象文件,具体代码如下:

```scala
import org.apache.log4j.{Level, Logger}
import org.apache.spark.sql.SQLContext
import org.apache.spark.storage.StorageLevel
import org.apache.spark.{SparkContext, SparkConf}

//定义用户类和网购交易明细类
case class User(userID: String, gender: String, age: Int, registerDate: String,
   provice:String, career: String)
case class TradeDetail(tradeID: String, tradeDate: String, productID: Int, amount:
   Int, userID: String)

object OnLineTradeStatistics {
  def main(args: Array[String]) {
    if (args.length < 1) {
      println("Usage:OnLineTradeStatistics consumingDataFilePath")
      System.exit(1)
    }
    Logger.getLogger("org.apache.hadoop").setLevel(Level.ERROR)
    Logger.getLogger("org.apache.spark").setLevel(Level.ERROR)
    Logger.getLogger("org.eclipse.jetty.server").setLevel(Level.OFF)

    //设置应用程序
    val conf = new SparkConf().setAppName("On Line Trade Data
        Statistics").setMaster("local")
    val ctx = new SparkContext(conf)
    val sqlCtx = new SQLContext(ctx)
    import sqlCtx.implicits._

    //读取用户文件,然后把数据由 RDD 转换为 DataFrame,并为临时表 user
    val userDF = ctx.textFile(args(0)).map(_.split(" ")).map(u => User(u(0), u(1),
        u(2).toInt,u(3),u(4),u(5))).toDF()
    userDF.createOrReplaceTempView("user")
    userDF.persist(StorageLevel.MEMORY_ONLY_SER)

    //读取网购交易明细文件,然后把数据由 RDD 转换为 DataFrame,并为临时表 trade
    val tradeDF = ctx.textFile(args(1)).map(_.split(" ")).map(u =>
        TradeDetail(u(0), u(1), u(2).toInt, u(3).toInt, u(4))).toDF()
    tradeDF.createOrReplaceTempView("trade")
    tradeDF.persist(StorageLevel.MEMORY_ONLY_SER)
```

```scala
    //获取2016年度网购总额
    val countOfTrade2016 = sqlCtx.sql("SELECT * FROM trade where tradeDate like '
        2016%'").count()
    println("2016年度网购总额:" + countOfTrade2016)

    //获取2016-1-1至2016-6-30用户编号为100网购总额
    println("获取2016-1-1至2016-7-1间用户编号为100网购总额: ")
    val  amountOfTradeForUser100  =  sqlCtx.sql("SELECT u.userID, t.tradeID,
        t.tradeDate, t.productID, t.amount FROM trade t,user u where u.userID =
        100 and u.userID = t.userID and t.tradeDate >= '2016-1-1' and t.tradeDate
        < '2016-7-1'").show()

    //获取2016年度交易额最高的5个省份,并显示该省交易最高、最低和平均价格
    println("获取2016年度交易额最高的5个省份,并显示该省交易最高、最低和平均价格:")
    val  orderStatsForUser5 = sqlCtx.sql("SELECT  max(amount)  as  maxAmount,
        min(amount)  as  minAmount,  avg(amount)  as  avgAmount,  sum(amount)  as
        sumAmount, provice FROM trade t, user u where u.userID = t.userID and
        t.tradeDate like '2016%' group by u.provice order by sumAmount desc")
    orderStatsForUser5.collect().take(5).map(order =>
      "Provice: " + order.getAs("provice")
        + " Sum =" + order.getAs("sumAmount")
        + ";Minimum =" + order.getAs("minAmount")
        + ";Maximum =" + order.getAs("maxAmount")
        + ";Avg =" + order.getAs("avgAmount")
    ).foreach(result => println(result))
  }
}
```

2. 运行查看结果

生成打包文件 sparklearning.jar 并移动到 Spark 根目录下,通过如下命令调用打包中的 OnLineTradeStatistics 方法:

```
$cd /app/spark/spark-2.0.0
$bin/spark-submit --class chapter7.OnLineTradeStatistics \
  --master spark://master:7077 \
  --executor-memory 1g \
  sparklearning.jar \
  hdfs://master:9000/chapter7/on_line_trade_user.txt \
  hdfs://master:9000/chapter7/on_line_trade_detail.txt
```

执行统计结果如下:
2016年度网购总额:58715

获取2016-1-1至2016-7-1间用户编号为100网购总额:

```
+------+-------+---------+---------+------+
|userID|tradeID|tradeDate|productID|amount|
+------+-------+---------+---------+------+
|   100| 495922|2016-6-15|      205|  2530|
|   100| 685706| 2016-2-4|      974| 12114|
+------+-------+---------+---------+------+
```

获取 2016 年度交易额最高的 5 个省份，并显示该省交易最高、最低和平均价格：

```
Provice: GZ Sum =21454360;Minimum =12;Maximum =19992;Avg =10344.435872709739
Provice: HL Sum =20584899;Minimum =10;Maximum =19942;Avg =10135.351550960118
Provice: GX Sum =20557688;Minimum =11;Maximum =19964;Avg =10187.159563924677
Provice: SH Sum =20282695;Minimum =12;Maximum =19978;Avg =9817.374152952565
Provice: AH Sum =20274977;Minimum =13;Maximum =19995;Avg =10260.61589068826
```

7.8 小结

在本章中先介绍了 Spark SQL 的基本情况。介绍了 Spark SQL 的发展历史，它前身是 Shark，Shark 的前身是 Hive，Spark SQL 相比较这两个工具在性能上有较大的提升；同时介绍 Spark1.6 引入的 DataFrame/Dataset，区别于 RDD 是其带有 Schema 元数据，类似于传统数据库中的二维表格。

然后本章介绍了 Spark SQL 运行原理。在该小结中先介绍原理是先讲解了通用 SQL 执行原理，然后介绍 Spark SQL 种由 Core、Catalyst、Hive 和 Hive-Thriftserver 四个部分组成，并介绍了 Tree 和 Rules 两个概念，接着对 Spark SQL 运行原理进行了深入分析，介绍了 Spark SQL 中从输入查询语句到输出数据执行过程。输入的 SQL 语句通过 Antlr 进行词法和语法解析，默认 Antlr 会构建出一棵语法树；在 Analyzer 中结合 SessionCatalog 元数据，对未绑定的逻辑计划进行解析，生成了已绑定的逻辑计划；而 Optimizer 利用 Rule 对逻辑计划和 Expression 进行迭代处理，从而达到树的节点进行合并和优化；SparkPlanner 使用 Planning Strategies 对优化的逻辑计划进行转换，生成可以执行的物理计划；SparkPlanner.preparations 对物理计划进行准备工作，规范返回数据行的格式等，然后调用 SparkPlan.execute 执行物理计划，从数据库中查询数据并生成 RDD。

接着介绍 Spark 两种工具：Hive-Console 和 SQLConsole，在 Hive-Console 使用 hive/console 工具，该工具可以查询 hive 数据并详细查看 Spark SQL 的运行计划，而利用 SQLConsole 可以读取不同格式数据并进行缓存、DSL 操作。再接着介绍了 SparkSQL CLI 和 Thrift Server 两种工具，SparkSQL CLI 允许用户使用命令界面直接输入 SQL 命令，然后发送到 Spark 集群进行执行，在界面中显示运行过程和最终的结果，而 ThriftServer 是一个 JDBC/ODBC 接口，用户可以通过 JDBC/ODBC 连接 ThriftServer 来访问 Spark SQL 的数据。

最后使用销售数据分类和网店销售数据统计两个实例，在销售数据分类中介绍 Spark SQL 对数据的操作同时，结合了 Hive、MLlib 和 SparkR 处理。

第 8 章

Spark Streaming

8.1 Spark Streaming 简介

Spark Streaming 是 Spark 核心 API 的一个扩展，具有吞吐量高、容错能力强的实时流数据处理系统，支持包括 Kafka、Flume、HFDS/S3、Twitter、ZeroMQ 以及 TCP Sockets 等数据源，获取数据后可以使用 Map、Reduce、Join 和 Window 等高级函数进行复杂算法的处理，处理结果存储到文件系统、数据库或展示到仪表盘等，其中 Spark Streaming 数据处理流程如图 8-1 所示。另外，Spark Streaming 也能和和 Spark 其他的组件，如 MLlib（机器学习）以及 Graphx 等融合，对实时数据进行更加复杂的处理。

图 8-1 Spark Streaming 数据处理流程

Spark 的各个子组件都是基于 Spark 核心，Spark Streaming 在内部的处理机制是：先接收实时流的数据，并根据一定的时间间隔拆分成一批批的数据，这些批数据在 Spark 内核对应一个 RDD 实例，因此，流数据的 DStream 可以看成是一组 RDDs，然后通过调用 Spark 核心的作业处理这些批数据，最终得到处理后的一批批结果数据。通俗点理解的话，在流数据分成一批一批后，通过一个先进先出的队列，然后 Spark 核心的作业从该队列中依次取出一个个批数据，把批数据封装成一个 RDD，然后进行处理，这是一个典型的生产者消费者模型。

作为构建于 Spark 之上的应用框架，Spark Streaming 承袭了 Spark 的编程风格，对于已经了解 Spark 的用户来说能够快速地上手。下面将以 Spark Streaming 官方提供的单词计数器代码为例来分析 Spark Streaming 相关内容：

```scala
import org.apache.spark._
import org.apache.spark.streaming._

//在本地启动名为 NetworkWordCount 的 Spark Streaming 应用
//该应用拥有两个线程，其批处理间隔时间为 1s
val conf = new SparkConf().setMaster("local[2]").setAppName
  ("NetworkWordCount")
val ssc = new StreamingContext(conf, Seconds(1))

//创建 SocketInputDStream，该 InputDStream 的 Receiver 监听本地机器的 9999 端口
val lines = ssc.socketTextStream("localhost", 9999)  //类型是 SocketInputDStream

//进行单词计数，利用 DStream 的 Transformation 构造出了
//lines -> words -> pairs -> wordCounts -> .print() 这样一个 DStreamGraph
val words = lines.flatMap(_.split(" "))              //类型是 FlatMappedDStream
val pairs = words.map(word => (word, 1))             //类型是 MappedDStream
val wordCounts = pairs.reduceByKey(_ + _)            //类型是 ShuffledDStream
wordCounts.print()                                   //类型是 ForeachDStream

//下面这行 start() 将在幕后启动 JobScheduler，进而启动 JobGenerator 和 ReceiverTracker
// ssc.start()
//   -> JobScheduler.start()
//     -> JobGenerator.start();不断生成一个一个 Batch
//     -> ReceiverTracker.start();在 Executor 上启动 ReceiverSupervisor，
//        进而创建和启动 Receiver
ssc.start()
ssc.awaitTermination()
```

8.1.1 术语定义

1. 离散流（Discretized Stream）或 DStream

DStream 作为 Spark Streaming 的基础抽象，它代表持续性的数据流。这些数据流既可以通过外部输入源来获取，也可以通过现有的 DStream 的转换操作来获得。在内部实现上，DStream 由一组时间序列上连续的 RDD 来表示，如图 8-2 所示。

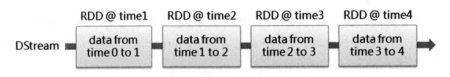

图 8-2　DStream 与 RDD 关系

在 DStream 中定义了名为 generatedRDDs 离散数据流，它是以时间为键、RDD 为值的哈希列表。在流数据接收过程中，源源不断地把接收到的数据放入到该列表中，而对于不需要的旧 RDD 从该列表中删除。

```
private[streaming] var generatedRDDs = new HashMap[Time, RDD[T]] ()
```

在单词计数例子中，如图 8-3 所示，在图中每一个椭圆形表示一个 RDD，椭圆形中的每个圆形代表一个 RDD 中的一个 Partition，图中的每一列的多个 RDD 表示一个 DStream（图中有 5 个 DStream），其中第一个列为监听 Socket 端口接收的数据形成 SocketInputDStream，而每一行最后一个 RDD，则表示每一个批处理所产生的中间结果 RDD。

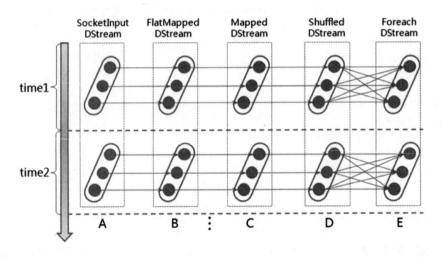

图 8-3　单词计数例子中 DStream 依赖关系

对 DStream 中数据的各种操作也是映射到内部的 RDD 上来进行的，对 DStream 的操作可以通过 RDD 的转换操作生成新的 DStream（具体操作介绍参见 7.2.2 节的内容）。如图 8-3 所示，在单词计数例子中，DStream 转换顺序依次为 SocketInputDStream→FlatMappedDStream→MappedDStream→ShuffledDStream→ForeachDStream。

2. DStreamGraph

在 Spark 核心中，作业是由一系列具有依赖关系的 RDD 及作用于这些 RDD 上的函数所组成的操作链，在遇到行动操作时触发运行，向 DAGScheduler 提交并运行作业。Spark Streaming 中作业的生成与 Spark 核心类似，对 DStream 进行的各种操作让它们之间构建起依赖关系。当遇到 DStream 使用输出操作时，这些依赖关系以及它们之间的操作会被记录到名为 DStreamGraph 的对象中表示一个作业。这些作业注册到 DStreamGraph 并不会立即运行，而是等到 Spark Streaming 启动后，到达批处理时间时，才根据 DStreamGraph 生成作业处理该批处理时间内接收的数据。在 Spark Streaming 如果应用程序中存在多个输出操作，那么在批处理中会产生多个作业。这些作业的调度阶段有可能有依赖，也可能依赖，需要根据实际情况进行判断。

在前面单词计数的例子中，每一个批处理内，RDD 之间存在依赖关系，其关系为 RddA<-RddB<-RddC<-RddD<-RddE，同样地，DStream 之间也存在依赖关系，其依赖关系为 DStreamA<-DStreamB<-DStreamC<-DStreamD<-DStreamE，这些 DStream 的依赖关系保存在 DStreamGraph 中。当对数据进行 print 时，DStreamGraph 会根据它所包含的 DStream 之间的依赖关系，生成打印作业输出结果，它们之间的关系图如图 8-4 所示。

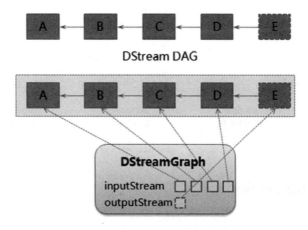

图 8-4　DStream 的依赖关系

3．批处理间隔（Batch Duration）

在 Spark Streaming 中，数据采集是逐条进行的，而数据处理是按批进行的，因此在 Spark Streaming 中会先设置好批处理间隔（batch duration）。当超过批处理间隔的时候就会把采集到的数据汇总起来成为一批数据交给系统去处理。

4．窗口间隔（Window Duration）和滑动间隔（Slide Duration）

对于窗口操作，批处理间隔、窗口间隔和滑动间隔是非常重要的 3 个时间概念，是理解窗口操作的关键所在。

对于窗口操作而言，在其窗口内部会有 N 个批处理数据，批处理数据的个数由窗口间隔（Window Duration）决定，其为窗口持续的时间，在窗口操作中只有窗口间隔满足了才会触发批数据的处理。除了窗口的长度，另一个重要的参数就是滑动间隔（Slide Duration），它指的是经过多长时间窗口滑动一次形成新的窗口，滑动窗口默认情况下和批次间隔的相同，而窗口间隔一般设置得要比它们俩都大。特别要注意的是，窗口间隔和滑动间隔的大小一定得设置为批处理间隔的整数倍。

如图 8-5 所示，批处理间隔是 1 个时间单位，窗口间隔是 3 个时间单位，滑动间隔是 2 个时间单位。对于初始的窗口 time 1~time 3，只有窗口间隔满足了才触发数据的处理。这里需要注意的一点是，初始的窗口有可能流入的数据没有撑满，但是随着时间的推进，窗口最终会被撑满。当每 2 个时间单位，窗口滑动一次后，会有新的数据流入窗口，这时窗口会移去最早的两个时间单位的数据，而与最新的两个时间单位的数据进行汇总形成新的窗口(time3~time5)。

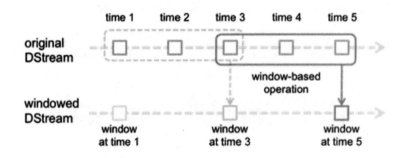

图 8-5 批处理间隔示意图

8.1.2 Spark Streaming 特点

1．流式处理

Spark Streaming 是将流式计算分解成一系列短小的批处理作业。这里的批处理引擎是 Spark Core，也就是把 Spark Streaming 的输入数据按照批处理间隔（如 1s）分成一段一段的数据（Discretized Stream），每一段数据都转换成 Spark 中的 RDD（Resilient Distributed Dataset），然后将 Spark Streaming 中对 DStream 的 Transformation 操作变为针对 Spark 中对 RDD 的 Transformation 操作，将 RDD 经过操作变成中间结果保存在内存中。整个流式计算根据业务的需求可以对中间的结果进行叠加或者存储到外部设备（参见运行架构图）。

2. 高容错

对于流式计算来说，容错性至关重要。首先我们要明确一下 Spark 中 RDD 的容错机制。每一个 RDD 都是一个不可变的分布式可重算的数据集，其记录着确定性的操作"血统"（lineage），所以只要输入数据是可容错的，那么任意一个 RDD 的分区（Partition）出错或不可用，都是可以利用原始输入数据通过转换操作而重新算出的。

对于 Spark Streaming 来说，其 RDD 的传承关系如图 8-6 所示，我们可以看到图中的每一个 RDD 都是通过"血统"相连接的，由于 Spark Streaming 输入数据可以来自于磁盘，例如 HDFS（多份副本）或是来自于网络的数据流（Spark Streaming 会将网络输入数据的每一个数据流复制两份到其他的机器）都能保证容错性，所以 RDD 中任意的 Partition 出错，都可以并行地在其他机器上将缺失的 Partition 计算出来。这个容错恢复方式比连续计算模型（如 Storm）的效率更高。

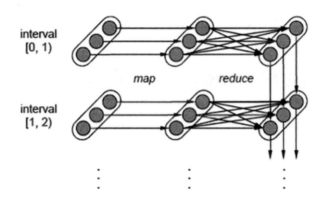

图 8-6　Spark Streaming 中 RDD 的"血统"关系图

3. 低延迟

对于实时性的讨论，会牵涉流式处理框架的应用场景。Spark Streaming 将流式计算分解成多个 Spark Job，对于每一段数据的处理都会经过 Spark DAG 图分解以及 Spark 的任务集的调度过程。对于目前版本的 Spark Streaming 而言，其最小的 Batch Size 的选取在 0.5～2s（Storm 目前最小的延迟是 100ms 左右），所以 Spark Streaming 能够满足除对实时性要求非常高（如高频实时交易）之外的所有流式准实时计算场景。

4. 吞吐量高

Spark 目前在 EC2 上已能够线性扩展到 100 个节点（每个节点 4Core），可以以数秒的延迟处理 6GB/s 的数据量（60M records/s），其吞吐量也比流行的 Storm 高 2～5 倍。图 8-7 是 Berkeley 利用 WordCount 和 Grep 两个用例所做的测试，在 Grep 这个测试中，Spark Streaming 中的每个节点的吞吐量是 670k records/s，而 Storm 是 115k records/s。

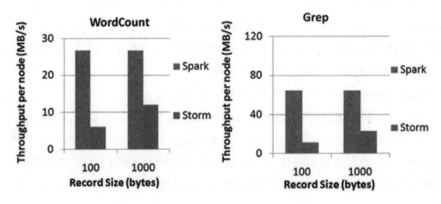

图 8-7　Spark Streaming 与 Storm 吞吐量比较

8.2　Spark Streaming 编程模型

8.2.1　DStream 的输入源

在 Spark Streaming 中所有的操作都是基于流，而输入源是这一系列操作的起点，Spark Streaming 提供基础和高级两种数据流来源。

1．基础来源

在前面的例子中我们已看到 ssc.socketTextStream()方法可以通过 TCP 套接字连接，从文本数据中创建了一个 DStream。除了套接字，StreamingContext 的 API 还提供了方法从文件系统中创建 DStreams 作为输入源。

StreamingContext.fileStream（dataDirectory）方法可以从多种文件系统（如 HDFS、S3 和 NFS 等）的文件中读取数据，然后创建一个 DStream。Spark Streaming 监控 dataDirectory 目录和在该目录下任何文件被创建处理（不支持在嵌套目录下写文件）。需要注意的是，读取的必须是具有相同的数据格式的文件，而且创建的文件必须在 dataDirectory 目录下，并通过自动移动或重命名成数据目录，如果文件一旦移动就不能被改变。

另外 Spark Streaming 也可以使用 streamingContext.queueStream(queueOfRDDs)方法创建基于 RDD 队列的 DStream，每个 RDD 队列将被视为 DStream 中一块数据流进行加工处理。

2．高级来源

这一类的来源需要外部接口，其中一些有复杂的依赖关系（如 Kafka 和 Flume），因此通过这些来源创建 DStreams 需要明确其依赖。例如，如果想创建一个使用 Twitter tweets 的数据的 DStream 流，必须按以下步骤来做：

（1）在 SBT 或 Maven 工程里添加 spark-streaming-twitter_2.11 依赖。

（2）导入 TwitterUtils 包，通过 TwitterUtils.createStream 方法创建一个 DStream。

（3）添加所有依赖的 jar 包（包括依赖的 spark-streaming-twitter_2.11 及其依赖），然后部署应用程序。

需要注意的是，这些高级的来源一般在 Spark Shell 中不可用，因此基于这些高级来源的应用不能在 Spark Shell 中进行测试。如果必须在 Spark Shell 中使用，需要下载相应的 Maven 工程的 Jar 依赖并添加到类路径中。常见的输入源和对应的项目如表 8-1 所示。

表 8-1 常见的输入源和对应的项目

输入源名称	项目名称	描述
Kafka	spark-streaming-kafka_2.11	可以从 Kafka 中接收流数据
Flume	spark-streaming-flume_2.11	可以从 Flume 中接收流数据
Twitter	spark-streaming-twitter_2.11	通过 TwitterUtils 工具类调用 Twitter4j，可以得到公众的流或得到基于关键词过滤流
MQTT	spark-streaming-mqtt_2.11	可以从 MQTT 消息队列中接收流数据
ZeroMQ	spark-streaming-zeromq_2.11	可以从 ZeroMQ 消息队列中接收流数据

8.2.2 DStream 的操作

与 RDD 类似的 DStream 也提供了一系列操作方法，这些操作可以分成 3 类：普通的转换操作、窗口转换操作和输出操作。

1．普通的转换操作

普通的转换操作如表 8-2 所示。

表 8-2 普通的转换操作

转换	描述
map(*func*)	源 DStream 的每个元素通过函数 func 返回一个新的 DStream
flatMap(*func*)	类似与 map 操作，不同的是每个输入元素可以被映射出 0 或者更多的输出元素
filter(*func*)	在源 DStream 上选择 Func 函数返回仅为 true 的元素,最终返回一个新的 DStream

续表

转　换	描　述
repartition(*numPartitions*)	通过输入的参数 numPartitions 的值来改变 DStream 的分区大小
union(*otherStream*)	返回一个包含源 DStream 与其他 DStream 的元素合并后的新 DStream
count()	对源 DStream 内部的所含有的 RDD 的元素数量进行计数，返回一个内部的 RDD 只包含一个元素的 DStreaam
reduce(*func*)	使用函数 func（有两个参数并返回一个结果）将源 DStream 中每个 RDD 的元素进行聚合操作，返回一个内部所包含的 RDD 只有一个元素的新 DStream
countByValue()	计算 DStream 中每个 RDD 内的元素出现的频次并返回新的 DStream[(K,Long)]，其中 K 是 RDD 中元素的类型，Long 是元素出现的频次
reduceByKey(*func*, [*numTasks*])	当一个类型为（K，V）键值对的 DStream 被调用的时候，返回类型为类型为（K，V）键值对的新 DStream,其中每个键的值 V 都是使用聚合函数 func 汇总。注意：默认情况下，使用 Spark 的默认并行度提交任务（本地模式下并行度为 2，集群模式下为 8），可以通过配置 numTasks 设置不同的并行任务数
join(*otherStream*, [*numTasks*])	当被调用类型分别为（K，V）和（K，W）键值对的 2 个 DStream 时，返回类型为（K，（V，W））键值对的一个新 DStream
cogroup(*otherStream*, [*numTasks*])	当被调用的两个 DStream 分别含有(K，V) 和(K，W)键值对时,返回一个(K，Seq[V], Seq[W])类型的新的 DStream
transform(*func*)	通过对源 DStream 的每个 RDD 应用 RDD-to-RDD 函数返回一个新的 DStream，这可以用来在 DStream 做任意 RDD 操作
updateStateByKey(*func*)	返回一个新状态的 DStream，其中每个键的状态是根据键的前一个状态和键的新值应用给定函数 func 后的更新。这个方法可以被用来维持每个键的任何状态数据

下面分析一下 ransform()和 updateStateByKey()两个操作。

（1）transform(func)操作。

transform 操作（转换操作）允许 DStream 上应用任意 RDD-to-RDD 函数。它可以被应用于未在 DStream API 中暴露任何的 RDD 操作。例如，在每批次的数据流与另一数据集的连接功能不直接暴露在 DStream API 中，但可以轻松地使用 transform 操作来做到这一点，这使得 DStream 的功能非常强大。例如，可以通过连接预先计算的垃圾邮件信息的输入数据流（可能也有 Spark 生成的），然后基于此做实时数据清理的筛选，如下面官方提供的伪代码所示。事实上，也可以在 transform 方法中使用机器学习和图形计算的算法。

```
//获取包含垃圾邮件信息的 RDD 数据流
val spamInfRDD = ssc.sparkContext.newAPIHadoopRDD(……)
val cleanedDStream = wordCounts.transform(rdd => {
```

```
//过滤邮件中垃圾的信息
rdd.join(spamInfoRDD).filter(……)
……
})
Some(newCount)
```

（2）updateStateByKey 操作。

该 updateStateByKey 操作可以保持任意状态，同时进行信息更新。要使用此功能，必须进行两个步骤。

1）定义状态：状态可以是任意的数据类型。

2）定义状态更新函数：用一个函数指定如何使用先前的状态和从输入流中获取的新值更新状态。

让我们用一个例子来说明，假设你要进行文本数据流中的单词来计数。在这里，正在运行的计数是状态，而且它是一个整数。我们定义了更新功能如下：

```
def updateFunction(newValues: Seq[Int], runningCount: Option[Int]): Option[Int] 
  = {
  //通过原来的计数信息得到新的技术信息
  val newCount = ……
  Some(newCount)
}
```

此函数应用于含有键值对的 DStream 中，它会针对里面的每个元素调用更新函数，其中，newValues 是最新的值，runningCount 是之前的值。

```
val runningCounts = pairs.updateStateByKey[Int](updateFunction _)
```

2. 窗口转换操作

Spark Streaming 提供了窗口的计算，允许通过滑动窗口对数据进行转换，窗口转换操作如表 8-3 所示。

表 8-3　窗口转换操作

转　　换	描　　述
window(*windowLength, slideInterval*)	返回一个基于源 DStream 的窗口批次计算后得到新的 DStream
countByWindow(*windowLength,slideInterval*)	返回基于滑动窗口的 DStream 中的元素的数量
reduceByWindow(*func, windowLength,slideInterval*)	基于滑动窗口对源 DStream 中的元素进行聚合操作，得到一个新的 DStream
reduceByKeyAndWindow(*func,windowLength, slideInterval,* [*numTasks*])	基于滑动窗口对（K，V）键值对类型的 DStream 中的值按 K 使用聚合函数 func 进行聚合操作，得到一个新的 DStream

续表

转换	描述
reduceByKeyAndWindow(*func, invFunc,windowLength, slideInterval, [numTasks]*)	一个更高效的 reduceByKkeyAndWindow() 的实现版本，先对滑动窗口中新的时间间隔内数据增量聚合并移去最早的与新增数据量的时间间隔内的数据统计量。例如，计算 t+4 秒时刻过去 5s 窗口的 WordCount，那么可以将 t+3 时刻过去 5s 的统计量加上[t+3, t+4]的统计量，再减去[t-2, t-1]的统计量，这种方法可以复用中间 3s 的统计量，提高统计的效率
countByValueAndWindow(*windowLength,slideInterval, [numTasks]*)	基于滑动窗口计算源 DStream 中每个 RDD 内每个元素出现的频次并返回 DStream[(K,Long)]，其中 K 是 RDD 中元素的类型，Long 是元素频次。与 countByValue 一样，reduce 任务的数量可以通过一个可选参数进行配置

3. 输出操作

Spark Streaming 允许 DStream 的数据输出到外部系统，如数据库或文件系统，输出的数据可以被外部系统所使用，该操作类似于 RDD 的输出操作。表 8-4 列出了目前主要的输出操作。

表 8-4　输出操作

转换	描述
print()	在 Driver 中打印 DStream 中数据的前 10 个元素
saveAsTextFiles(*prefix, [suffix]*)	将 DStream 中的内容以文本的形式保存为文本文件，其中每次批处理间隔内产生的文件以 prefix-TIME_IN_MS[.suffix]的方式命名
saveAsObjectFiles(*prefix, [suffix]*)	将 DStream 中的内容按对象序列化并且以 SequenceFile 的格式保存。其中每次批处理间隔内产生的文件以 prefix-TIME_IN_MS[.suffix]的方式命名
saveAsHadoopFiles(*prefix, [suffix]*)	将 DStream 中的内容以文本的形式保存为 Hadoop 文件，其中每次批处理间隔内产生的文件以 prefix-TIME_IN_MS[.suffix]的方式命名
foreachRDD(*func*)	最基本的输出操作，将 func 函数应用于 DStream 中的 RDD 上，这个操作会输出数据到外部系统，比如保存 RDD 到文件或者网络数据库等。需要注意的是，func 函数是在运行该 streaming 应用的 Driver 进程里执行的。

8.3 Spark Streaming 运行架构

8.3.1 运行架构

前面我们知道，Spark Streaming 相对其他流处理系统最大的优势在于流处理引擎和数据处理在同一个软件栈，其中 Spark Streaming 功能主要包括流处理引擎的流数据接收与存储以及批处理作业的生成与管理，而 Spark 核心负责处理 Spark Streaming 发送过来的作业。Spark Streaming 分为 Driver 端和 Client 端，运行在 Driver 端为 StreamingContext 实例。该实例包括 DStreamGraph 和 JobScheduler（包括 ReceiverTracker 和 JobGenerator）等，而 Client 包括 ReciverSupervisor 和 Reciver 等。

Spark Streaming 进行流数据处理大致可以分为：启动流处理引擎、接收及存储流数据、处理流数据和输出处理结果等 4 个步骤，其运行架构如图 8-8 所示。

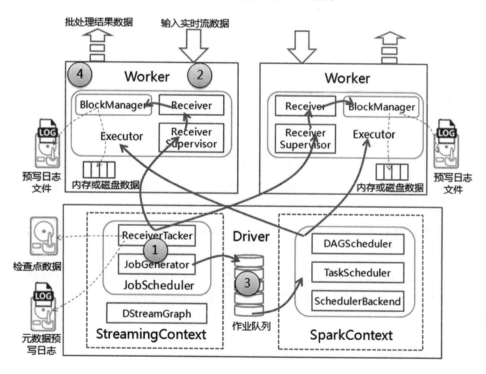

图 8-8　Spark Streaming 运行架构

（1）初始化 StreamingContext 对象，在该对象启动过程中实例化 DStreamGraph 和 JobScheduler，其中 DStreamGraph 用于存放 DStream 以及 DStream 之间的依赖关系等信息，而 JobScheduler

中包括ReceiverTracker和JobGenerator。其中ReceiverTracker为Driver端流数据接收器(Receiver)的管理者，JobGenerator为批处理作业生成器。在ReceiverTracker启动过程中，根据流数据接收器分发策略通知对应的Executor中的流数据接收管理器（ReciverSupervisor）启动，再由ReciverSupervisor启动流数据接收器。

（2）当流数据接收器Receiver启动后，持续不断地接收实时流数据，根据传过来数据的大小进行判断，如果数据量很小，则攒多条数据成一块，然后再进行块存储；如果数据量大，则直接进行块存储。对于这些数据Receiver直接交给ReciverSupervisor，由其进行数据转储操作。块存储根据设置是否预写日志分为两种，一种是使用非预写日志BlockManagerBasedBlockHandler方法直接写到Worker的内存或磁盘中，另一种是进行预写日志WriteAheadLogBasedBlockHandler方法，即在预写日志同时把数据写入到Worker的内存或磁盘中。数据存储完毕后，ReciverSupervisor会把数据存储的元信息上报给ReceiverTracker，ReceiverTracker再把这些信息转发给ReceivedBlockTracker，由它负责管理收到数据块的元信息。

（3）在StreamingContext的JobGenerator中维护一个定时器，该定时器在批处理时间到来时会进行生成作业的操作。在该操作中进行如下操作：

1）通知ReceiverTracker将接收到的数据进行提交，在提交时采用synchronized关键字进行处理，保证每条数据被划入一个且只被划入一个批中。

2）要求DStreamGraph根据DStream依赖关系生成作业序列Seq[Job]。

3）从第一步中ReceiverTracker获取本批次数据的元数据。

4）把批处理时间time、作业序列Seq[Job]和本批次数据的元数据包装为JobSet，调用JobScheduler.submitJobSet(JobSet)提交给JobScheduler，JobScheduler将把这些作业发送给Spark核心进行处理，由于该执行为异步，因此本步执行速度将非常快。

5）只要提交结束（不管作业是否被执行），Spark Streaming对整个系统做一个检查点（Checkpoint）。

（4）在Spark核心的作业对数据进行处理，处理完毕后输出到外部系统，如数据库或文件系统，输出的数据可以被外部系统所使用。由于实时流数据的数据源源不断地流入，Spark会周而复始地进行数据处理，相应也会持续不断地输出结果。

8.3.2 消息通信

上面我们对Spark Streaming的运行架构进行了分析，在这里将对StreamingContext启动流处理引擎和接收存储流数据中进行的消息通信进行详细分析，如图8-9所示。在启动流处理引擎过程中，将进行启动所有流数据接收器Receiver和注册流数据接收器Receiver两个消息通信；

在接收存储流数据中,当数据块存储完成后发送添加数据块消息;而当 Spark Streaming 停止时,需要发送关闭所有所有流数据接收器 Receiver 消息。

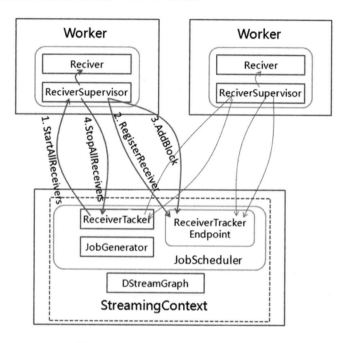

图 8-9　Spark Streaming 消息通信过程

具体过程如下:

(1) 在启动流处理引擎过程中,JobScheduler 在内部启动 ReceiverTracker 和 ReceiverTrackerEndPoint 终端点,当 ReceiverTracker 准备完毕后向终端点发送 StartAllReceivers 消息,通知其分发并启动所有流数据接收器 Receiver,其代码如下:

```
override def receive: PartialFunction[Any, Unit] = {
  case StartAllReceivers(receivers) =>
    //根据流数据接收器分发策略,匹配流数据接收器 Receiver 和 Executor,
    //其分发过程参见下节的 Receiver 分布内容
    val scheduledExecutors = schedulingPolicy.scheduleReceivers(receivers,
      getExecutors)
    for (receiver <- receivers) {
      val executors = scheduledExecutors(receiver.streamId)
      updateReceiverScheduledExecutors(receiver.streamId, executors)
      //在 HashMap 中保存流数据接收器 Receiver 首选位置
      receiverPreferredLocations(receiver.streamId) = receiver.
        preferredLocation
```

```
        //在指定的Executor中启动流数据接收器Receiver
        startReceiver(receiver, executors)
    }
    ......
}
```

（2）启动流数据接收器Receiver前，ReceiverSupervisor会向ReceiverTrackerEndPoint终端点发送RegisterReceiver注册消息，当注册成功后才继续进行流数据接收器Receiver的启动。其中注册过程代码位于ReceiverSupervisor的startReceiver方法中，具体如下：

```
def startReceiver(): Unit = synchronized {
  try {
    //先调用ReceiverSupervisorImpl类的onReceiverStart方法进行注册，
    //如果注册成功，则继续进行流数据接收器Receiver启动
    if (onReceiverStart()) {
      logInfo("Starting receiver")
      receiverState = Started
      receiver.onStart()
      logInfo("Called receiver onStart")
    } else {
      //如果Driver端的TrackerReceiver拒绝注册或注册失败，则停止流数据接收器，
      //并发送注销流数据接收器DeregisterReceiver消息
      stop("Registered unsuccessfully because Driver refused to start receiver
        " + streamId, None)
    }
  } catch { …… }
}
```

发送注册流数据接收器消息RegisterReceiver代码如下：

```
override protected def onReceiverStart(): Boolean = {
  val msg = RegisterReceiver(streamId, receiver.getClass.getSimpleName,
    hostPort, endpoint)
  trackerEndpoint.askWithRetry[Boolean](msg)
}
```

（3）在流数据接收器Receiver接收数据的过程中，当保存完一个数据块时，作为数据转储的管理者ReceiverSupervisor会把数据块的元数据发送给ReceiverTrackerEndPoint终端点，ReceiverTracker再把这些信息转发给ReceivedBlockTracker，由它负责管理收到数据块的元信息。其中发送增加数据块消息代码位于ReceiverSupervisorImpl的pushAndReportBlock方法中，具体代码如下：

```
def pushAndReportBlock(
    receivedBlock: ReceivedBlock,
    metadataOption: Option[Any],
    blockIdOption: Option[StreamBlockId]
```

```
) {
    val blockId = blockIdOption.getOrElse(nextBlockId)
    val time = System.currentTimeMillis
    //调用 ReceivedBlockHandler 的 storeBlock 方法进行保存数据块
    val blockStoreResult=receivedBlockHandler.storeBlock(blockId,receivedBlock)
    val numRecords = blockStoreResult.numRecords
    //把数据块的元信息发送给 ReceiverTrackerEndPoint 终端点
    val blockInfo = ReceivedBlockInfo(streamId, numRecords, metadataOption,
        blockStoreResult)
    trackerEndpoint.askWithRetry[Boolean](AddBlock(blockInfo))
}
```

（4）当 Spark Streaming 停止时，ReceiverTracker 发送注销所有流数据接收器 Receiver 消息，ReceiverTrackerEndPoint 终端点接到该消息会调用 ReceiverTracker.stop 方法注销。在停止操作过程中，ReceiverTracker 会发送两次注销消息，发送两次消息之间的间隔为 10s，用于等待流数据接收器 Receiver。

```
def stop(graceful: Boolean): Unit = synchronized {
    if (isTrackerStarted) {
        if (!skipReceiverLaunch) {
            //第一次发送注销所有流数据接收器 Receiver 消息并等待 10s
            endpoint.askWithRetry[Boolean](StopAllReceivers)
            receiverJobExitLatch.await(10, TimeUnit.SECONDS)
            ......
            //第二次发送注销所有所有流数据接收器 Receiver 消息，并返回注销结果
            val receivers = endpoint.askWithRetry[Seq[Int]](AllReceiverIds)
            if (receivers.nonEmpty) {
                logWarning("Not all of the receivers have deregistered, " + receivers)
            } else {
                logInfo("All of the receivers have deregistered successfully")
            }
    ......
}
```

8.3.3 Receiver 分发

Spark Streaming 中处理的数据一方面通过内部接口获取，另一方面来自于 Kafka、Flume、Twitter、ZeroMQ 以及 TCP Sockets 等外部系统源，获取后对这些输入的数据源进行 Map、Reduce、Join 和 Window 等操作，完成较为复杂的数据加工和处理，这些数据均抽象于 DStream，如图 8-10 所示。

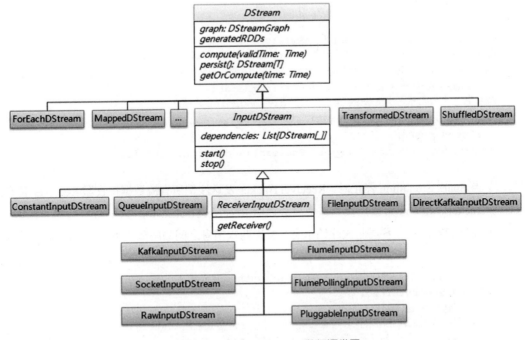

图 8-10　Spark Streaming 数据源类图

直接继承于 DStream 可以分为输入 DStream 和操作 DStream 等类别，而根据输入 DStream 来源不同可以分为 ReceiverInputDStream（通过流接收器获取输入数据）、QueueInputDStream（使用队列获取输入数据）、FileInputDStream（通过读取文件方式获取输入数据）和 DirectKafkaInputDStream（直接读取 Kafka 获取输入数据）等方式。根据流接收器类型又分为 SocketInputDStream、KafkaInputDStream、FlumeInputDStream 和 MQTTInputDStream 等类型。另外也支持通过继承 PluggableInputDStream 自定义流接收器获取流数据。

不同的 ReceiverInputDStream 包含不同的流数据接收器，而这些接收器继承于 Receiver，其类关系如图 8-11 所示。

以 Socket 为例，SocketInputDStream 对应的流数据接收器为 SocketReceiver，在 SocketInputDStream 中通过 getReceiver 方法获取该流数据接收器，其类关系如图 8-12 所示。

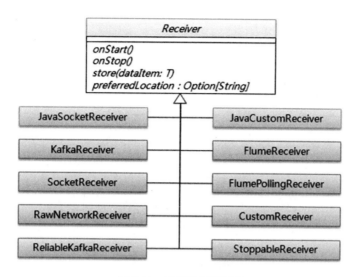

图 8-11　数据接收器类图

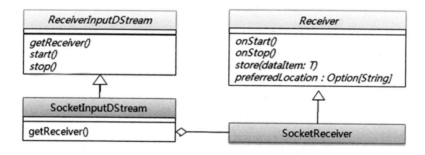

图 8-12　Socket 中 DStream 与流数据接收器类图

在 StreamingContext 启动过程中，ReceiverTracker 会把流数据接收器 Receiver 分发到 Executor 上，在每个 Executor 上，由 ReceiverSupervisor 启动对应的 Receiver。在 Spark 1.4 及以前的版本中根据 N 个 Receiver 实例，在 StreamingContext 中创建一个作业，该作业包含 N 个任务，其创建结构如图 8-13 所示。

其创建过程具体如下。

（1）先遍历 ReceiverInputStream，通过其 getReceiver 获取需要启动的 N 个 Receiver 实例，然后把这些实例作为 N 份数据，在 StreamingContext 创建一个 RDD 实例，该 RDD 分为 N 个 partition，每个 partition 对应包含一个 Receiver 数据（即 Receiver 实例）。

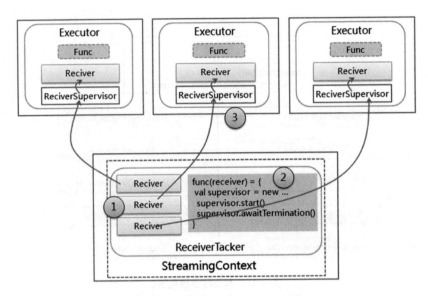

图 8-13 Spark 1.4 及以前版本 StreamingContext 启动流数据接收器

（2）在这里我们把 Receiver 所进行的计算定义为 func 函数，该函数以 Receiver 实例为参数构造 ReceiverSupervisorImpl 实例 supervisor，构造完毕后使用新线程启动该 supervisor 并阻塞该线程，其代码如下：

```
val supervisor = new ReceiverSupervisorImpl(receiver, SparkEnv.get,
    serializableHadoopConf.value, checkpointDirOption)
supervisor.start()
supervisor.awaitTermination()
```

（3）把 ReceiverTracker 尽可能地按照 Receiver 的首选位置分发到集群并启动，启动完毕后 Receiver 会处于阻塞状态，持续不断地接入流数据。

该 Receiver 分发方式在长时间的运行过程中，如果出现某个任务失败，则 Spark 会重新发送该任务到其他 Executor 进行重跑，但由于该分发过程属于随机分配，无法实现集群的负载均衡，有可能出现某 Worker 节点运行多个任务，而某些 Worker 节点却是空闲。而当该任务失败次数超过规定的上限 spark.task.maxFailures（默认=4），则会导致该 Receiver 无法启动。针对这些问题在 Spark 1.5 及以后的版本中，在 StreamingContext 中会根据 N 个 Receiver 实例创建 N 个作业，各个作业中只包含一个任务，并加入了可插拔的 Receiver 分发策略，其结构如图 8-14 所示。

这样在 Spark Streaming 中每个 Receiver 都有一个作业来分发（该作业只包含一个任务），而且对于这仅有的一个任务只在第一次启动的时，才尝试启动 Receiver。如果该任务失败了，则不再尝试启动 Receiver，对应的作业可以置为完成状态。此时，ReceiverTracker 会新生成一个作业，在其他 Executor 尝试启动，直到成功为止，这样 Receiver 就不会受到任务失败上限而无法启动。通过这种方式，Spark Streaming 中的所有 Receiver 总是保持活性，并不会随 Executor 的失败而停止。

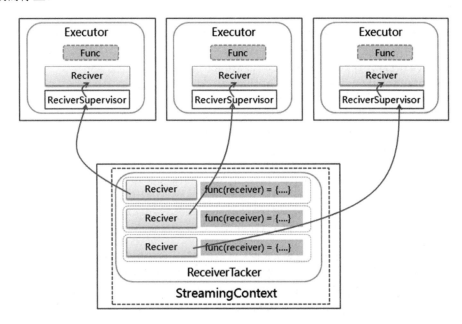

图 8-14　Spark 1.5 及以后版本 StreamingContext 启动流数据接收器

可插拔的 Receiver 分发策略在 ReceiverSchedulingPolicy 类中定义，在 Receiver 分发之前会收集所有的 InputDStream 包含的所有 Receiver 实例和 Executor，然后调用该类中的 scheduleReceivers 方法计算每个 Receiver 对应的 Executor。在该方法中以轮询调度（Round-Robin）方式进行分配，首先对存在首选位置的 Receiver 进行处理，尽可能地把 Receiver 运行在首选位置机器运行 Receiver 个数最少的 Executor 中，接着对于没有首选位置的 Receiver，则优先分配到运行 Receiver 个数最少的 Executor 中，分配完成后返回调度好的 Executor 列表。

```
//该方法中传入 InputDStream 包含的所有 Receiver 实例和 Executor 序列
  def scheduleReceivers(receivers: Seq[Receiver[_]],executors: Seq[String]):
    Map[Int, Seq[String]] = {
  ……
  //Executor 格式为 host:port，获取冒号的前半部分信息，即 IP 地址或机器名
  val hostToExecutors = executors.groupBy(_.split(":")(0))

  //定义 Receivers 个数的数组用于存放 Executors 列表，该数组和 Receiver 序列一一对应
```

```scala
    val scheduledExecutors = Array.fill(receivers.length)(new
      mutable.ArrayBuffer[String])
    //定义Executor，运行Receiver个数的哈希列表
    val numReceiversOnExecutor = mutable.HashMap[String, Int]()
    //设置初始值为0
    executors.foreach(e => numReceiversOnExecutor(e) = 0)

    //首先，如果Receiver存在首选位置，则把该首选位置放在候选调度Executor列表中
    for (i <- 0 until receivers.length) {
      receivers(i).preferredLocation.foreach { host =>
        hostToExecutors.get(host) match {
    //如果Receiver的首选位置在传入的Executor序列中,则在scheduledExecutors的Receiver
    //加入对应的Executor,在numReceiversOnExecutor对应Executor的Receiver个数加1，
    //需要注意的是，获取该首选位置机器运行Receiver个数最少的Executor
          case Some(executorsOnHost) =>
            val leastScheduledExecutor =
              executorsOnHost.minBy(executor => numReceiversOnExecutor(executor))
            scheduledExecutors(i) += leastScheduledExecutor
            numReceiversOnExecutor(leastScheduledExecutor) =
              numReceiversOnExecutor(leastScheduledExecutor) + 1

    //如果Receiver的首选位置不在传入的Executor序列中，则该Executor存在两种可能，
    //一种是还未启动完毕，另一种是Executor已死或是所在主机不在集群中，对于这些情况
    //则先把该主机加到scheduledExecutors中
          case None =>scheduledExecutors(i) += host
        }
      }
    }

    //对于那些不存在首选位置的Receiver（即scheduledExecutors数组对应为空的元素），
    //则在numReceiversOnExecutor获取Receiver计数最少的元素，把该Receiver放到
    //其Excutor进行运行，并更新numReceiversOnExecutor列表信息
    for (scheduledExecutorsForOneReceiver <- scheduledExecutors.filter
      (_.isEmpty)) {
      val (leastScheduledExecutor, numReceivers) = numReceiversOnExecutor.minBy
        (_._2)
      scheduledExecutorsForOneReceiver += leastScheduledExecutor
      numReceiversOnExecutor(leastScheduledExecutor) = numReceivers + 1
    }

    //返回调度好的Executor列表，该列表和Receiver序列一一对应
    receivers.map(_.streamId).zip(scheduledExecutors).toMap
  }
```

8.3.4 容错性

实时流处理系统需要长时间接收并处理数据，在这个过程中出现异常是难以避免的，需要流程系统具备高容错性。Spark Streaming 在设计之初考虑了这种情况：一方面利用 Spark 自身的容错设计、存储级别（如设置存储级别为 MEMORY_AND_DISK_2）和 RDD 抽象设计能够处理集群中任何 Worker 节点的故障；另一方面由于 Spark 运行多种运行模式，其 Driver 端可能运行在 Master 节点或者在集群中的任意节点，这样让 Driver 端具备容错能力是一个很大的挑战，但是由于 Spark Streaming 接收的数据是按照批进行存储和处理，这些批次数据的元数据可以通过执行检查点的方式定期写入到可靠的存储中，在 Driver 端重新启动中恢复这些状态。

虽然 Spark Streaming 在绝大多数情况下能够处理并恢复运行的状态和数据，但是对于接收数据存在于内存中存在丢失的风险。例如对于 Kafka 和 Flume 等其他数据源，由于接收到的数据还只缓存在 Executor 的内存中，尚未及时被处理，当 Executor 出现异常时会丢失这些数据；如果集群处于独立运行模式或 YARN 运行模式或 Mesos 运行模式，当 Driver 端失败时该 Driver 端所管理的 Executor 及内存中数据将终止，即使 Driver 端重新启动这些缓存的数据也不能被恢复。为了避免这种数据损失，从 Spark 1.2 版本起引进了预写日志（WriteAheadLogs）功能。

预写日志通常被用于数据库和文件系统中，保证数据操作的持久性。预写日志通常是先将操作写入到一个持久可靠的日志文件中，然后才对数据施加该操作，当加入施加该操作中出现了异常，可以通过读取日志文件并重新施加该操作，从而恢复系统。当启动预写日志以后，所有收到的数据同时还保存到了容错文件系统的日志文件中，当 Spark Streaming 失败，这些接收到的数据也不会丢失。另外，接收数据的正确性只在数据被预写到日志以后接收器才会确认，已经缓存但还没有保存的数据可以在 Driver 重新启动之后由数据源再发送一次。这两个机制确保了零数据丢失，即所有的数据或者从日志中恢复，或者由数据源重发。

在一个 Spark Streaming 应用开始时（也就是 Driver 开始时），相关的 StreamingContext（所有流功能的基础）使用 SparkContext 启动接收器成为长驻运行任务。这些接收器接收并保存流数据到 Spark 内存中以供处理。用户传送数据的生命周期如图 8-15 所示。

（1）接收数据：接收器将数据流分成一系列小块，存储到 Executor 内存或磁盘中。如果启用预写日志，数据同时还写入到容错文件系统的预写日志文件中。

（2）通知 StreamingContext：接收块中的元数据（Metadata）被发送到 Driver 的 StreamingContext。这个元数据包括：①定位其在 Executor 内存或磁盘中数据位置的块编号；②块数据在日志中的偏移信息（如果启用了预写日志）。

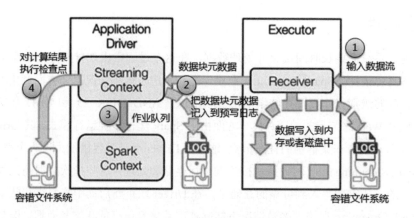

图 8-15　用户传送数据的生命周期

（3）处理数据：每批数据的间隔，流上下文使用块信息产生弹性分布数据集 RDD 和它们的作业(Job)，StreamingContext 通过运行任务处理 Executor 内存或磁盘中的数据块来执行作业。

（4）周期性地设置检查点：为了恢复的需要，流计算（换句话说，即 StreamingContext 提供的 DStreams）周期性地设置检查点，并保存到同一个容错文件系统中另外的一组文件中。

当一个失败的 Driver 端重启时，会进行如下处理，如图 8-16 所示。

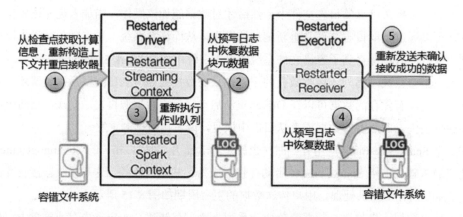

图 8-16　Driver 端异常容错处理

（1）恢复计算：使用检查点信息重启 Driver 端，重新构造上下文并重启接收器。

（2）恢复元数据块：为了保证能够继续下去所必备的全部元数据块都被恢复。

（3）未完成作业的重新形成：由于失败而没有处理完成的批处理，将使用恢复的元数据再次产生 RDD 和对应的作业。

（4）读取保存在日志中的块数据：在这些作业执行时，块数据直接从预写日志中读出，这将恢复在日志中可靠地保存所有必要的数据。

（5）重发尚未确认的数据：失败时没有保存到日志中的缓存数据将由数据源再次发送。

因此通过预写日志和可靠的接收器，Spark Streaming 就可以保证没有输入数据会由于 Driver 端的失败（或换言之，任何失败）而丢失。但需要注意的是，在启用了预写日志以后，数据接收吞吐率会有轻微的降低，因为所有数据都被写入容错文件系统，文件系统的写入吞吐率和用于数据复制的网络带宽，可能就是潜在的瓶颈。

8.4 Spark Streaming 运行原理

8.4.1 启动流处理引擎

在运行架构分析中，我们了解到在启动流处理引擎阶段需要初始化 StreamingContext 中的 DStreamGraph 和 JobScheduler，进而启动 JobScheduler 中的 ReceiverTracker 和 JobGenerator，其启动时序图如图 8-17 所示。

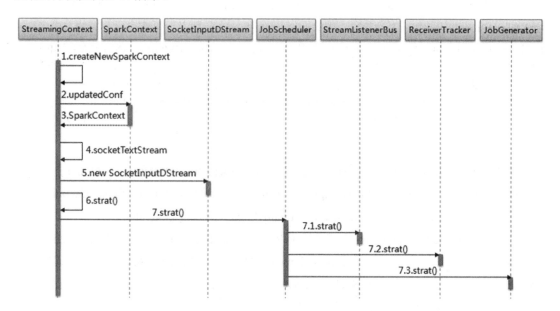

图 8-17 StreamingContext 启动时序图

1．初始化 StreamingContext

对于一个 Spark Streaming 应用程序，首先要做的事情就是初始化 StreamingContext，对

Streaming 初始化的可以有多个重载方法,在这里使用了 SparkConf 的对象和 Duration(批处理的时间间隔)作为参数传入 StreamingContext 的构造函数中,需要注意的是,在这里 Checkpoint 默认被设置为 null,代码如下:

```
def this(conf: SparkConf, batchDuration: Duration) = {
  this(StreamingContext.createNewSparkContext(conf), null, batchDuration)
}
```

结合 8.1 节中演示的例子,SparkConf 初始化中设置了运行模式 Local[2],即以本地模式的两个线程进行运行,应用程序的名称为 NetworkWordCount。

与 SparkContext 类似,StreamingContext 在初始化的过程中也会进行成员变量的初始化,在这些成员变量中重要的有 DStreamGraph、JobScheduler 和 StreamingTab 等。其中 DStreamGraph 跟 RDD 的有向无环图类似,包含 DStream 之间相互依赖的有向无环图;JobScheduler 的作用是定时查看 DStreamGraph,然后根据流入的数据生成运行作业;StreamingTab 是在 Spark Streaming 的作业运行的时候,提供对流数据处理的监控。

2. 创建 InputDStream

在例子中接着调用 StreamingContext 的 socketTextStream 方法生成具体的 InputDStream。在 socketTextStream 方法中有 3 个参数,其中 hostname 和 port 分别表示要连接服务端的主机名和端口号,而 storageLevel 是数据的存储等级,默认值是 StorageLevel.MEMORY_AND_DISK_SER_2。

```
def socketTextStream(hostname: String,port: Int,
    storageLevel: StorageLevel = StorageLevel.MEMORY_AND_DISK_SER_2
  ): ReceiverInputDStream[String] = withNamedScope("socket text stream") {
  socketStream[String](hostname, port, SocketReceiver.bytesToLines,
    storageLevel)
}
```

继续跟踪 socketStream 方法,在其中创建了一个 SocketInputDStream。通过 8.3.3 节中的 DStream 类图可以看到 SocketInputDStream 类继承自 ReceiverInputDStream,而 ReceiverInputDStream 继承自 InputDStream,InputDStream 继承自 DStream。

另外,在 SocketInputDStream 内部重写了 ReceiverInputDStream 中的 getReceiver 方法。该方法是用来生成接收器的,当 Spark Streaming 进行分发流接收器 Receiver 进行调用。在 getReceiver 方法内部创建了一个 SocketReceiver 实例,并在该实例中启动线程接收数据。其中 SocketInputDStream 代码如下:

```
class SocketInputDStream[T: ClassTag]( @transient ssc_ : StreamingContext,host:
    String,port: Int,bytesToObjects: InputStream => Iterator[T],storageLevel:
    StorageLevel ) extends ReceiverInputDStream[T](ssc_) {
```

```
def getReceiver(): Receiver[T] = {
  //创建 SocketReceiver 实例，在该实例中启动线程接收 Socket 数据
  new SocketReceiver(host, port, bytesToObjects, storageLevel)
  }
}
```

通过以上步骤创建了一个 InputDStream 对象，接着就是对 InputDStream 进行 flatMap、map、reduceByKey 和 print 等连续操作，类似于 RDD 的转换操作。

3．启动 JobScheduler

创建完 InputDStream 后，调用 StreamingConntext 的 start 方法进行 Spark Streaming 应用程序的启动，其最重要的就是启动 JobScheduler。在 JobScheduler 启动过程中，实例化并启动 ReceiverTracker 和 JobGenerator。其中 JobScheduler 启动代码如下：

```
def start(): Unit = synchronized {
  //如果 JobScheduler 已经启动，则退出
  if (eventLoop != null) return

  logDebug("Starting JobScheduler")
  eventLoop = new EventLoop[JobSchedulerEvent]("JobScheduler") {
    override protected def onReceive(event: JobSchedulerEvent): Unit =
      processEvent(event)
    override protected def onError(e: Throwable): Unit = reportError("Error in
      job scheduler", e)
  }
  eventLoop.start()

  for {
   inputDStream <- ssc.graph.getInputStreams
   rateController <- inputDStream.rateController
  } ssc.addStreamingListener(rateController)

  //启动 Spark Streaming 的消息总线
  listenerBus.start(ssc.sparkContext)
  receiverTracker = new ReceiverTracker(ssc)
  inputInfoTracker = new InputInfoTracker(ssc)

  //启动 ReceiverTracker 和 JobGenerator
  receiverTracker.start()
  jobGenerator.start()
  logInfo("Started JobScheduler")
}
```

4. 启动 JobGenerator

启动 JobGenerator 需要判断是否第一次运行，如果不是第一次运行需要进行上次检查点的恢复，如果是第一次运行，则调用了 JobGenerator 的 startFirstTime 方法。在该方法中初始化了定时器的开启时间，并启动 DStreamGraph 和定时器 timer，其代码如下：

```
private def startFirstTime() {
  val startTime = new Time(timer.getStartTime())
  graph.start(startTime - graph.batchDuration)
  timer.start(startTime.milliseconds)
}
```

其中 timer.getStartTime 方法，它会计算出来下一个周期的到期时间，计算公式：以当前的时间/除以间隔时间，再用 math.floor 求出它的上一个整数（即上一个周期的到期时间点），加上 1，再乘以周期就等于下一个周期的到期时间，其代码如下：

```
def getStartTime(): Long = {
  (math.floor(clock.getTimeMillis().toDouble / period) + 1).toLong * period
}
```

对于 DStreamGraph 启动时间为：启动时间 = startTime - graph.batchDuration，这里可以看出它的启动时间比定时器要早一个时间间隔。而对于它的 start 方法，主要是对向它注册过的 ForEachDStream 进行操作。

8.4.2　接收及存储流数据

1. 启动 ReceiverTracker

启动 ReceiverTracker 先调用 ReceiverTracker.launchReceivers 方法。该方法会向 ReceiverTrackerEndPoint 终端点发送分发，并启动所有流数据接收器的消息（参见 8.3.2 中的第一个消息分析），在 ReceiverTracker.startReceiver 的方法中进行所有流数据接收器 Receiver 的分发并启动（参见 8.3.3 节中 Receiver 的分发内容），其中 launchReceivers 代码如下：

```
private def launchReceivers(): Unit = {
  //获取所有 InputDStream 中定义的流数据接收器
  val receivers = receiverInputStreams.map(nis => {
    val rcvr = nis.getReceiver()
    rcvr.setReceiverId(nis.id)
    rcvr
  })

  runDummySparkJob()
  logInfo("Starting " + receivers.length + " receivers")
  //向终端点 ReceiverTrackerEndPoint 发送分发并启动所有流数据接收器的消息
```

```
      endpoint.send(StartAllReceivers(receivers))
}
```

ReceiverTracker 启动过程时序图调用如图 8-18 所示。

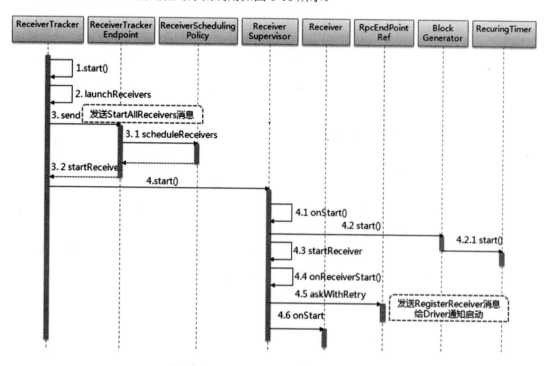

图 8-18 ReceiverTracker 启动过程时序图

分发到 Executor 的流数据接收器由 ReceiverTrackerEndpoint.startReceiver 方法进行启动，在该方法中开启了一个 ReceiverSupervisor 对象来管理该流数据接收器，其代码如下：

```
  private def startReceiver(receiver: Receiver[_], scheduledExecutors:
    Seq[String]): Unit = {
  ......
    val startReceiverFunc: Iterator[Receiver[_]] => Unit =
      (iterator: Iterator[Receiver[_]]) => {
        if (!iterator.hasNext) {
          throw new SparkException("Could not start receiver as object not found.")
        }
        if (TaskContext.get().attemptNumber() == 0) {
          val receiver = iterator.next()
          //创建流数据接收器管理器，用于监管该流数据接收器
          val supervisor = new ReceiverSupervisorImpl(
            receiver, SparkEnv.get, serializableHadoopConf.value, checkpointDirOption)
```

```
        supervisor.start()
        supervisor.awaitTermination()
      } else {}
    }
    ……
}
```

在 ReceiverSupervisor 的 start 方法中调用了其自身的 onStart 和 startReceiver 两个方法，在 onStart 方法中启动了 BlockGenerator，而在 startReceiver 方法中完成了流数据接收器的注册和启动。

2．启动流数据接收器并接收数据

在 ReceiverSupervisor.startReceiver 方法进行流数据接收器的启动，先调用 ReceiverSupervisorImpl 的 onReceiverStart 方法向 ReceiverTrackerEndPoint 终端点发送 RegisterReceiver 注册消息（参见 8.3.2 中发送注册流数据接收器消息内容），如果注册成功了，则调用 SocketReceiver.onStart 方法开始接收数据。

在 SocketReceiver.onStart 中接收数据调用了 receive 方法，该方法中通过监听 Socket 端口传递过来的流数据，并把接收到的数据转化为对象进行保存。

```
def receive() {
  var socket: Socket = null
  try {
    //根据机器名和端口建立 Socket 连接
    socket = new Socket(host, port)
    //把接收到的数据转化为对象进行保存
    val iterator = bytesToObjects(socket.getInputStream())
    while(!isStopped && iterator.hasNext) {
      store(iterator.next)
    }
    ……
  } finally {
    //关闭 Socket 连接
    if (socket != null) {
      socket.close()
    }
  }
}
```

3．启动 BlockGenerator 并生成数据块

在 ReceiverSupervisorImpl.onStart 方法中调用 BlockGenerator.start 启动 BlockGenerator。而 BlockGenerator.start 方法主要完成两件事情：第一是启动一个数据块生成定时器，将当前

currentBuffer 缓存中的数据按照用户在 Spark Streaming 应用程序里定义的批处理时间间隔封装成一个 Block 数据块，然后存放到 BlockGenerator 的 blocksForPushing 队列中；第二是启动一个 blockPushingThread 线程，不断地将 BlockForPush 队列中的数据块传递给 BlockManager。

```
private val blockIntervalTimer = new RecurringTimer(clock, blockIntervalMs,
    updateCurrentBuffer, "BlockGenerator")
private val blockQueueSize = conf.getInt("spark.streaming.blockQueueSize", 10)
private val blocksForPushing = new ArrayBlockingQueue[Block](blockQueueSize)
private val blockPushingThread = new Thread() { override def run()
    { keepPushingBlocks() } }

def start(): Unit = synchronized {
  if (state == Initialized) {
    state = Active
    //开启一个定时器，定期地把缓存中的数据封装成数据块
    blockIntervalTimer.start()
    //开始一个线程，不断地将封装好的数据推送给 BlockManager
    blockPushingThread.start()
    logInfo("Started BlockGenerator")
  } else {
    throw new SparkException(
     s"Cannot start BlockGenerator as its not in the Initialized state [state =
        $state]")
  }
}
```

跟进 blockIntervalTimer.start 方法，看一下定时器的执行流程为 start(getStartTime())，而 start(getStartTime()) 方法中开启了一个线程来执行 RecurringTimer.loop 方法，loop 方法代码如下：

```
private def loop() {
  try {
    while (!stopped) {
      triggerActionForNextInterval()
    }
    triggerActionForNextInterval()
  } catch {
    case e: InterruptedException =>
  }
}

private def triggerActionForNextInterval(): Unit = {
  clock.waitTillTime(nextTime)
  //定时执行回调函数，该函数为 updateCurrentBuffer
  callback(nextTime)
```

```
        prevTime = nextTime
        nextTime += period
        logDebug("Callback for " + name + " called at time " + prevTime)
      }
```

在 loop 方法中是数据块生成定时器真正处理的工作。该工作内容是定时地执行回调函数，其为初始化数据块生成定时器传入的 updateCurrentBuffer 参数。在该回调函数中，先把内存中的数据 currentBuffer 赋值给 newBlockBuffer，然后把 newBlockBuffer 封装成一个数据块，最后把这个数据块放进 blockForPushing 队列中，该队列默认情况下数目为 10 个。其中 BlockGenerator.updateCurrentBuffer 代码如下所示：

```
  private val blockQueueSize = conf.getInt("spark.streaming.blockQueueSize", 10)
  private val blocksForPushing = new ArrayBlockingQueue[Block](blockQueueSize)
  private val blockPushingThread = new Thread() { override def run()
    { keepPushingBlocks() } }
  private def updateCurrentBuffer(time: Long): Unit = {
    try {
      var newBlock: Block = null
      synchronized {
        if (currentBuffer.nonEmpty) {
          val newBlockBuffer = currentBuffer
          currentBuffer = new ArrayBuffer[Any]
          val blockId = StreamBlockId(receiverId, time - blockIntervalMs)
          listener.onGenerateBlock(blockId)
          newBlock = new Block(blockId, newBlockBuffer)
        }
      }

      if (newBlock != null) {
        blocksForPushing.put(newBlock)
      }
    } catch { …… }
  }
```

这里需要非常注意的就是，在加入 currentBuffer 数组时会先由 RateLimiter 检查一下速率，是否加入的频率已经太高。如果太高的话，就需要阻塞等到下一秒再开始添加。这里的最高频率是由 spark.streaming.receiver.maxRate(default=Long.MaxValue)控制的，是单个流数据接收器每秒钟允许添加的条数。控制了这个速率，就控制了整个 SparkStreaming 系统每个批处理需要处理的最大数据量。Spark1.5 版本开始 Spark Streaming 加入了分别动态控制每个流数据接收器速率的特性。

BlockGenerator 生成的数据块使用 keepPushingBlocks 方法传递给 BlockManager，该操作由

pushBlock 方法进行具体实现，而在 pushBlock 中继续调用 listener.onPushBlock，其中 listener 是在 ReceiversupervisorImpl 中初始化 BlockGenerator 时传给它的参数 BlockGeneratorListener 对象。在 BlockGeneratorListener.onPushBlock 方法里会继续调用 ReceiversupervisorImpl.pushArrayBuffer 方法，在 pushArrayBuffer 方法中则调用该类下的 pushAndReportBlock 方法发送 AddBlock 消息给 ReceiverTracker，进行数据块的保存。其中 keepPushingBlocks 代码如下：

```
private def keepPushingBlocks() {
  def areBlocksBeingGenerated: Boolean = synchronized {
    state != StoppedGeneratingBlocks
  }

  try {
    //不断地从blocksForPushing队列中取出数据块，时间间隔为10ms
    while (areBlocksBeingGenerated) {
      Option(blocksForPushing.poll(10, TimeUnit.MILLISECONDS)) match {
        case Some(block) => pushBlock(block)
        case None =>
      }
    }

    //在退出之前会判断blocksForPushing是否为空，如果不为空，则会在退出之前把剩下的
    //数据块一起输出
    while (!blocksForPushing.isEmpty) {
      val block = blocksForPushing.take()
      pushBlock(block)
    }
  } catch {……}
}
```

4．数据存储

作为流数据接收器调用 Receiver.store 方式进行数据存储，该方法有多个重载方法，如果数据量很小，则攒多条数据成数据块再进行块存储；如果数据量大，则直接进行块存储。对于需要攒成数据块的操作，在 Receiver.store 方法中调用 ReceiverSupervisorImpl 的 pushSingle 方法进行处理。在 pushSingle 方法中，通过调用 BlockGenerator.addData 把数据保存到内存 currentBuffer 中，这些内存的数据被数据块生成定时器定时封装加入队列并调用 ReceiverSupervisor 的 pushArrayBuffer 方法进行处理。而对于直接存储的操作，则调用 ReceiverSupervisor 的 pushIterator 或 pushBytes 方法进行处理。

这两种情况下均是调用 pushAndReportBlock 方法进行数据存储，该方法一方面会调用

ReceivedBlockHandler 的 storeBlock 方法保存数据并根据配置进行预写日志，另一方面会处理好的数据块的元信息发送给 ReceiverTrackerEndPoint 终端点（发送数据块的元信息内容参见 8.3.2 节），ReceiverTracker 再把这些信息转发给 ReceivedBlockTracker，由它负责管理收到数据块的元信息。

其中 ReceivedBlockHandler 根据 SparkEnv 中配置项的实际值来决定，如果设置了预写日志，则实例化 WriteAheadLogBasedBlockHandler；如果未设置预写日志，则实例化 BlockManagerBasedBlockHandler，其在 ReceiverSupervisorImpl 类中代码如下：

```
private val receivedBlockHandler: ReceivedBlockHandler = {
  if (WriteAheadLogUtils.enableReceiverLog(env.conf)) {
    if (checkpointDirOption.isEmpty) {
      throw new SparkException("……")
    }
    new WriteAheadLogBasedBlockHandler(env.blockManager, receiver.streamId,
      receiver.storageLevel, env.conf, hadoopConf, checkpointDirOption.get)
  } else {
    new BlockManagerBasedBlockHandler(env.blockManager, receiver.storageLevel)
  }
}
```

其中在 WriteAheadLogBasedBlockHandler.storeBlock 方法中，先使用 BlockManager 的 putBytes 方法，把数据保存到 Executor 的内存或磁盘中，然后把数据写入到预写日志文件中，最后当两个写入过程完成后返回写入数据块的元数据中，代码如下：

```
def storeBlock(blockId: StreamBlockId, block: ReceivedBlock):
  ReceivedBlockStoreResult = {
  ……
  //使用 BlockManager,把数据保存在内存或者磁盘中
  val storeInBlockManagerFuture = Future {
    val putSucceeded = blockManager.putBytes(blockId,
      serializedBlock, effectiveStorageLevel, tellMaster = true)
    if (!putSucceeded) {throw new SparkException(
      s"Could not store $blockId to block manager with storage level $storageLevel")
    }
  }

  //调用 WriteAheadLogRecordHandle 中的 write 方法在预写日志文件中写入数据
  val storeInWriteAheadLogFuture = Future {
    writeAheadLog.write(serializedBlock, clock.getTimeMillis())
  }

  //等待写入日志和数据完成，完成后返回存储块数据的元数据
```

```
    val combinedFuture = storeInBlockManagerFuture.zip
      (storeInWriteAheadLogFuture).map(_._2)
    val walRecordHandle = Await.result(combinedFuture, blockStoreTimeout)
    WriteAheadLogBasedStoreResult(blockId, numRecords, walRecordHandle)
  }
```

数据存储类时序图如图 8-19 所示。

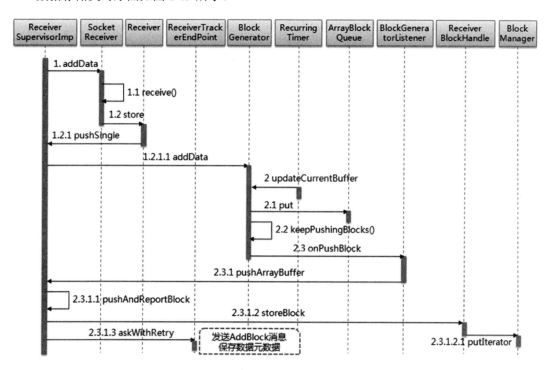

图 8-19 数据存储类时序图

8.4.3 数据处理

前面介绍了数据的接收和保存，这些数据必须经过处理才有意义，那么已经存储的数据真正处理是什么触发的呢？在这里我们把视线转移到一开始提供的 Spark Streaming 的官方示例代码中，我们知道在生成 SocketInputStream 对象后，它会进行 flatMap、map、reduceByKey 和 print 等一系列操作，在这里看一下 DStream 的 print() 方法。

```
  def print(num: Int): Unit = ssc.withScope {
    def foreachFunc: (RDD[T], Time) => Unit = {
      (rdd: RDD[T], time: Time) => {
        val firstNum = rdd.take(num + 1)
```

```
    println("-------------------------------------------")
    println("Time: " + time)
    println("-------------------------------------------")
    firstNum.take(num).foreach(println)
    if (firstNum.length > num) println("……")
    println()
  }
}
new ForEachDStream(this, context.sparkContext.clean(foreachFunc)).register()
}
```

在这里 print 方法是 DStreamGraph 的最后一个操作,很像 RDD 的行动操作。在这个方法中生成了一个 ForEachDStream,并定义了一个作用于它的函数 foreachFunc,同时在最后还调用了 ForEachDStream 的 register 方法向 DStreamGraph 注册。在 ForEachDStream 的 register 方法里(这里需要在它的父类 DStream 中查找),调用了 DStreamGraph 的 addOutputStream 方法,可以看到新初始化的 ForEachDStream 已经被添加到 DStreamGraph 的一个 ArrayBuffer[DStream[_]]中。

在前面我们介绍在启动 JobGenerator 中启动了作业生成器的定时器 timer,其定义如下:

```
val clock = {
  val clockClass = ssc.sc.conf.get("spark.streaming.clock",
    "org.apache.spark.util.SystemClock")
  try {
    Utils.classForName(clockClass).newInstance().asInstanceOf[Clock]
  } catch { …… }
}

private val timer = new RecurringTimer(clock, ssc.graph.batchDuration.
    milliseconds,
    longTime => eventLoop.post(GenerateJobs(new Time(longTime))),
      "JobGenerator")
```

在定时器 timer 中会定时地调用 processEvent 方法,对于接收到的 GenerateJobs 消息会调用 **JobGenerator.generateJobs** 方法继续处理。

```
    private def generateJobs(time: Time) {
      SparkEnv.set(ssc.env)
      ssc.sparkContext.setLocalProperty(RDD.CHECKPOINT_ALL_MARKED_ANCESTORS,
        "true")
      Try {
        jobScheduler.receiverTracker.allocateBlocksToBatch(time)
        graph.generateJobs(time)
      } match {
        case Success(jobs) =>
          val streamIdToInputInfos = jobScheduler.inputInfoTracker.getInfo(time)
```

```
      jobScheduler.submitJobSet(JobSet(time, jobs, streamIdToInputInfos))
    case Failure(e) =>jobScheduler.reportError("Error generating jobs for time
      " + time, e)
  }
  eventLoop.post(DoCheckpoint(time, clearCheckpointDataLater = false))
}
```

在 JobGenerator 的 generateJobs 方法中，主要进行了 5 个步骤：

（1）要求 ReceiverTracker 将目前已收到的数据进行一次提交，即将上次批处理切分后未处理的数据切分到到本次批处理中。RecevierTracker 的 allocateBlocksToBatch 方法继续调用 ReceivedBlockTracker 的 allocateBlocksToBatch 方法对本次批处理进行分配数据块，处理之前把批处理时间和对应的数据块的元信息记录到预写日志文件中，具体代码如下：

```
def allocateBlocksToBatch(batchTime: Time): Unit = synchronized {
  if (lastAllocatedBatchTime == null || batchTime > lastAllocatedBatchTime) {
    val streamIdToBlocks = streamIds.map { streamId =>
        (streamId, getReceivedBlockQueue(streamId).dequeueAll(x => true))
    }.toMap
    val allocatedBlocks = AllocatedBlocks(streamIdToBlocks)
    writeToLog(BatchAllocationEvent(batchTime, allocatedBlocks))
    timeToAllocatedBlocks(batchTime) = allocatedBlocks
    lastAllocatedBatchTime = batchTime
    allocatedBlocks
  } else {
    logInfo(s"Possibly processed batch $batchTime need to be processed again in
      WAL recovery")
  }
}
```

（2）调用 DStreamGraph 的 generateJobs 方法生成作业。在这个方法内部会通过 outputStream 的 generateJob 方法生成作业。

```
def generateJobs(time: Time): Seq[Job] = {
  val jobs = this.synchronized {
    outputStreams.flatMap(outputStream => outputStream.generateJob(time))
  }
  jobs
}
```

这里的 outputStream 是我们前面提到的 ForEachDStream，ForEachDStream 的 generateJob 方法会调用 DStream 的 getOrComputer 方法生成 RDD，然后再定义一个作用于作业的 jobFunc，最后会初始化一个作业，并把定义好的 jobFunc 函数作为参数传给作业。

```
private[streaming] def generateJob(time: Time): Option[Job] = {
  getOrCompute(time) match {
```

```
    case Some(rdd) => {
      val jobFunc = () => {
        val emptyFunc = { (iterator: Iterator[T]) => {} }
        context.sparkContext.runJob(rdd, emptyFunc)
      }
      Some(new Job(time, jobFunc))
    }
    case None => None
  }
}
```

继续看 DStream 的 getOrComputer 方法看它如何生成 RDD：

```
private[streaming] final def getOrCompute(time: Time): Option[RDD[T]] = {
  generatedRDDs.get(time).orElse {
    if (isTimeValid(time)) {
      //每个 DStream 在自身实现的 compute 中会进行重写
      val rddOption = createRDDWithLocalProperties(time) {
        PairRDDFunctions.disableOutputSpecValidation.withValue(true) {
          compute(time)
        }
      }

      rddOption.foreach { case newRDD =>
        if (storageLevel != StorageLevel.NONE) {
          newRDD.persist(storageLevel)
        }
        if (checkpointDuration != null && (time - zeroTime).isMultipleOf
          (checkpointDuration)) {
          newRDD.checkpoint()
        }
        //把生成的 RDD 保存到 generatedRDDs 哈希 Map 中，便于下次使用
        generatedRDDs.put(time, newRDD)
      }
      rddOption
    } else {
      None
    }
  }
}
```

（3）获取接收到的 Block 信息，通过调用 InputInfoTracker 的 getInfo 方法把 Receivertracker 中接收到的数据块元数据，保存到 batchTimeToInputInfos 中这个哈希 Map 中，然后作为参数传给 JobSet。

```
def getInfo(batchTime: Time): Map[Int, StreamInputInfo] = synchronized {
```

```
val inputInfos = batchTimeToInputInfos.get(batchTime)
// Convert mutable HashMap to immutable Map for the caller
inputInfos.map(_.toMap).getOrElse(Map[Int, StreamInputInfo]())
}
```

（4）调用 JobScheduler 的 submitJobSet 方法提交作业，在这个方法里，最重要的一行代码是 jobSet.jobs.foreach(job => jobExecutor.execute(new JobHandler(job)))，它会遍历 jobSet 里所有的作业，然后通过 jobExecutor 这个线程池把所有的作业进行提交。JobExecutor 作为 JobScheduler 的成员变量进行初始化，可以看到它是一个线程池。JobHandler 类主要做两件事情：第一是在作业运行前后分别发送 JobStarted 消息和 JobCompleted 消息给 JobScheduler；第二是进行作业的运行。

```
private val numConcurrentJobs = ssc.conf.getInt("spark.streaming.concurrentJobs", 1)
private val jobExecutor =ThreadUtils.newDaemonFixedThreadPool(
  numConcurrentJobs, "streaming-job-executor")
def submitJobSet(jobSet: JobSet) {
  if (jobSet.jobs.isEmpty) {
    logInfo("No jobs added for time " + jobSet.time)
  } else {
    listenerBus.post(StreamingListenerBatchSubmitted(jobSet.toBatchInfo))
    jobSets.put(jobSet.time, jobSet)
    jobSet.jobs.foreach(job => jobExecutor.execute(new JobHandler(job)))
    logInfo("Added jobs for time " + jobSet.time)
  }
}
```

（5）提交完作业后，发送一个 DoCheckpoint 消息给 JobGenerator，然后调用 JobGenerator 的 doCheckpoint 进行检查点操作。这里做检查点是异步的，不用等检查点真正写完成即可返回。

数据处理类时序图如图 8-20 所示。

通过以上步骤生成作业并提交运行，Spark 核心会对提交的作业进行处理，处理完毕后把处理按照要求进行输出。

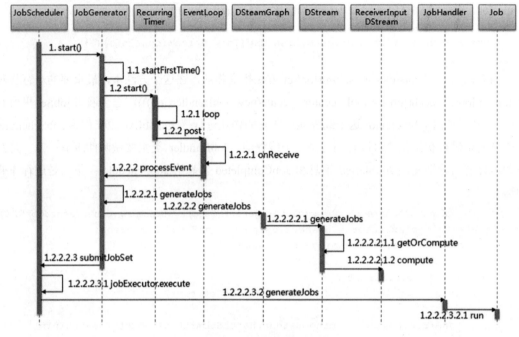

图 8-20　数据处理类时序图

8.5　实例演示

8.5.1　流数据模拟器

1．流数据说明

在实例演示中模拟实际情况，需要源源不断地接入流数据，为了在演示过程中更接近真实环境将定义流数据模拟器。该模拟器主要功能是通过 Socket 方式监听指定的端口号，当外部程序通过该端口连接并请求数据时，模拟器将定时将指定的文件数据随机获取发送给外部程序。

2．模拟器代码

模拟器的代码如下：

```
import java.io.{PrintWriter}
import java.net.ServerSocket
import scala.io.Source

object StreamingSimulation {
```

```scala
//定义随机获取整数的方法
def index(length: Int) = {
  import java.util.Random
  val rdm = new Random
  rdm.nextInt(length)
}

def main(args: Array[String]) {
  //调用该模拟器需要3个参数,分别为文件路径、端口号和间隔时间(单位:ms)
  if (args.length != 3) {
    System.err.println("Usage: <filename><port><millisecond>")
    System.exit(1)
  }

  //获取指定文件总的行数
  val filename = args(0)
  val lines = Source.fromFile(filename).getLines.toList
  val filerow = lines.length

  //指定监听某端口,当外部程序请求时建立连接
  val listener = new ServerSocket(args(1).toInt)
  while (true) {
    val socket = listener.accept()
    new Thread() {
      override def run = {
        println("Got client connected from: " + socket.getInetAddress)
        val out = new PrintWriter(socket.getOutputStream(), true)
        while (true) {
          Thread.sleep(args(2).toLong)
          //当该端口接受请求时,随机获取某行数据发送给对方
          val content = lines(index(filerow))
          println(content)
          out.write(content + '\n')
          out.flush()
        }
        socket.close()
      }
    }.start()
  }
}
```

3. 生成打包文件

在打包配置界面中，需要在 Class Path 加入：/app/soft/scala-2.11.8/lib/scala-swing.jar /app/soft/scala-2.11.8/lib/scala-library.jar /app/soft/scala-2.11.8/lib/scala-actors.jar，各个 jar 包之间用空格分开，如图 8-21 所示。

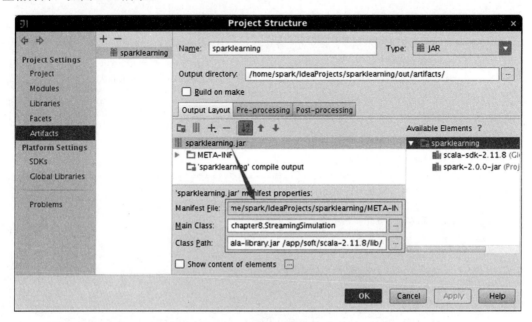

图 8-21　打包文件配置界面

单击菜单 Build→Build Artifacts，弹出选择动作，选择 Build 或者 Rebuild 动作，使用如下命令复制打包文件到 Spark 根目录下。

```
$cd /home/spark/IdeaProjects/sparklearning/out/artifacts/
$cp sparklearning.jar /app/spark/spark-2.0.0/
$ll /app/spark/spark-2.0.0/
```

8.5.2　销售数据统计实例

1. 演示说明

在该实例中将由 8.5.1 节流数据模拟器以 1s 的频度发送模拟数据（销售数据），Spark Streaming 通过 Socket 接收流数据并每 5s 运行一次用来处理接收到数据，处理完毕后打印该时间段内销售数据总和。需要注意的是，各处理段时间之间状态并无关系。

2. 演示代码

本实例的代码如下:

```scala
import org.apache.log4j.{Level, Logger}
import org.apache.spark.{SparkContext, SparkConf}
import org.apache.spark.streaming.{Milliseconds, Seconds, StreamingContext}
import org.apache.spark.streaming.StreamingContext._
import org.apache.spark.storage.StorageLevel

object SaleStatics {
  def main(args: Array[String]) {
if (args.length != 2) {
    System.err.println("Usage: SaleStatics<hostname><port>")
    System.exit(1)
    }
  Logger.getLogger("org.apache.spark").setLevel(Level.ERROR)
  Logger.getLogger("org.eclipse.jetty.server").setLevel(Level.OFF)

  val conf = new SparkConf().setAppName("SaleStatics").setMaster("local[2]")
  val sc = new SparkContext(conf)
  val ssc = new StreamingContext(sc, Seconds(5))

  //通过 Socket 获取数据,需要提供 Socket 的主机名和端口号,数据保存在内存和硬盘中
  val lines = ssc.socketTextStream(args(0), args(1).toInt,
      StorageLevel.MEMORY_AND_DISK_SER)
  val words = lines.map(_.split(",")).filter(_.length == 6)
  val wordCounts = words.map(x=>(1, x(5).toDouble)).reduceByKey(_ + _)

  wordCounts.print()
  ssc.start()
  ssc.awaitTermination()
  }
}
```

3. 运行代码

启动 8.5.1 节打包好的流数据模拟器,在该实例中将定时发送/home/spark/work/saledata 目录下的 tbStockDetail.txt 数据文件(参见附录 B 中对该数据描述,该文件可以在本系列配套资源目录/saledata 中找到),其中表 tbStockDetail 字段分别为订单号、行号、货品、数量、金额,数据内容如图 8-22 所示。

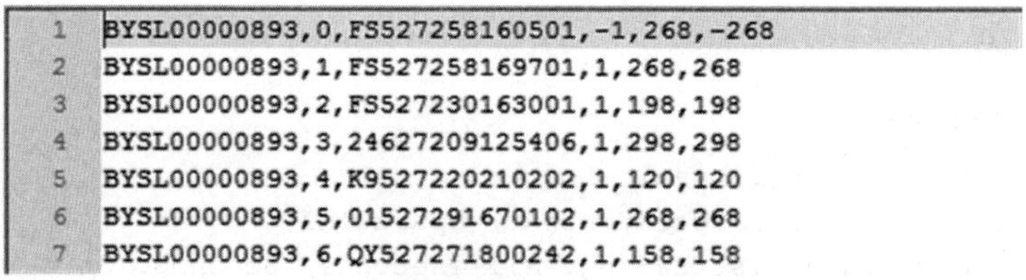

```
1  BYSL00000893,0,FS527258160501,-1,268,-268
2  BYSL00000893,1,FS527258169701,1,268,268
3  BYSL00000893,2,FS527230163001,1,198,198
4  BYSL00000893,3,24627209125406,1,298,298
5  BYSL00000893,4,K9527220210202,1,120,120
6  BYSL00000893,5,01527291670102,1,268,268
7  BYSL00000893,6,QY527271800242,1,158,158
```

图 8-22　数据文件中的数据内容

模拟器 Socket 端口号为 9999，频度为 1s，如图 8-23 所示。

```
$cd /app/spark/spark-2.0.0
$java -cp sparklearning.jar chapter8.StreamingSimulation /home/spark/work/
saledata/tbStockDetail.txt 9999 1000
```

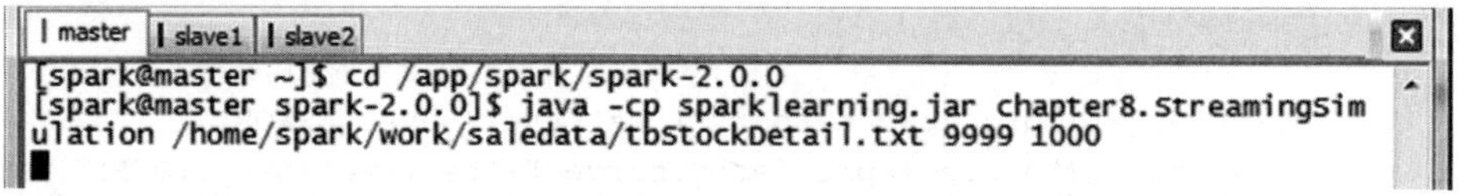

```
[spark@master ~]$ cd /app/spark/spark-2.0.0
[spark@master spark-2.0.0]$ java -cp sparklearning.jar chapter8.StreamingSim
ulation /home/spark/work/saledata/tbStockDetail.txt 9999 1000
```

图 8-23　启动流数据模拟器

在 IDEA 中运行该实例，该实例需要配置连接 Socket 主机名和端口号，在这里配置参数机器名为 master 和端口号为 9999

4．查看结果

（1）观察模拟器发送情况。

IDEA 中的 Spark Streaming 程序运行时与模拟器建立连接，当模拟器检测到外部连接时开始发送销售数据，时间间隔为 1s，如图 8-24 所示。

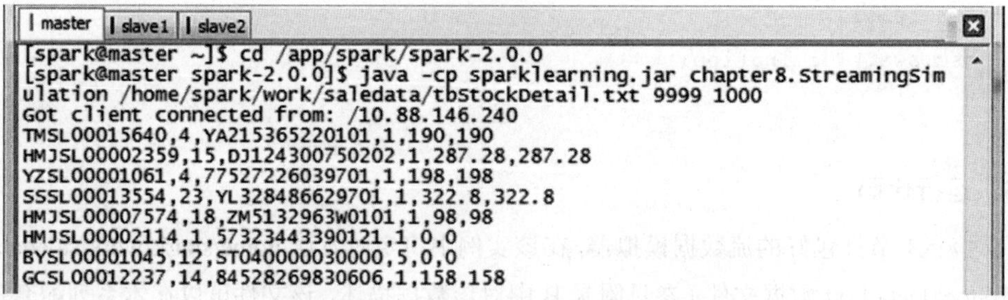

```
[spark@master ~]$ cd /app/spark/spark-2.0.0
[spark@master spark-2.0.0]$ java -cp sparklearning.jar chapter8.StreamingSim
ulation /home/spark/work/saledata/tbStockDetail.txt 9999 1000
Got client connected from: /10.88.146.240
TMSL00015640,4,YA215365220101,1,190,190
HMJSL00002359,15,DJ124300750202,1,287.28,287.28
YZSL00001061,4,775272226039701,1,198,198
SSSL00013554,23,YL328486629701,1,322.8,322.8
HMJSL00007574,18,ZM5132963W0101,1,98,98
HMJSL00002114,1,57323443390121,1,0,0
BYSL00001045,12,ST040000030000,5,0,0
GCSL00012237,14,84528269830606,1,158,158
```

图 8-24　流数据模拟器以 1s 间隔发送数据

（2）IDEA 运行情况。

在 IDEA 的运行窗口中，可以观察到每 5s 运行一次作业（两次运行间隔为 5000ms），运行完毕后打印该时间段内销售数据总和，如图 8-25 所示。

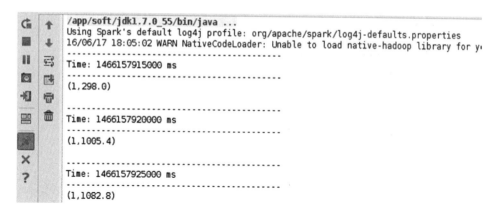

图 8-25　IDEA 统计销售结果

（3）在监控页面观察执行情况。

在 WebUI 上监控作业运行情况，可以观察到每 5s 运行一次作业，如图 8-26 所示。

Active Jobs (1)

Job Id	Description	Submitted	Duration	Stages: Succeeded/Total	Tasks (for all stages): Succeeded/Total
0	Streaming job running receiver 0 start at SaleAmount.scala:28	2016/04/21 14:58:54	2.3 min	0/1	0/1

Completed Jobs (54)

Job Id	Description	Submitted	Duration	Stages: Succeeded/Total	Tasks (for all stages): Succeeded/Total
54	Streaming job from [output operation 0, batch time 15:01:05] print at SaleAmount.scala:27	2016/04/21 15:01:05	31 ms	1/1 (1 skipped)	1/1 (5 skipped)
53	Streaming job from [output operation 0, batch time 15:01:05] print at SaleAmount.scala:27	2016/04/21 15:01:05	92 ms	2/2	6/6
52	Streaming job from [output operation 0, batch time 15:01:00] print at SaleAmount.scala:27	2016/04/21 15:01:00	27 ms	1/1 (1 skipped)	1/1 (5 skipped)
51	Streaming job from [output operation 0, batch time 15:01:00] print at SaleAmount.scala:27	2016/04/21 15:01:00	96 ms	2/2	6/6

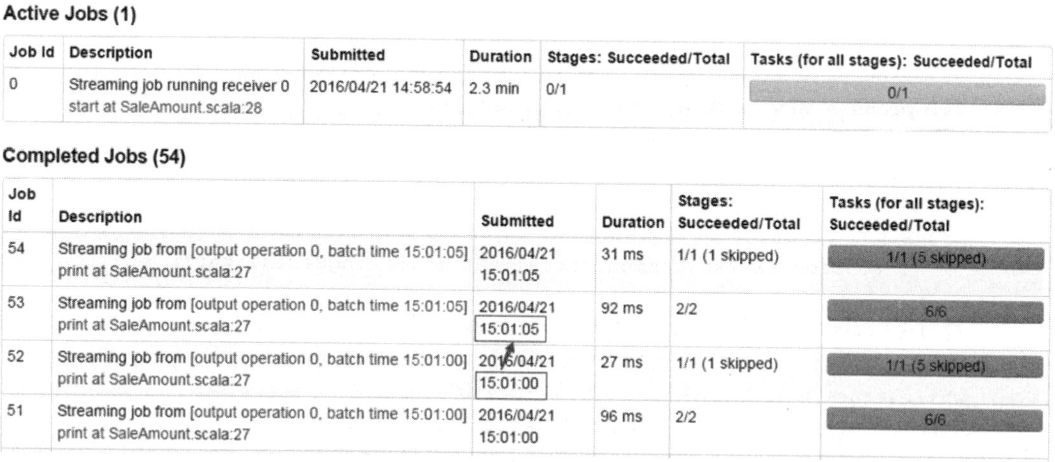

图 8-26　WebUI 上作业运行情况

8.5.3　Spark Streaming+Kafka 实例

1．演示说明

该实例演示 Spark Streaming 从 Kafka 获取数据并进行实时统计。该实例是典型的生产者—消费者模型，作为生产者 Kafka 以每秒发送 3 条消息，每个消息包含 5 个 0～9 随机数，而作为消费者，Spark Streaming 使用 reduceByKeyAndWindow()方法进行统计并显示结果，在该方法中需要指定窗口时间长度和滑动时间间隔。

2. 演示代码

（1）KafkaWordCountProducer.scala 的代码如下：

```scala
import java.util.HashMap
import org.apache.kafka.clients.producer.KafkaProducer
import org.apache.kafka.clients.producer.ProducerConfig
import org.apache.kafka.clients.producer.ProducerRecord

object KafkaWordCountProducer {
  def main(args: Array[String]) {
    //对输入参数进行验证，如果少于4个则退出
    if (args.length < 4) {
      System.err.println("Usage: KafkaWordCountProducer <metadataBrokerList> " +
        "<topic> " +
        "<messagesPerSec><wordsPerMessage>")
      System.exit(1)
    }
    val Array(brokers, topic, messagesPerSec, wordsPerMessage) = args

    //Zookeeper连接属性，通过第一个传入参数获取brokers地址信息
    val props = new HashMap[String, Object]()
    props.put(ProducerConfig.BOOTSTRAP_SERVERS_CONFIG, brokers)
    props.put(ProducerConfig.VALUE_SERIALIZER_CLASS_CONFIG,
      "org.apache.kafka.common.serialization.StringSerializer")
    props.put(ProducerConfig.KEY_SERIALIZER_CLASS_CONFIG,
      "org.apache.kafka.common.serialization.StringSerializer")

    //创建Kafka实例，该实例通过参数确定每秒发送消息的条数，及每个消息包含0~9随机数的个数
    val producer = new KafkaProducer[String, String](props)
    while(true) {
      (1 to messagesPerSec.toInt).foreach { messageNum =>
        val str = (1 to wordsPerMessage.toInt).map(x => scala.util.Random.
          nextInt(10).toString).mkString(" ")
        val message = new ProducerRecord[String, String](topic, null, str)
        //调用Kafka发送消息方法
        producer.send(message)
      }

      Thread.sleep(1000)
    }
  }
}
```

（2）KafkaWordCount.scala 的代码如下：

```
import org.apache.log4j.{Level, Logger}
import org.apache.spark.SparkConf
import org.apache.spark.streaming._
import org.apache.spark.streaming.kafka._

object KafkaWordCount  {
  def main(args: Array[String]) {
    //对输入参数进行验证，如果少于 4 个则退出
    if (args.length < 4) {
      System.err.println("Usage: KafkaWordCount <zkQuorum><group><topics>
        <numThreads>")
      System.exit(1)
    }
    Logger.getRootLogger.setLevel(Level.WARN)
    val Array(zkQuorum, group, topics, numThreads) = args

    //初始化 Spark Streaming 环境
    val sparkConf = new SparkConf().setAppName("KafkaWordCount")
    val ssc = new StreamingContext(sparkConf, Seconds(2))
    ssc.checkpoint("checkpoint")

    //通过 Zookeeper 连接属性，获取 Kafka 的组和主题信息，并创建连接获取数据流
    val topicMap = topics.split(",").map((_, numThreads.toInt)).toMap
    val lines = KafkaUtils.createStream(ssc, zkQuorum, group,
      topicMap).map(_._2)

    //对获取的数据流进行分词，使用 reduceByKeyAndWindow 进行统计并打印
    val words = lines.flatMap(_.split(" "))
    val wordCounts = words.map(x => (x, 1L)).reduceByKeyAndWindow(_ + _, _ - _,
      Minutes(10), Seconds(2), 2)
    wordCounts.print()

    ssc.start()
    ssc.awaitTermination()
  }
}
```

3. 运行代码

（1）打包代码

参考 2.4.2 节的内容对运行代码进行打包，由于在程序中使用了 Zookeeper 和 Kafka，所以需要在打包中加入其依赖包，这些依赖包除了 scala-swing.jar、scala-library.jar 和 scala-actors.jar，还需要如下 Jar 包，如图 8-27 所示。

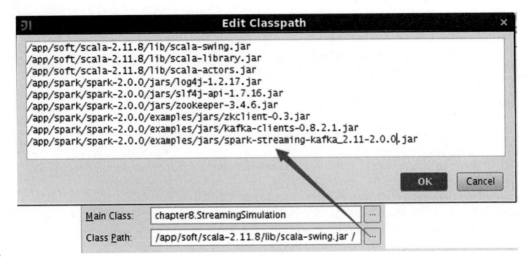

图 8-27　打包配置界面

```
/app/spark/spark-2.0.0/jars/log4j-1.2.17.jar
/app/spark/spark-2.0.0/jars/slf4j-api-1.7.16.jar
/app/spark/spark-2.0.0/jars/zookeeper-3.4.6.jar
/app/spark/spark-2.0.0/examples/jars/zkclient-0.3.jar
/app/spark/spark-2.0.0/examples/jars/kafka-clients-0.8.2.1.jar
/app/spark/spark-2.0.0/examples/jars/spark-streaming-kafka_2.11-2.0.0.jar
```

（2）启动数据生产者

参见附录 C 和附录 D 在各个节点中启动 Zookeeper 和 Kafka，然后使用如下命令启动打包好的数据生产者，该生产者 Kafka 以每秒发送 3 条消息，每个消息包含 5 个 0～9 随机数。

```
$cd /app/spark/spark-2.0.0
$java -cp sparklearning.jar chapter8.KafkaWordCountProducer master:9092 test 3 5
```

其中，参数 master:9092 表示 producer 的地址和端口，test 表示 Kafka 主题名，3 表示每秒发多少条消息，5 表示每条消息中包含几个单词，运行截图如图 8-28 所示。

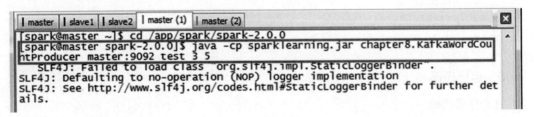

图 8-28　启动数据生产者

(3) 启动数据消费者

在启动数据消费者前需要启动 Spark 集群，然后使用 spark-submit 脚本提交程序到集群进行运行，其运行命令和截图如图 8-29 所示。

```
$cd /app/spark/spark-2.0.0
$./bin/spark-submit --class chapter8.KafkaWordCount \
  --master spark://master:7077 \
  --executor-memory 1024m \
  sparklearning.jar \
  master:2181 test-consumer-group test 1
```

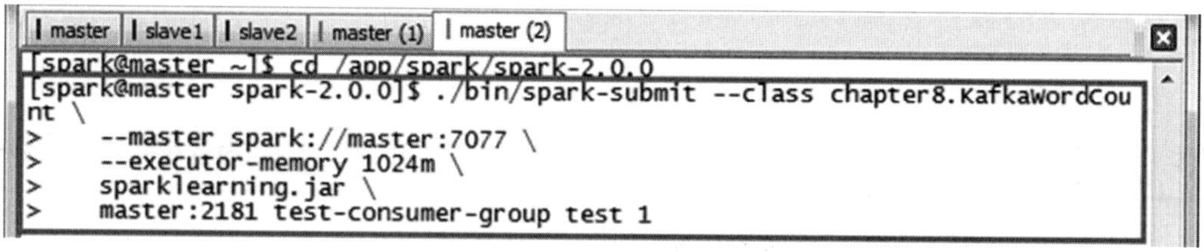

图 8-29　启动数据消费者

4. 查看结果

在运行窗口中，可以观察到每 2s 输出一次结果，结果为 10min 内 0～9 数字出现的次数列表，其结果如图 8-30 所示。

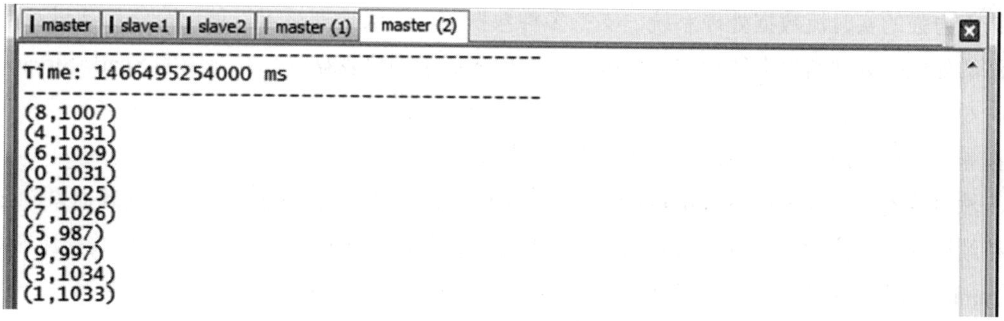

图 8-30　查看运行结果

在 WebUI 上监控作业运行情况，可以观察到每 2s 运行一次作业，另外也通过 Streaming 页面查看其数据接收的速率和每个批次处理的时间等信息，如图 8-31 所示。

Active Jobs (1)

Job Id	Description	Submitted	Duration	Stages: Succeeded/Total	Tasks (for all stages): Succeeded/Total
11	Streaming job running receiver 0 start at KafkaWordCount.scala:29	2016/06/21 15:38:27	1.6 min	0/1	0/1

Completed Jobs (115)

Job Id	Description	Submitted	Duration	Stages: Succeeded/Total	Tasks (for all stages): Succeeded/Total
115	Streaming job from [output operation 0, batch time 15:40:00] print at KafkaWordCount.scala:27	2016/06/21 15:40:00	75 ms	1/1 (1 skipped)	1/1 (2 skipped)
114	Streaming job from [output operation 0, batch time 15:40:00] print at KafkaWordCount.scala:27	2016/06/21 15:40:00	0.2 s	2/2	3/3
113	Streaming job from [output operation 0, batch time 15:39:58] print at KafkaWordCount.scala:27	2016/06/21 15:39:59	41 ms	1/1	1/1
112	Streaming job from [output operation 0, batch time 15:39:58] print at KafkaWordCount.scala:27	2016/06/21 15:39:58	0.4 s	1/1 (5 skipped)	2/2 (10 skipped)
111	Streaming job from [output operation 0, batch time 15:39:58] print at KafkaWordCount.scala:27	2016/06/21 15:39:58	0.3 s	2/2 (4 skipped)	3/3 (8 skipped)

图 8-31　查询 WebUI 作业运行情况

8.6　小结

在本章先介绍 Spark Streaming 基本情况，它是 Spark 核心 API 的一个扩展，具有吞吐量高、容错能力强的实时流数据处理系统，支持多种数据源，获取数据后可以使用高级函数进行复杂算法的处理，处理结果存储到文件系统、数据库或展示到仪表盘等。在介绍的 Spark Streaming 同时对比了流处理系统 Storm，它们之间在扩展性和可容错性模型不同、在实现方式和编程 API 存在差异、在生成支持和集群管理集成有所区别。

然后接着介绍 Spark Streaming 编程模型，DStream 作为 Spark Streaming 的基础抽象，它代表持续性的数据流。在 Spark Streaming 提供基础和高级两种数据流来源，对于这些数据流 Spark Streaming 提供了对 DStream 一系列操作方法，这些操作可以分成三类：普通的转换操作、窗口转换操作和输出操作。

接着介绍 Spark Streaming 的运行架构，Spark Streaming 分为 Driver 端和 Client 端，运行在 Driver 端为 StreamingContext 实例，该实例包括了 DStreamGraph 和 JobScheduler（包括了 ReceiverTracker 和 JobGenerator）等，而 Client 包括了 ReciverSupervisor 和 Reciver 等。Spark Streaming 进行流数据处理大致可以分为：启动流处理引擎、接收及存储流数据、处理流数据和输出处理结果等四个步骤。再接着详细介绍了 Spark Streaming 运行原理，着重介绍了 Spark Streaming 启动流处理引擎、接收及存储流数据、数据处理三个过程。

最后通过三个实例演示了 Spark Streaming 的应用，在前两个实例中构建了一个流数据模拟器，通过该模拟器演示真实环境的流数据。在销售数据统计实例中实时接收这些数据，以每 5 秒运行一次用来处理接收到数据，处理完毕后打印该时间段内销售数据总和；在 Stateful 实例中，同样每 5 秒运行一次用来处理接收到数据，区别前一个实例各时间段之间状态是相关的；最后一个实例演示 Spark Streaming 从 Kafka 获取数据并进行实时统计，使用 reduceByKeyAndWindow 方法进行统计并显示结果。

第 9 章

Spark MLlib

9.1 Spark MLlib 简介

9.1.1 Spark MLlib 介绍

MLlib 是 Spark 机器学习库，它的目标是让机器学习更加容易和可伸缩性。MLlib 是 MLBase 的一部分，其中 MLBase 分为 4 个部分，分别是 MLlib、MLI、ML Optimizer 和 MLRuntime，如图 9-1 所示。

- ML Optimizer 会选择它认为最适合的已经在内部实现好了的机器学习算法和相关参数，来处理用户输入的数据，并返回模型或其他的帮助分析的结果。
- MLI 是一个进行特征抽取和高级 ML 编程抽象的算法实现的 API 或平台。
- MLlib 是 Spark 实现一些常见的机器学习算法和实用程序。
- MLRuntime 基于 Spark 计算框架，将 Spark 的分布式计算应用到机器学习领域。

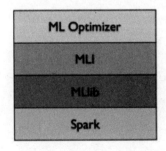

图 9-1 Spark MLlib 的架构分层

MLlib 提供了常用机器学习算法的实现，包括分类、回归、聚集、协同过滤和降维等。使用 MLlib 来做机器学习工作，可以说是非常简单，通常只需要在对原始数据进行处理后，然后直接调用相应的 API 就可以实现了。但是要想选择合适的算法，高效、准确地对数据进行分析，还需要深入了解一下算法原理，以及相应 MLlib API 实现的参数的意义。

需要提及的是，Spark 机器学习库从 1.2 版本以后被分为两个包。

- **spark.mllib**：Spark MLlib 历史比较长了，1.0 以前的版本中已经包含了这个包，提供的算法实现都是基于原始的 RDD，从学习角度上来讲，其实比较容易上手。如果读者已经有机器学习方面的经验，那么只需要熟悉一下 MLlib 的 API 就可以开始数据分析工作了。想要基于这个包提供的工具构建完整并且复杂的机器学习流水线是比较困难的。

- **spark.ml**：Spark ML Pipeline 从 Spark1.2 版本开始，目前已经从 Alpha 阶段毕业，成为可用并且较为稳定的新的机器学习库。ML Pipeline 弥补了原始 MLlib 库的不足，向用户提供了一个基于 DataFrame 的机器学习工作流式 API 套件，使用 ML Pipeline API，我们可以很方便地把数据处理、特征转换、正则化，以及多个机器学习算法联合起来，构建一个单一完整的机器学习流水线。显然，这种新的方式给我们提供了更灵活的方法，而且这也更符合机器学习过程的特点。

从官方文档来看，Spark ML Pipeline 虽然是被推荐的机器学习方式，但是并不会在短期内替代原始的 MLlib 库，因为 MLlib 已经包含了丰富稳定的算法实现，并且部分 ML Pipeline 实现基于 MLlib。而且就笔者看来，并不是所有的机器学习过程都需要被构建成一个流水线，有时候原始数据格式整齐且完整，而且使用单一的算法就能实现目标，我们就没有必要把事情复杂化，采用最简单且容易理解的方式才是正确的选择。

从目前来看，AMPLab 建议尽可能使用新引入的 spark.ml 包。因为使用 DataFrame 能够提供更加通用和灵活的接口，不过旧的 spark.mllib 包将继续更新开发。图 9-2 为 Spark MLlib 算法架构图。

从图 9-2 中可以看出，MLlib 主要包含两部分。

- **底层基础**：主要包括 Spark 的运行库、矩阵库和向量库，其中向量接口和矩阵接口基于 Netlib 和 BLAS/LAPACK 开发的线性代数库 Breeze。MLlib 支持本地的密集向量和稀疏向量，并且支持标量向量；它同时支持本地矩阵和分布式矩阵，支持的分布式矩阵分为 RowMatrix、IndexedRowMatrix 和 CoordinateMatrix 等。
- **算法库**：包含分类、回归、聚集、协同过滤、梯度下降和特征提取和变换等算法。

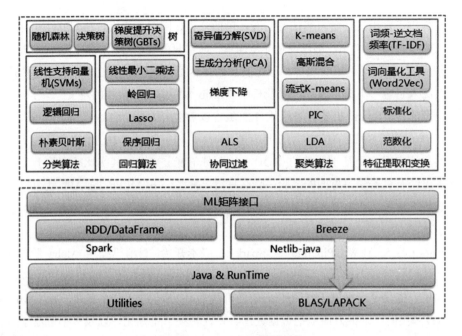

图 9-2　Spark MLlib 算法架构

表 9-1 为 Spark MLlib 算法的分类情况。

表 9-1　Spark MLlib 算法分类

分　类	具体算法
二元分类	线性支持向量机、逻辑回归、决策树、随机森林、梯度提升决策树、朴素贝叶斯
多类分类	逻辑回归、决策树、随机森林、朴素贝叶斯
回归	线性最小二乘法、Lasso、岭回归、决策树、随机森林、梯度提升决策树、保序回归

9.1.2　Spark MLlib 数据类型

1．本地向量（Local Vector）

本地向量是由整型索引和双精度浮点型数值组成的数据结构，本地向量的使用方便了 MLlib 对数据的描述和操作。其中本地向量分为密集向量和稀疏向量两种类型，密集向量通过一个浮点数组来表示，而稀疏向量通过索引和值两个并列的数组来表示。比如，向量(1.0, 0.0, 3.0)可以以密集格式[1.0, 0.0, 3.0]或稀疏格式(3, [0, 2], [1.0, 3.0])表示，后者第一个 3 表示向量长度，即列数。

关于密集型和稀疏型的向量的示例如下：

```
import org.apache.spark.mllib.linalg.{Vector, Vectors}

//创建密集向量 (1.0, 0.0, 3.0)
val dv: Vector = Vectors.dense(1.0, 0.0, 3.0)
//创建稀疏向量(1.0, 0.0, 3.0)，其中3表示向量长度，后面两个数组分别为对应索引和值数组
val sv1: Vector = Vectors.sparse(3, Array(0, 2), Array(1.0, 3.0))
//创建稀疏向量(1.0, 0.0, 3.0)，其中3表示向量长度，后面两个序列分别为具体索引和值组合
val sv2: Vector = Vectors.sparse(3, Seq((0, 1.0), (2, 3.0)))
```

稀疏向量在含有大量非零元素的计算中会节省大量的空间并提高计算速度，如下所示。

```
Training set:
                        dense   sparse
• 12 million examples
                storage  47GB    7GB
• 500 features
                time     240s    58s
• sparsity: 10%

40GB savings in storage, 4x speedup in computation
```

2．标签点（Labeled Point）

标签点由一个本地向量（密集或稀疏）和一个类标签组成。其中使用双精度浮点型来存储一个类标签，在二元分类中标签或为 0（负向）或为 1（正向），而在多元分类中标签应该是从 0 开始的索引，如 0，1，2，……

标签点在实际中大量使用，例如判断邮件是否为垃圾邮件时，可以把表示为 1.0 的判断为正常邮件，而表示为 0.0 则作为垃圾邮件来看待，类似于如下代码：

```
import org.apache.spark.mllib.linalg.Vectors
import org.apache.spark.mllib.regression.LabeledPoint

//使用正标记和密集向量来创建标记点
val pos = LabeledPoint(1.0, Vectors.dense(1.0, 0.0, 3.0))
//使用负标记和稀疏向量来创建标记点
val neg = LabeledPoint(0.0, Vectors.sparse(3, Array(0, 2), Array(1.0, 3.0)))
```

实际运用中，稀疏数据是很常见的，在 MLlib 可以读取以 libSVM 格式存储的训练实例。其中 libSVM 格式是 LIBSVM 和 LIBLINEAR 的默认格式，这是一种文本格式，每行代表一个含类标签的稀疏特征向量，格式如下：

```
label index1:value1 index2:value2 ……
```

索引是从 1 开始并且递增，加载完成后，索引被转换为从 0 开始，代码示例如下所示：

```
import org.apache.spark.mllib.regression.LabeledPoint
import org.apache.spark.mllib.util.MLUtils
import org.apache.spark.rdd.RDD
```

```
//调用 MLUtils.loadLibSVMFile 来读取稀疏的数据训练集
val examples: RDD[LabeledPoint] = MLUtils.loadLibSVMFile(sc, "data/mllib/
    sample_libsvm_data.txt")
```

3. 本地矩阵（Local Matrix）

MLlib 中的矩阵其实是向量型的 RDD，分为本地矩阵和分布式矩阵两种。本地矩阵由整型行列索引数据和对应的双精度浮点型值数据组成并存储。MLlib 支持密集矩阵，实体值以列优先的方式存储在一个双精度浮点型数组中。例如下面的矩阵，其存储方式是一个一维数组 [1.0, 3.0, 5.0, 2.0, 4.0, 6.0] 和矩阵大小为(3, 2)。

$$\begin{pmatrix} 1.0 & 2.0 \\ 3.0 & 4.0 \\ 5.0 & 6.0 \end{pmatrix}$$

本地矩阵的基类是 Matrix，提供了密集矩阵和稀疏矩阵两种对应实现方法，可以通过这些实现方法来创建本地矩阵，代码如下：

```
import org.apache.spark.mllib.linalg.{Matrix, Matrices}

//创建密集矩阵((1.0, 2.0), (3.0, 4.0), (5.0, 6.0))
val dm: Matrix = Matrices.dense(3, 2, Array(1.0, 3.0, 5.0, 2.0, 4.0, 6.0))

//创建稀疏矩阵((9.0, 0.0), (0.0, 8.0), (0.0, 6.0))
val sm: Matrix = Matrices.sparse(3,2,Array(0,1,3),Array(0,2,1),Array(9,6,8))
```

4. 分布式矩阵

分布式矩阵由长整型行列索引和双精度浮点型值数据组成，分布式存储在一个或多个 RDD 中，如下所示。对于巨大的分布式的矩阵来说，选择正确的存储格式非常重要，将一个分布式矩阵转换为另一个不同格式需要混洗（Shuffle），其代价很高。在 MLlib 实现了三类分布式矩阵存储格式，分别为行矩阵（RowMatrix）、行索引矩阵（IndexedRowMatrix）、三元组矩阵（CoordinateMatrix）和分块矩阵（BlockMatrix）等四种。

（1）行矩阵（RowMatrix）：行矩阵是一个面向行的分布式矩阵，行索引是没有具体含义的。比如一系列特征向量的一个集合，通过一个 RDD 来代表所有的行，每一行就是一个本地向量。行矩阵直接通过 RDD[Vector]来定义并可以用来统计平均数、方差、协同方差等，示例代码如下：

```
import org.apache.spark.mllib.linalg.Vector
import org.apache.spark.mllib.linalg.distributed.RowMatrix

//定义一个本地向量的 RDD
val rows: RDD[Vector] = ……
//使用 RDD[Vector]来创建行矩阵
val mat: RowMatrix = new RowMatrix(rows)

//分别获取行矩阵的行、列数
val m = mat.numRows()
val n = mat.numCols()
```

（2）行索引矩阵（IndexedRowMatrix）：行索引矩阵和行矩阵类似，但其行索引具有特定含义，本质上是一个含有索引信息的行数据集合，每一行由长整型索引和一个本地向量组成。行索引矩阵可从一个 RDD[IndexedRow]实例创建，这里的 IndexedRow 是(Long, Vector)的封装类，剔除行索引矩阵中的行索引信息就变成一个行矩阵，代码示例如下：

```
import org.apache.spark.mllib.linalg.distributed.{IndexedRow,
    IndexedRowMatrix, RowMatrix}

//定义一个含有索引信息的行数据集合的 RDD
val rows: RDD[IndexedRow] = ……
//由 RDD[IndexedRow]实例创建行索引矩阵
val mat: IndexedRowMatrix = new IndexedRowMatrix(rows)

//获取该行索引矩阵的行、列数
val m = mat.numRows()
val n = mat.numCols()

//通过提出行索引矩阵的行索引信息得到行矩阵
val rowMat: RowMatrix = mat.toRowMatrix()
```

（3）三元组矩阵（CoordinateMatrix）：三元组矩阵是一个分布式矩阵，其实体集合是一个 RDD，每一个实体是一个(i: Long, j: Long, value: Double)三元组，其中 i 代表行索引，j 代表列索引，value 代表实体的值。三元组矩阵常用于稀疏性比较高的计算中，是由 RDD[MatrixEntry]来构建的。MatrixEntry 是一个 Tuple 类型的元素，其中包含行、列和元素值，代码示例如下所示：

```scala
import org.apache.spark.mllib.linalg.distributed.{CoordinateMatrix,
    MatrixEntry}

//定义一个含有MatrixEntry集合的RDD
val entries: RDD[MatrixEntry] = ……
//通过RDD[MatrixEntry]三元组矩阵
val mat: CoordinateMatrix = new CoordinateMatrix(entries)

//获取该三元组矩阵的行、列数
val m = mat.numRows()
val n = mat.numCols()

//把三元组矩阵转换为行索引矩阵
val indexedRowMatrix = mat.toIndexedRowMatrix()
```

（4）分块矩阵（BlockMatrix）：分块矩阵是支持矩阵分块（MatrixBlocks）RDD 的分布式矩阵，其中矩阵分块（MatrixBlocks）由（(Int, Int), Matrix）元组所构成，(Int, Int) 表示该分块矩阵所处父矩阵的索引位置，Matrix 表示该索引位置上的子矩阵。分块矩阵支持矩阵的加法和乘法，并设有辅助函数验证用于检查分块矩阵是否设置正确。

分块矩阵可以使用 toBlockMatrix 很容易地从行索引矩阵或三元组矩阵转换，在该方法中默认创建的矩阵大小为 1024×1024，也可以通过带有参数的转换方法 toBlockMatrix(rowsPerBlock, colsPerBlock)来指定矩阵的大小，示例代码如下：

```scala
import org.apache.spark.mllib.linalg.distributed.{BlockMatrix,
    CoordinateMatrix, MatrixEntry}

//创建形如(i, j, v)矩阵实体的RDD
val entries: RDD[MatrixEntry] = …… // an RDD of (i, j, v) matrix entries
//通过RDD[MatrixEntry]创建三元组矩阵
val coordMat: CoordinateMatrix = new CoordinateMatrix(entries)
// Transform the CoordinateMatrix to a BlockMatrix
//使用toBlockMatrix把三元组矩阵转换为分块矩阵
val matA: BlockMatrix = coordMat.toBlockMatrix().cache()

//对该分块矩阵进行检验，确认该矩阵分块是否正确，如果不正确，则报异常
matA.validate()

//对分块矩阵进行乘法运算A^T A.
val ata = matA.transpose.multiply(matA)
```

9.1.3 Spark MLlib 基本统计方法

1．汇总统计（Summary Statistics）

在 Statistics 类中的 colStats 方法对 RDD[Vector]格式数据的列汇总统计。该方法返回一个 MultivariateStatisticalSummary 实例，里面包括面向列的最大值、最小值、均值、方差、非零值个数以及总数量，示例代码如下：

```
import org.apache.spark.mllib.linalg.Vector
import org.apache.spark.mllib.stat.{MultivariateStatisticalSummary, Statistics}

//创建一个向量集的 RDD
val observations: RDD[Vector] = ……

//使用 colStats 获取 MultivariateStatisticalSummary 实例
val summary: MultivariateStatisticalSummary =
    Statistics.colStats(observations)
println("均值:"+summary.mean)
println("方差:"+summary.variance)
println("非零统计量个数:"+summary.numNonzeros)
println("总数:"+summary.count)
println("最大值:"+summary.max)
println("最小值:"+summary.min)
```

2．相关性（Correlations）

MLlib 支持两种相关性系数：皮尔逊相关系数（Pearson）和斯皮尔曼等级相关系数（Spearsman）。相关系数是用以反映变量之间相关关系密切程度的统计指标，简单来说就是相关系数绝对值越大（值越接近 1 或者-1）则表示数据越可进行线性拟合，如图 9-3 所示。

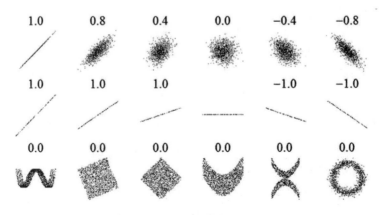

图 9-3　线性拟合图

皮尔逊相关系数计算公式为

$$r = \frac{n\sum xy - \sum x \sum y}{\sqrt{n\sum x^2 - (\sum x)^2}\sqrt{n\sum y^2 - (\sum y)^2}}$$

斯皮尔曼等级相关系数计算公式为

$$\rho = 1 - \frac{6\sqrt{\sum d_i^2}}{n(n^2-1)}$$ （其中，di = xi-yi）

Statistics 类的 corr 方法提供了计算系列间相关性的方法，输入参数的类型可以使 RDD[Double]或 RDD[Vector]或 JavaRDD[java.lang.Double]，输出将相应是一个双精度浮点型值或相关矩阵。

```
import org.apache.spark.SparkContext
import org.apache.spark.mllib.linalg._
import org.apache.spark.mllib.stat.Statistics

val sc: SparkContext = ……

//创建两个具有相同分区数和基数的RDD
val seriesX: RDD[Double] = ……
val seriesY: RDD[Double] = ……
//使用皮尔逊相关系数方法计算相关性，默认情况下使用皮尔逊相关系数，可以通过指定"spearman"
//使用斯皮尔曼等级相关系数方法进行计算
val correlation: Double = Statistics.corr(seriesX, seriesY, "pearson")

//定义一个只包含行向量的RDD
val data: RDD[Vector] = ……
//使用皮尔逊相关系数方法计算矩阵相关性，默认情况下使用皮尔逊相关系数，
//可以通过指定"spearman"使用斯皮尔曼等级相关系数方法进行计算
val correlMatrix: Matrix = Statistics.corr(data, "pearson")
```

3．分层抽样（Stratified Sampling）

不同于其他统计方法，分层抽样方法运行在键值对的 RDD 上，其中键是一个标签，值是特定的属性，比如，键可以是男人或女人、文档编号，其相应的值可以是人口数据中的年龄列表或者文档中的词列表等。分层抽样方法包括 sampleByKey 和 sampleByKeyExact 两种，sampleByKey 方法对每一个观测掷币决定是否抽中它，所以需要对数据进行一次遍历和输入期望抽样的大小。而 sampleByKeyExact 方法并不是简单地在每一层中使用 sampleByKey 方法随机抽样，它需要更多资源，但将提供置信度高达 99.99%的精确抽样大小。

sampleByKeyExact 方法允许使用者准确抽取[$f_k \cdot n_k$]∀k∈K 个元素，这里的 f_k 是从键 k 中

期望抽取的比例，nk 是从键 k 中抽取的键值对数量，而 K 是键集合。为了确保凑样大小，无放回抽样对数据会多一次遍历，而有放回的抽样会多两次遍历。

```
import org.apache.spark.SparkContext
import org.apache.spark.SparkContext._
import org.apache.spark.rdd.PairRDDFunctions

val sc: SparkContext = ……

//创建一个任意数据格式键值对的 RDD 和为每个键指定值的哈希列表
val data = ……
val fractions: Map[K, Double] = ……

//获得每层的精确样本
val approxSample = data.sampleByKey(withReplacement = false, fractions)
val exactSample = data.sampleByKeyExact(withReplacement = false, fractions)
```

4．假设检验（Hypothesis Testing）

在统计分析中，假设检验是一个强大的工具，用来判断结果是不是统计得充分以及结果是不是随机的。MLlib 支持皮尔森卡方检测（Pearson's chi-squared tests）是最著名的卡方检测方法之一，一般提到卡方检测时若无特殊说明，则代表使用的是皮尔森卡方检测。皮尔森卡方检测可以用来进行适配度检测和独立性检测。

- 适配度检测（Goodness of Fit test）：验证一组观察值的次数分配是否异于理论上的分配。其 H_0 假设（虚无假设，null hypothesis）为一个样本中已发生事件的次数分配会服从某个特定的理论分配。通常情况下这个特定的理论分配指的是均匀分配，目前 Spark 默认的是均匀分配。
- 独立性检测（Independence Test）：验证从两个变量抽出的配对观察值组是否互相独立，其虚无假设是两个变量呈统计独立性。

5．随机数据生成

随机数据生成对对随机的算法、原型和性能测试来说是有用的。MLlib 支持指定分布类型来生成随机 RDD，如均匀分布、标准正态分布和泊松分布。RandomRDDs 提供工厂方法来生成随机双精度浮点型 RDDs 或者向量 RDDs。下面示例生成一个随机的双精度浮点型 RDD，其值服从标准正态分布 N（0，1），然后将其映射为 N（1，4）。

```
import org.apache.spark.SparkContext
import org.apache.spark.mllib.random.RandomRDDs._
```

```
val sc: SparkContext = ……

//定义一个随机的双精度浮点型RDD，其值服从标准正态分布N(0,1)
val u = normalRDD(sc, 1000000L, 10)
//将RD映射为服从标准正态分布N(1,4)
val v = u.map(x => 1.0 + 2.0 * x)
```

6．核密度估计（Kernel Density Estimation）

核密度估计是在概率论中用来估计未知的密度函数，属于非参数检验方法之一。它由 Rosenblatt 和 Emanuel Parzen 提出，又名 Parzen 窗（Parzen window）。Ruppert 和 Cline 基于数据集密度函数聚类算法提出修订的核密度估计方法。由于该方法不利用有关数据分布的先验知识，对数据分布不附加任何假定，是一种从数据样本本身出发研究数据分布特征的方法，因此在统计学理论和应用领域均受到高度的重视。

```
import org.apache.spark.mllib.stat.KernelDensity
import org.apache.spark.rdd.RDD

//创建一个样本数据的RDD
val data: RDD[Double] = ……

//构建密度估计与样本数据和用于高斯核的标准偏差
val kd = new KernelDensity().setSample(data).setBandwidth(3.0)

//计算给定数据的密度估计值
val densities = kd.estimate(Array(-1.0, 2.0, 5.0))
```

7．流显著性验证（Streaming Significance Testing）

MLlib 提供了使用类似 A/B 验证的在线实现方法，这些方法可以提供对 Spark Streaming 中的数据对象 DStream[(Boolean,Double)]进行验证，其中第一个元素假表示对照组而真表示实验组，第二个元素表示观测值。流显著性验证支持如下两个参数。

- peacePeriod：指定在初始流中，需要忽略数据的时间段。
- windowSize：设置进行假设验证的时间窗口大小，设置为 0 表示使用所有批次数据进行计算。

```
val data = ssc.textFileStream(dataDir).map(line => line.split(",") match {
  case Array(label, value) => BinarySample(label.toBoolean, value.toDouble)
})

val streamingTest = new StreamingTest().setPeacePeriod(0)
  .setWindowSize(0).setTestMethod("welch")
```

```
val out = streamingTest.registerStream(data)
out.print()
```

9.1.4 预言模型标记语言

预测模型标记语言（Predictive Model Markup Language，PMML）是一种可以呈现预测分析模型的事实标准语言。它支持在 PMML 兼容应用程序之间轻松共享预测解决方案。借助预测分析，石油和化工业可以创建解决方案来预测机械故障，确保安全。PMML 受多种顶级统计工具的支持。因此，将预测分析模型应用于实践的过程非常简单，因为用户可以在一个工具中建立一个模型，然后立即在另一个工具中对其进行部署。在这个传感器和数据收集日益普遍深入的世界，像 PMML 这样的预测分析和标准使人们可以从智慧解决方案中受益，并真正改变他们的生活。

PMML 是一种事实标准语言，用于呈现数据挖掘模型。预测分析模型和数据挖掘模型是指代数学模型的术语，这些模型采用统计技术了解大量历史数据中隐藏的模式。预测分析模型采用定型过程中获取的知识来预测新数据中是否有已知模式。PMML 允许您在不同的应用程序之间轻松共享预测分析模型。因此，您可以在一个系统中定型一个模型，在 PMML 中对其进行表达，然后将其移动到另一个系统中，并在该系统中使用上述模型预测机器失效的可能性等。

PMML 是数据挖掘群组的产物。该群组是一个由供应商领导的委员会，由各种商业和开放源码分析公司组成。因此，现在的大部分领先数据挖掘工具都可以导出或导入 PMML。作为一个已发展 10 多年的成熟标准，PMML 既可以呈现用于从数据中了解模型的统计技术（如人工神经网络和决策树），也可以呈现原始输入数据的预处理以及模型输出的后处理（参见图 9-4）。

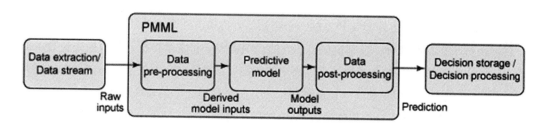

图 9-4　PMML 数据处理流程

MLlib 支持将模型导出为预测模型标记语言，表 9-2 列出了 MLlib 模型导出为 PMML 对应模型。

表 9-2　MLlib 模型导出为 PMML 对应模型

`spark.mllib`模型	PMML 模型
KMeansModel	ClusteringModel
LinearRegressionModel	RegressionModel (functionName="regression")
RidgeRegressionModel	RegressionModel (functionName="regression")
LassoModel	RegressionModel (functionName="regression")
SVMModel	RegressionModel (functionName="classification" normalizationMethod="none")
BinaryLogisticRegressionModel	RegressionModel (functionName="classification" normalizationMethod="logit")

9.2　线性模型

9.2.1　数学公式

许多标准机器学习方法可以转换为凸优化问题，即一个为凸函数 f 找到最小值的任务，这个函数 f 依赖于一个有 d 个值的向量变量 w。更正式一点地说，这是一个优化问题，其目标函数 f 具有下面的形式：

$$f(w) := \lambda R(w) + \frac{1}{n}\sum_{i=1}^{n} L(w, x_i, y_i)$$

向量 $x_i \in R^d$ 是训练数据样本，其中，$1 \leqslant i \leqslant n$，$y_i \in R$ 是相应的类标签，也是需要预测的目标。如果 L(w;x,y) 能被表述为 $w^T x$ 和 y 的一个函数，则称该方法是线性的。

目标函数 f 包括两部分：控制模型复杂度的正则化因子和度量模型误差的损失函数。损失函数 L(w;.) 是典型关于 w 的凸函数。事先锁定的正则化参数 $\lambda \geqslant 0$ 承载了我们在最小化损失量（训练误差）和最小化模型复杂度（避免过渡拟合）两个目标之间的权衡取舍。

表 9-3 概述了 MLlib 支持的损失函数及其梯度或子梯度，其中 hinge loss 表示用于软间隔支持向量机的损失函数，logistic loss 表示用于逻辑回归的损失函数，而 squared loss 表示用于最小二乘法的损失函数。

表 9-3　MLlib 支持的损失函数及其梯度或子梯度

	损失函数 $L(w;x_i,y_i)$	梯度或子梯度
hinge loss	$\max\{0, 1-yw^Tx\}, y \in \{-1,+1\}$	$\begin{cases} -y \cdot x & \text{if } yw^Tx < 1, \\ 0 & \text{otherwise.} \end{cases}$
logistic loss	$\log(1+\exp(-yw^Tx)), y \in \{-1,+1\}$	$-y(1-\dfrac{1}{1+\exp(-yw^Tx)}) \cdot x$
squared loss	$\dfrac{1}{2}(w^Tx-y)^2, y \in R$	$(w^Tx-y) \cdot x$

正则化因子的目标是获得简单模型和避免过度拟合，在 MLlib 中支持下面的正则化因子，如表 9-4 所示。

表 9-4　MLlib 中支持正则化因子

	正则化因子 R(w)	梯度或子梯度
零	0	O
L2 范数	$\dfrac{1}{2}\|w\|_2^2$	w
L1 范数	$\|w\|_1$	sign(w)
elastic net	$\alpha\|w\|_1 + (1-\alpha)\dfrac{1}{2}\|w\|_2^2$	$\alpha\, sign(w) + (1-\alpha)w$

在这里，sign(w)是一个代表 w 中所有实体的类标签（signs (± 1)）的向量。

与 L1 正则化问题比较，由于 L2 的平滑性，L2 正则化问题一般较容易解决。但是，由于可以强化权重的稀疏性，L1 正则化更能产生较小的和更容易解释的模型，而后者在特征选择是非常有用的。

9.2.2　线性回归

线性回归（Linear Regression）问题属于监督学习，又称为分类（Classification）或归纳学习（Inductive Learning）。线性回归是利用称为线性回归方程的函数对一个或多个自变量和因变量之间关系进行建模的一种回归分析方法，只有一个自变量的情况称为简单回归，大于一个自变量情况的叫做多元回归，在实际情况中大多数都是多元回归。线性回归数学表达如下：

$$h(x)=a_0+a_1x_1+a_2x_2+\cdots+a_nx_n+J(\theta)$$

其中，h(x)为预测函数，$a_i(i=1,2,\cdots,n)$为估计参数，模型训练的目的就是计算出这些参数的值。而线性回归分析的整个过程可以简单描述为如下 3 个步骤。

（1）寻找合适的预测函数，即上文中的 h(x)，用来预测输入数据的判断结果。这个过程是非常关键的，需要对数据有一定的了解或分析，知道或者猜测预测函数的"大概"形式，比如是线性函数还是非线性函数，若是非线性的，则无法用线性回归来得出高质量的结果。

（2）构造一个 Loss 函数（损失函数），该函数表示预测的输出（h）与训练数据标签之间的偏差，可以是二者之间的差（h-y）或者是其他的形式（如平方差开方）。综合考虑所有训练数据的"损失"，将 Loss 求和或者求平均，记为 J(θ) 函数，表示所有训练数据预测值与实际类别的偏差。

（3）显然，J(θ) 函数的值越小，预测函数就越准确（即 h 函数越准确），所以这一步需要做的是找到 J(θ) 函数的最小值。找函数的最小值有不同的方法，Spark 中采用的是梯度下降法（stochastic gradient descent, SGD）。

线性回归同样可以采用正则化手段，其主要目的就是防止过拟合。

当采用 L1 正则化时，则变成了 Lasso 回归；当采用 L2 正则化时，则变成了岭回归；线性回归未采用正则化手段。通常来说，在训练模型时是建议采用正则化手段的，特别是在训练数据的量特别少的时候，若不采用正则化手段，过拟合现象会非常严重。L2 正则化相比 L1 而言会更容易收敛（迭代次数少），但 L1 可以解决训练数据量小于维度的问题（也就是 n 元一次方程只有不到 n 个表达式，这种情况下是多解或无穷解的）。

MLlib 提供 L1、L2 和无正则化 3 种方法，如表 9-5 所示。

表 9-5　MLlib 提供 L1、L2 和无正则化方法

	正则化因子	梯度或子梯度
零（未正则化）	0	O
L2 范数	$\frac{1}{2}\|w\|_2^2$	w
L1 范数	$\alpha\|w\|_1$	sign(w)

9.2.3　线性支持向量机

二元分类将数据项划分为两类：正例和反例。MLlib 支持两种二元分类的线性方法：线性支持向量机和逻辑回归。在 MLlib 中，MLlib 都支持 L1、L2 正则化，并且训练数据集用 RDD[LabeledPoint] 格式来表示，为了与多类标签保持一致，反例标签是 0，正例标签是+1。

线性支持向量机 SVMs（Linear Support Vector Machines）对于大规模的分类任务来说，线性支持向量机是标准方法。在前面"数学公式"一节中所描述的线性方法，其损失函数是 hinge loss：$\max\{0, 1-yw^Tx\}, y \in \{-1, +1\}$。默认配置下，线性 SVM 使用 L2 正则化训练，该算法产出

是一个 SVM 模型。给定新数据点 X，该模型基于 w^Tx 的值来预测，默认情形下，$w^Tx \geqslant 0$ 时为正例，否则为反例。

9.2.4 逻辑回归

逻辑回归广泛运用于二元因变量预测。它是 9.3.1 节"数学公式"中所描述的线性方法，其损失函数是 $\log(1+\exp(-yw^Tx))$，$y \in \{-1,+1\}$。逻辑回归算法的产出是一个逻辑回归模型。给定新数据点 X，该模型运用下面逻辑函数来预测：

$$f(z) = \frac{1}{1+e^{-z}}$$

在这里，$z=w^Tx$。默认情况下，若 $f(w^Tx)>0.5$，输出是正例，否则是反例。与线性支持向量机不同之处在于，线性回归模型 $f(z)$ 的输出含有一个概率解释（即 x 是正例的概率）。

9.2.5 线性最小二乘法、Lasso 和岭回归

线性最小二乘法是回归问题中最常用的公式，它是 9.3.1 节"数学公式"中所描述的线性方法，其损失函数是平方损失（squared loss）：$L(w;x,y) = \frac{1}{2}(w^Tx-y)^2$。

根据正则化参数类型的不同，将相关算法分为不同回归算法：
- 普通最小二乘法或线性最小二乘法：未正则化。
- 岭回归算法：使用 L2 正则化。
- Lasso 算法：使用 L1 正则化。

所有相关模型，其平均损失或训练误差计算公式为：$\frac{1}{n}\sum_{t=1}^{n}(w^Tx_i-y_i)^2$，即均方误差。

9.2.6 流式线性回归

当数据以流的形式传入，当收到新数据时更新模型参数，在线拟合回归模型是有用的。MLlib 目前使用普通最小二乘法实现流的线性回归。这种拟合的处理机制与离线方法相似，但其拟合发生于每一数据块到达时之外，目的是为了持续更新以反应流中数据。

9.3 决策树

决策树（Decision Tree）是一个树结构（可以是二叉树或非二叉树）。其每个非叶节点表示一个特征属性上的测试，每个分支代表这个特征属性在某个值域上的输出，而每个叶节点存放一个类别。使用决策树进行决策的过程就是从根节点开始，测试待分类项中相应的特征属性，并按照其值选择输出分支，直到到达叶子节点，将叶子节点存放的类别作为决策结果。

决策树的构造过程不依赖领域知识，它使用属性选择度量来选择将元组最好地划分成不同的类的属性。所谓决策树的构造就是进行属性选择度量确定各个特征属性之间的拓扑结构。构造决策树的关键步骤是分裂属性。所谓分裂属性就是在某个节点处按照某一特征属性的不同划分构造不同的分支，其目标是让各个分裂子集尽可能地"纯"。尽可能"纯"就是尽量让一个分裂子集中待分类项属于同一类别。分裂属性分为 3 种不同的情况：

- 属性是离散值且不要求生成二叉决策树，此时用属性的每一个划分作为一个分支。
- 属性是离散值且要求生成二叉决策树，此时使用属性划分的一个子集进行测试，按照"属于此子集"和"不属于此子集"分成两个分支。
- 属性是连续值，此时确定一个值作为分裂点，按照大于分裂点和小于等于分裂点生成两个分支。

构造决策树的关键性内容是进行属性选择度量，属性选择度量是一种选择分裂准则，是将给定的类标记的训练集合的数据划分 D "最好"地分成个体类的启发式方法，它决定了拓扑结构及分裂点的选择。属性选择度量算法有很多，一般使用自顶向下递归分治法，并采用不回溯的贪心策略。这里介绍 ID3 和 C4.5 两种常用算法。

1．ID3 算法

ID3 算法的核心思想就是以信息增益度量属性选择，选择分裂后信息增益最大的属性进行分裂，下面先定义几个要用到的概念。

设 D 为用类别对训练元组进行的划分，则 D 的熵（entropy）表示为

$$info(D) = -\sum_{i=1}^{m} p_i log_2(p_i)$$

其中，pi 表示第 i 个类别在整个训练元组中出现的概率，可以用属于此类别元素的数量除以训练元组元素总数量作为估计。熵的实际意义表示是 D 中元组的类标号所需要的平均信息量。

现在我们假设将训练元组 D 按属性 A 进行划分，则 A 对 D 划分的期望信息为

$$info_A(D) = \sum_{j=1}^{v} \frac{|D_j|}{|D|} info(D_j)$$

而信息增益即为两者的差值：

$$gain(A) = info(D) - info_A(D)$$

ID3 算法就是在每次需要分裂时，计算每个属性的增益率，然后选择增益率最大的属性进行分裂。

2．ID4.5 算法

ID3 算法存在一个问题，就是偏向于多值属性，例如，如果存在唯一标识属性 ID，则 ID3 会选择它作为分裂属性，这样虽然使得划分充分纯净，但这种划分对分类几乎毫无用处。ID3 的后继算法 C4.5 使用增益率（gain ratio）的信息增益扩充，试图克服这个偏倚。

C4.5 算法首先定义了"分裂信息"，其定义可以表示为

$$split\_info_A(D) = -\sum_{j=1}^{v} \frac{|D_j|}{|D|} log_2(\frac{|D_j|}{|D|})$$

其中各符号意义与 ID3 算法相同，然后，增益率被定义为

$$gain\_ratio(A) = \frac{gain(A)}{split\_info(A)}$$

C4.5 选择具有最大增益率的属性作为分裂属性，其具体应用与 ID3 类似，不再赘述。

9.4 决策模型组合

模型组合（比如说有 Boosting、Bagging 等）与决策树相关的算法比较多，这些算法最终的结果是生成 N（可能会有几百棵以上）棵树，这样可以大大地减少单决策树带来的问题，有点类似于"三个臭皮匠顶一个诸葛亮"的做法。虽然这几百棵决策树中的每一棵都很简单（相对于 C4.5 这种单决策树来说），但是它们组合起来确是很强大。

模型组合+决策树相关的算法有两种比较基本的形式——随机森林与 GBDT(Gradient Boost Decision Tree)。随机森林和梯度提升树（GBTs）两者之间主要的差别在于每棵树训练的顺序。随机森林通过对数据随机采样来单独训练每一棵树。这种随机性也使得模型相对于单决策树更健壮，且不易在训练集上产生过拟合。而 GBTs 则一次只训练一棵树，后面每一棵新的决策树逐步矫正前面决策树产生的误差。随着树的添加，模型的表达力也越强。最后，两种方法都生成了一个决策树的权重集合。该集成模型通过组合每棵独立树的结果来进行预测。图 9-5 显示了一个由 3 棵决策树集成的简单实例。

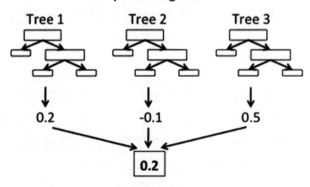

图 9-5　决策树集成的简单实例图

在上述例子的回归集合中，每棵树都预测出一个实值。这些预测值被组合起来产生最终集成的预测结果。这里，我们通过取均值的方法来取得最终的预测结果（当然不同的预测任务需要用到不同的组合算法）。

9.4.1　随机森林

随机森林顾名思义，是用随机的方式建立一个森林，森林里面有很多的决策树组成，随机森林的每一棵决策树之间是没有关联的。在得到森林之后，当有一个新的输入样本进入的时候，就让森林中的每一棵决策树分别进行一下判断，看看这个样本应该属于哪一类（对于分类算法），然后看看哪一类被选择最多，就预测这个样本为那一类。

在建立每一棵决策树的过程中，有两点需要注意，即采样与完全分裂。首先是两个随机采样的过程，随机森林对输入的数据要进行行、列的采样。对于行采样，采用有放回的方式，也就是在采样得到的样本集合中，可能有重复的样本。假设输入样本为 N 个，那么采样的样本也为 N 个。这样使得在训练的时候，每一棵树的输入样本都不是全部的样本，使得相对不容易出现过拟合。然后进行列采样，从 M 个 feature 中，选择 m 个（m << M）。之后就是对采样之后的数据使用完全分裂的方式建立出决策树，这样决策树的某一个叶子节点要么是无法继续分裂的，要么里面的所有样本的都是指向的同一个分类。一般很多的决策树算法都有一个重要的步骤，即剪枝，但是这里不这样操作。由于之前的两个随机采样的过程保证了随机性，所以就算不剪枝，也不会出现过拟合。

按这种算法得到的随机森林中的每一棵都是很弱的，但是组合起来就很厉害了。可以这样比喻随机森林算法：每一棵决策树就是一个精通于某一个窄领域的专家，这样在随机森林中就有了很多个精通不同领域的专家，对一个新的问题（新的输入数据），可以用不同的角度去看待它，最终由各个专家，投票得到结果。

9.4.2 梯度提升决策树

梯度提升决策树（Gradient Boosting Decision Tree，GBDT）又叫 MART（Multiple Additive Regression Tree），是一种迭代的决策树算法。该算法由多棵决策树组成，所有树的结论累加起来做最终答案。

Gradient Boost 其实是一个框架，里面可以套入很多不同的算法，其中 Boost 是"提升"的意思，一般 Boosting 算法都是一个迭代的过程，每一次新的训练都是为了改进上一次的结果。原始的 Boost 算法是在算法开始的时候，为每一个样本赋上一个权重值，初始的时候，大家都是一样重要的。在每一步训练中得到的模型，会使得数据点的估计有对有错，我们就在每一步结束后，增加分错的点的权重，减少分对的点的权重，这样使得某些点如果老是被分错，那么就会被"严重关注"，也就被赋上一个很高的权重。然后等进行了 N 次迭代（由用户指定），将会得到 N 个简单的分类器（Basic Learner），然后我们将它们组合起来（比如说可以对它们进行加权、或者让它们进行投票等），得到一个最终的模型。

而 Gradient Boost 与传统的 Boost 的区别是，每一次的计算是为了减少上一次的残差（Residual），而为了消除残差，可以在残差减少的梯度（Gradient）方向上建立一个新的模型。所以说，在 Gradient Boost 中，每个新的模型的简历是为了使得之前模型的残差往梯度方向减少，与传统 Boost 对正确、错误的样本进行加权有着很大的区别。

9.5 朴素贝叶斯

贝叶斯公式又叫做贝叶斯定理，是贝叶斯分类的基础，目前研究较多的 4 种贝叶斯分类算法，即 Naive Bayes、TAN、BAN 和 GBN。如何在已知事件 A 和 B 分别发生的概率，和事件 B 发生时事件 A 发生的概率，来求得事件 A 发生时事件 B 发生的概率，这就是贝叶斯公式的作用。其表述如下：

$$P(B|A) = \frac{P(A|B) \times P(B)}{P(A)}$$

朴素贝叶斯分类（Naive Bayes），其核心思想非常简单：对于某一预测项，分别计算该预测

项为各个分类的概率,然后选择概率最大的分类为其预测分类。朴素贝叶斯分类的数学定义如下:

(1) 设 $x=\{a_1,a_2,\cdots,a_m\}$ 为一个待分类项,而每个 a_i 为 x 的一个特征属性。

(2) 已知类别集合 $C=\{y_1,y_2,\cdots,y_n\}$。

(3) 计算 xx 为各个类别的概率:$P(y_1|x),P(y_2|x),\cdots,P(y_n|x)$。

(4) 如果 $P(y_k|x)=\max\{P(y_1|x),P(y_2|x),\cdots,P(y_n|x)\}$,则 x 的类别为 y_k。

如何获取第四步中的最大值,也就是如何计算第三步中的各个条件概率最为重要。用户可以采用如下做法:

(1) 获取训练数据集,即分类已知的数据集。

(2) 统计得到在各类别下各个特征属性的条件概率估计,即 $P(a_1|y_1),P(a_2|y_1),\cdots,P(a_m|y_1)$;$P(a_1|y_2),P(a_2|y_2),\cdots,P(a_m|y_2)$;$\cdots$;$P(a_1|y_n),P(a_2|y_n),\cdots,P(a_m|y_n)$,其中数据既可以是离散的,也可以是连续的。

(3) 如果各个特征属性是条件独立的,则根据贝叶斯定理有如下推导:$P(yi|x)=P(x|yi)P(yi)$ 对于某 x 来说,分母是固定的,所以只要找出分子最大的即为条件概率最大的,又因为各特征属性是条件独立的,所以有

$$P(x|y_i)P(y_i) = P(a_1|y_i),P(a_2|yi)\ldots,P(a_m|y_i)P(y_i) = P(y_i)\prod_{j=1}^{m}P(a_j|y_i)$$

9.6 协同过滤

协同过滤常被应用于推荐系统,旨在补充用户—商品关联矩阵中所缺失的部分。MLlib 当前支持基于模型的协同过滤,其中用户和商品通过小组隐语义因子进行表达,并且这些因子也用于预测缺失的元素。Spark MLlib 已实现了交替最小二乘法(Alternating Least Squares,ALS),它通过观察到的所有用户给产品打分,来推断每个用户的喜好,并向用户推荐适合的产品。

用户对物品或者信息的偏好,根据应用本身的不同,可能包括用户对物品的评分、用户查看物品的记录、用户的购买记录等。其实这些用户的偏好信息可以分为两类。

- 显式的用户反馈:这类是用户在网站上自然浏览或者使用网站以外显式地提供反馈信息,例如用户对物品的评分或者对物品的评论。
- 隐式的用户反馈:这类是用户在使用网站时产生的数据,隐式地反映了用户对物品的喜好,例如用户购买了某物品,用户查看了某物品的信息,等等。

显式的用户反馈能准确地反映用户对物品的真实喜好,但需要用户付出额外的代价;而隐

式的用户行为，通过一些分析和处理，也能反映用户的喜好，只是数据不是很精确，有些行为的分析存在较大的噪音。但只要选择正确的行为特征，隐式的用户反馈也能得到很好的效果，只是行为特征的选择可能在不同的应用中有很大的不同，例如在电子商务的网站上，购买行为其实就是一个能很好表现用户喜好的隐式反馈。

推荐引擎根据不同的推荐机制可能用到数据源中的一部分，然后根据这些数据，分析出一定的规则或者直接对用户对其他物品的喜好进行预测计算。这样推荐引擎可以在用户进入时给他推荐他可能感兴趣的物品。

MLlib 目前支持基于协同过滤的模型，在这个模型里，用户和产品被一组可以用来预测缺失项目的潜在因子来描述。特别是我们实现交替最小二乘（ALS）算法来学习这些潜在的因子，在 MLlib 中的实现有如下参数。

- numBlocks：用于并行化计算的分块个数（设置为-1 时为自动配置）。
- rank：表示模型中隐性因子的个数。
- iterations：表示迭代的次数。
- lambda：表示 ALS 的正则化参数。
- implicitPrefs：决定了是用显性反馈 ALS 的版本还是用隐性反馈数据集的版本。
- alpha：是一个针对于隐性反馈 ALS 版本的参数，这个参数决定了偏好行为强度的基准。

协同过滤的算法如图 9-6 所示。

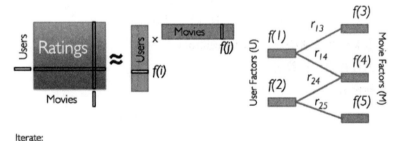

图 9-6　协同过滤的算法图

9.7 聚类

聚类（Cluster Analysis）也称为簇类，其定义为：聚类问题就是给定一个元素集合 D，其中每个元素具有 n 个可观察属性，使用某种算法将 D 划分成 k 个子集，要求每个子集内部的元素之间相异度尽可能低，而不同子集的元素相异度尽可能高，其中每个子集叫做一个簇。其核心任务是：将一组目标对象划分为若干个簇，每个簇之间的对象尽可能地相似，簇与簇之间的对象尽可能地相异。与分类不同，分类是示例式学习，要求分类前明确各个类别，并断言每个元素映射到一个类别，而聚类是观察式学习，在聚类前可以不知道类别甚至不给定类别数量，是无监督学习的一种。目前聚类广泛应用于统计学、生物学、数据库技术和市场营销等领域，相应的算法也非常多。

聚类算法是机器学习中重要的一部分，除了最为简单的 K-means 聚类算法外，MLlib 还支持如下聚类算法：

- K-means。
- 高斯混合（Gaussian Mixture）。
- 快速迭代聚类 PIC（Power Iteration Clustering）。
- LDA（Latent Dirichlet Allocation）。
- 二分 K-means（Bisecting K-means）。
- 流式 K-means（Streaming K-means）。

9.7.1 K-means

K-means 算法是将样本聚类成 k 个簇，具体算法描述如下：

（1）随机选取 k 个聚类质心点（cluster centroids）为 $\mu_1, \mu_2, \ldots, \mu_k \in \mathbb{R}^n$。

（2）重复下面过程直到收敛 {

对于每一个样例 i，计算其应该属于的类

$$c^{(i)} := \arg\min_j \|x^{(i)} - \mu_j\|^2.$$

对于每一个类 j，重新计算该类的质心

$$\mu_j := \frac{\sum_{i=1}^m 1\{c^{(i)} = j\} x^{(i)}}{\sum_{i=1}^m 1\{c^{(i)} = j\}}.$$

}

K 是我们事先给定的聚类数，$c^{(i)}$ 代表样例 i 与 k 个类中距离最近的那个类，$c^{(i)}$ 的值是 1～

k 中的一个。质心μ_j代表我们对属于同一个类的样本中心点的猜测,拿星团模型来解释就是要将所有的星星聚成 k 个星团,首先随机选取 k 个宇宙中的点(或者 k 个星星)作为 k 个星团的质心,然后第一步对于每一个星星计算其到 k 个质心中每一个的距离,然后选取距离最近的那个星团作为$c^{(i)}$,这样经过第一步每一个星星都有了所属的星团;第二步对于每一个星团,重新计算它的质心μ_j(对里面所有的星星坐标求平均)。重复迭代第一步和第二步直到质心不变或者变化很小。

图 9-7 展示了对 n 个样本点进行 K-means 聚类的效果,这里 k 取 2。

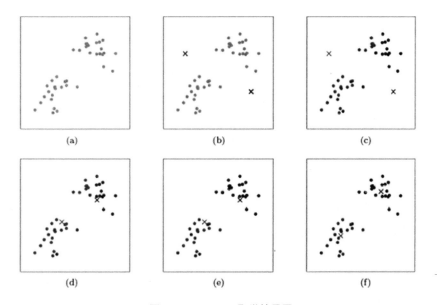

图 9-7 K-means 聚类效果图

K-means 面对的第一个问题是如何保证收敛,前面的算法中强调结束条件就是收敛,可以证明的是 K-means 完全可以保证收敛性。下面我们定性的描述一下收敛性,我们定义畸变函数(distortion function)如下:

$$J(c,\mu) = \sum_{i=1}^{m} ||x^{(i)} - \mu_{c^{(i)}}||^2$$

J 函数表示每个样本点到其质心的距离平方和。K-means 是要将 J 调整到最小。假设当前 J 没有达到最小值,那么首先可以固定每个类的质心μ_j,调整每个样例的所属的类别$c^{(i)}$来让 J 函数减少,同样,固定$c^{(i)}$,调整每个类的质心μ_j也可以使 J 减小。这两个过程就是内循环中使 J 单调递减的过程。当 J 递减到最小时,μ 和 c 也同时收敛。(在理论上,可以有多组不同的μ和 c

值能够使得 J 取得最小值，但这种现象实际上很少见）。

由于畸变函数 J 是非凸函数，意味着我们不能保证取得的最小值是全局最小值，也就是说 K-means 对质心初始位置的选取比较感冒，但一般情况下 k-means 达到的局部最优已经满足需求。但如果你怕陷入局部最优，那么可以选取不同的初始值跑多遍 k-Means，然后取其中最小的 J 对应的 μ 和 c 输出。

MLlib 中 K-means 实现包含一个 K-means++方法的并行化变体 K-means||，实现有如下的参数。

- k：是所需的类簇的个数。
- maxIterations：是最大的迭代次数。
- initializationMode：这个参数决定了是用随机初始化还是通过 k-means|| 进行初始化；
- runs：是跑 k-means 算法的次数（k-mean 算法不能保证能找出最优解，如果在给定的数据集上运行多次，算法将会返回最佳的结果）。
- initializiationSteps：决定了 k-means|| 算法的步数。
- epsilon：决定了判断 k-means 是否收敛的距离阈值。

9.7.2 高斯混合

高斯混合（Gaussian Mixture）与 K-means 一样，给定的训练样本是 $\{x^{(1)}, ..., x^{(m)}\}$，将隐含类别标签用 $z^{(i)}$ 表示。与 K-means 的硬指定不同，首先认为 $z^{(i)}$ 是满足一定的概率分布的，这里认为满足多项式分布，$z^{(i)} \sim \text{Multinomial}(\emptyset)$，其中 $p(z^{(i)} = j) = \emptyset_j$, $\emptyset_j \geq 0$, $\sum_{j=1}^{k} \emptyset_j = 1$，$z^{(i)}$ 有 k 个值 $\{1,...,k\}$ 可以选取。而且在给定 $z^{(i)}$ 后，$x^{(i)}$ 满足多值高斯分布，即 $(x^{(i)} | z^{(i)} = j) \sim N(\mu_j, \Sigma_j)$，由此可以得到联合分布 $p(x^{(i)}, z^{(i)}) = p(x^{(i)} | z^{(i)}) p(z^{(i)})$。

整个模型简单描述为对于每个样例 $x^{(i)}$，先从 k 个类别中按多项式分布抽取一个 $z^{(i)}$，然后根据所对应的 k 个多值高斯分布中的一个生成样例 $x^{(i)}$，整个过程称作混合高斯模型。需要注意的是，这里的 $z^{(i)}$ 仍然是隐含随机变量，模型中还有 3 个变量 $\emptyset$，μ 和 Σ，最大似然估计为 $p(x, z)$。对数化后如下：

$$\begin{aligned}\ell(\phi, \mu, \Sigma) &= \sum_{i=1}^{m} \log p(x^{(i)}; \phi, \mu, \Sigma) \\ &= \sum_{i=1}^{m} \log \sum_{z^{(i)}=1}^{k} p(x^{(i)} | z^{(i)}; \mu, \Sigma) p(z^{(i)}; \phi).\end{aligned}$$

这个式子的最大值是不能通过前面使用的求导数为 0 的方法来解决的，因为求的结果不是 close form。但是假设知道了每个样例的 $z^{(i)}$，那么上式可以简化为

$$\ell(\phi,\mu,\Sigma) = \sum_{i=1}^{m} \log p(x^{(i)}|z^{(i)};\mu,\Sigma) + \log p(z^{(i)};\phi).$$

这时候再来对ϕ，μ和Σ进行求导得到

$$\phi_j = \frac{1}{m}\sum_{i=1}^{m} 1\{z^{(i)} = j\},$$

$$\mu_j = \frac{\sum_{i=1}^{m} 1\{z^{(i)} = j\}x^{(i)}}{\sum_{i=1}^{m} 1\{z^{(i)} = j\}},$$

$$\Sigma_j = \frac{\sum_{i=1}^{m} 1\{z^{(i)} = j\}(x^{(i)} - \mu_j)(x^{(i)} - \mu_j)^T}{\sum_{i=1}^{m} 1\{z^{(i)} = j\}}.$$

ϕ_j就是样本类别中$z^{(i)} = j$的比率。μ_j是类别为 j 的样本特征均值，Σ_j是类别为 j 的样例的特征的协方差矩阵。实际上，当知道$z^{(i)}$后，最大似然估计就近似于高斯判别分析模型（Gaussian discriminant analysis model）了，所不同的是 GDA 中类别 y 是伯努利分布，而这里的 z 是多项式分布，还有这里的每个样例都有不同的协方差矩阵，而 GDA 中认为只有一个。

之前假设给定了$z^{(i)}$，实际上$z^{(i)}$是不知道的。那么怎么办呢？考虑之前提到的 EM 的思想，第一步是猜测隐含类别变量 z，第二步是更新其他参数，以获得最大的最大似然估计。用到这里就是循环下面的步骤，直到收敛：{

（E 步）对于每一个 i 和 j，计算

$$w_j^{(i)} := p(z^{(i)} = j|x^{(i)};\Phi,\mu,\Sigma)$$

（M 步），更新参数：

$$\phi_j := \frac{1}{m}\sum_{i=1}^{m} w_j^{(i)},$$

$$\mu_j := \frac{\sum_{i=1}^{m} w_j^{(i)} x^{(i)}}{\sum_{i=1}^{m} w_j^{(i)}},$$

$$\Sigma_j := \frac{\sum_{i=1}^{m} w_j^{(i)}(x^{(i)} - \mu_j)(x^{(i)} - \mu_j)^T}{\sum_{i=1}^{m} w_j^{(i)}}$$

}

在 E 步中，将其他参数Φ, μ, Σ看作是常量，计算$z^{(i)}$的后验概率，也就是估计隐含类别变量。

9.7.3 快速迭代聚类

快速迭代聚类是一种简单可扩展的图聚类方法，它利用数据归一化的逐对相似度矩阵，采用截断的快速迭代法寻找数据集的一个超低维嵌入。这种嵌入恰好是很有效的聚类指标，使它在真实的数据集上总是好于广泛使用的谱聚类方法，比如 NCut。在大规模数据集上，PIC 非常快，比基于最好的特征计算技术实现的 Ncut 快 1000 倍。

9.7.4 LDA

LDA（Latent Dirichlet Allocation）是一种文档主题生成模型，也称为一个三层贝x叶斯概率模型，包含词、主题和文档三层结构。所谓生成模型，就是说，我们认为一篇文章的每个词都是通过"以一定概率选择了某个主题，并从这个主题中以一定概率选择某个词语"这样一个过程得到。文档到主题服从多项式分布，主题到词服从多项式分布。

LDA 是一种非监督机器学习技术，可以用来识别大规模文档集（document collection）或语料库（corpus）中潜藏的主题信息。它采用了词袋（bag of words）的方法，这种方法将每一篇文档视为一个词频向量，从而将文本信息转化为了易于建模的数字信息。但是词袋方法没有考虑词与词之间的顺序，这简化了问题的复杂性，同时也为模型的改进提供了契机。每一篇文档代表了一些主题所构成的一个概率分布，而每一个主题又代表了很多单词所构成的一个概率分布。

对于语料库中的每篇文档，LDA 定义了如下生成过程（generativeprocess）：

（1）对每一篇文档，从主题分布中抽取一个主题。

（2）从上述被抽到的主题所对应的单词分布中抽取一个单词。

（3）重复上述过程直至遍历文档中的每一个单词。

更形式化一点说，语料库中的每一篇文档与 T（通过反复试验等方法事先给定）个主题的一个多项分布相对应，将该多项分布记为 θ。每个主题又与词汇表（vocabulary）中的 V 个单词的一个多项分布相对应，将这个多项分布记为 φ。上述词汇表是由语料库中所有文档中的所有互异单词组成，但实际建模的时候要剔除一些停用词（stopword），还要进行一些词干化（stemming）处理等。θ 和 φ 分别有一个带有超参数（hyperparameter）α 和 β 的 Dirichlet 先验分布。对于一篇文档 d 中的每一个单词，我们从该文档所对应的多项分布 θ 中抽取一个主题 z，然后我们再从主题 z 所对应的多项分布 φ 中抽取一个单词 w。将这个过程重复 N_d 次，就产生了文档 d，这里的 N_d 是文档 d 的单词总数。这个生成过程可以用如图 9-8 所示的模型来表示。

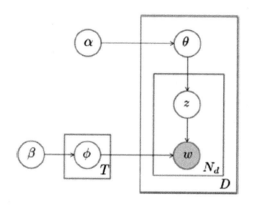

图 9-8　LDA 生成过程模型图

这个图模型表示法也称作"盘子表示法"（plate notation）。图中的阴影圆圈表示可观测变量（observed variable），非阴影圆圈表示潜在变量（latent variable），箭头表示两变量间的条件依赖性（conditional dependency），方框表示重复抽样，重复次数在方框的右下角。

该模型有两个参数需要推断（infer）：一个是"文档-主题"分布 θ，另外是 T 个"主题-单词"分布 ϕ。通过学习这两个参数，我们可以知道文档作者感兴趣的主题，以及每篇文档所涵盖的主题比例等。推断方法主要有 LDA 模型作者提出的变分-EM 算法，还有现在常用的 Gibbs 抽样法。

9.7.5　二分 K-means

二分 K-means（Bisecting K-means）算法的思想是：首先将所有点作为一个簇，然后将该簇一分为二，之后选择能最大程度降低聚类代价函数（也就是误差平方和）的簇划分为两个簇（或者选择最大的簇等，选择方法多种）。以此进行下去，直到簇的数目等于用户给定的数目 k 为止。

以上隐含着一个原则是：因为聚类的误差平方和能够衡量聚类性能，该值越小表示数据点越接近于它们的质心，聚类效果就越好。所以就需要对误差平方和最大的簇进行再一次的划分。因为误差平方和越大，表示该簇聚类越不好，越有可能是多个簇被当成一个簇了，所以首先需要对这个簇进行划分。

二分 K-means 聚类的优点如下：
- 二分 K 均值算法可以加速 K-means 算法的执行速度，因为它的相似度计算少了。
- 不受初始化问题的影响，因为这里不存在随机点的选取，且每一步都保证了误差最小。

9.7.6 流式 K-means

流式 K-means（Streaming k-means）算法是基于流式的，文件中的数据只参与计算一次，算法对加入一篇文档到新的类别和产生新类的代价分别进行评估，当满足一定的阈值才会将文档添加到聚类中去或者产生新的聚类。对每一批数据，将所有的点分配到最近的族簇，并计算最新的族簇中心，然后更新每个族簇的公式为

$$c_{t+1} = \frac{c_t n_t a + x_t m_t}{n_t a + m_t}$$

$$n_{t+1} = n_t + m_t$$

c_t 是前一次计算得到的族簇中心，n_t 是已经分配到族簇的点数，X_t 是从当前批次得到新的族簇中心，m_t 是当前批次加入族簇的点数。流式 K-means 算法和常见的 K-means 算法主要有如下区别：

- 所有文档只参与计算一次，不会重复参与计算。
- 聚类的个数是变化的，但是最终产生的聚类数目是小于等于 k 的。
- 添加到新类和产生新类需要满足一定条件的，不一定会选择最相似的类加入。
- 聚类的数目是动态变化的，而且文档的总数需要预估，文档的数量可以无限大，预估主要是为了对训练参数进行计算评估。

9.8 降维

9.8.1 奇异值分解降维

奇异值分解（Singular Value Decomposition，SVD）是线性代数中一种重要的矩阵分解。为了实现类似特征值分解的计算，可以使用奇异值分解。奇异值分解适用于任何矩阵，如下所示，其中 A 是一个 m*n 的矩阵：

$$A = U_{m*m} \sum\nolimits_{m*n} V^T_{n*n}$$

其中，

（1）U 是一个 m*m 的正交矩阵，其向量被称为左奇异向量。

（2）V 也是一个 n*n 的正交矩阵，其向量被称为右奇异向量。

（3）Σ 是一个 m*n 的矩阵，其对角线上的元素为奇异值，其余元素皆为 0。

当选取 top k 个奇异值时，可以将矩阵降维成为

$$A_{m*n} \approx U_{m*k} \sum\nolimits_{k*k} V_{k*n}^T$$

奇异值可以通过特征值来得出：

（1）求出 A^TA 的特征值和特征向量，$(A^TA)v_i = \lambda_i v_i$。

（2）计算奇异值 $\delta_i = \sqrt{\lambda_i}$。

（3）右奇异向量等于 v_i。

（4）左奇异向量等于 $\frac{1}{\delta_i} Av_i$。

9.8.2 主成分分析降维

主成分分析（Principal Components Analysis，PCA）是一种分析、简化数据集的技术。其主成分分析经常用于减少数据集的维数，同时保持数据集中的对方差贡献最大的特征。其方法主要是通过对协方差矩阵进行特征分解，以得出数据的主成分（即特征向量）与它们的权值（即特征值）。

PCA 的数学定义是：一个正交化线性变换，把数据变换到一个新的坐标系统中，使得这一数据的任何投影的第一大方差在第一个坐标（称为第一主成分）上，第二大方差在第二个坐标（第二主成分）上，依次类推。如图 9-9 所示，通过变化将 X-Y 坐标系映射到 signal 和 noise 上。

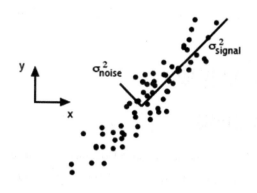

图 9-9　主成分分析

图 9-9 中通过坐标变化后，找出方差最大的方向为第一个坐标（signal），然后在其正交的平面上找出方差最大的方向为第二个坐标（noise）。这样就可以通过选取 top k 个坐标方向来达到对维度（特征）的提炼。其数学表达如下，其中，A 矩阵表示 m 行 n 维的数据，P 表示坐标系的变换：

$$A_{m*n}P_{m*n} = \tilde{A}_{m*n}$$

PCA 将上述中的维度 n 进行提炼，降维成 k（k<n），数学表达如下：

$$A_{m*n}P_{m*k} = \tilde{A}_{m*k}$$

前文中提到 SVD 的降维公式为

$$A_{m*n} \approx U_{m*k} \sum_{k*k} V_{k*n}^T$$

若将两边同时乘以 V_{n*k}，则得到

$$A_{m*n}V_{n*k} \approx U_{m*k} \sum_{k*k} V_{k*n}^T V_{n*k}$$

由于 V_{n*k} 为正交矩阵，所以，

$$A_{m*n}V_{n*k} \approx U_{m*k} \sum_{k*k} \approx \tilde{A}_{m*k}$$

就这样通过 SVD 神奇地实现了 PCA 的坐标系变换和特征的提炼。此外，SVD 还可以进行数据的压缩，如下：

$$U_{k*m}^T A_{m*n} \approx U_{k*m}^T U_{m*k} \sum_{k*k} V_{k*n}^T$$

同样由于 U_{n*k} 为正交矩阵，所以，

$$U_{k*m}^T A_{m*n} \approx \sum_{k*k} V_{k*n}^T \approx \tilde{A}_{k*k}$$

如此则将 m 行数据压缩成 k 行数据，其含义就是去除那些十分相近的数据。

9.9 特征提取和变换

9.9.1 词频—逆文档频率

词频—逆文档频率（Term Frequency–Inverse Document Frequency，TF-IDF）是一种用于资讯检索与资讯探勘的常用加权技术。TF-IDF 是一种统计方法，用以评估一个字词对于一个文件集或一个语料库中的其中一份文件的重要程度。字词的重要性随着它在文件中出现的次数成正比增加，但同时会随着它在语料库中出现的频率成反比下降。TF-IDF 加权的各种形式常被搜寻引擎应用，作为文件与用户查询之间相关程度的度量或评级。除了 TF-IDF 以外，互联网上的搜寻引擎还会使用基于连接分析的评级方法，以确定文件在搜寻结果中出现的顺序。

在一份给定的文件里，词频（Term Frequency，TF）指的是某一个给定的词语在该文件中出现的次数。这个数字通常会被正规化，以防止它偏向长的文件。同一个词语在长文件里可能

会比短文件有更高的词频，而不管该词语重要与否，对于在某一特定文件里的词语 t_i 来说，它的重要性可表示为

$$tf_{i,j} = \frac{n_{i,j}}{\sum_k n_{k,j}}$$

以上式子中，$n_{i,j}$ 是该词在文件 d_j 中的出现次数，而分母则是在文件 d_j 中所有字词的出现次数之和。

逆向文件频率(Inverse Document Frequency，IDF)是一个词语普遍重要性的度量。某一特定词语的 IDF，可以由总文件数目除以包含该词语之文件的数目，再将得到的商取对数得到

$$idf_i = \log \frac{|D|}{|\{d:d \ni t_i\}|}$$

其中，

$|D|$：语料库中的文件总数。

$|\{d:d \ni t_i\}|$：包含词语 t_i 的文件数目（即 $n_i \neq 0$ 的文件数目）。

然后，

$$tfidf_{j,j} = tf_{i,j} \cdot idf_i$$

某一特定文件内的高词语频率，以及该词语在整个文件集合中的低文件频率，可以产生出高权重的 TF-IDF。因此，TF-IDF 倾向于过滤掉常见的词语，保留重要的词语。

有很多不同的数学公式可以用来计算 TF-IDF。词频（TF）是一词语出现的次数除以该文件的总词语数。假如一篇文件的总词语数是 100 个，而词语"母牛"出现了 3 次，那么"母牛"一词在该文件中的词频就是 0.03 (3/100)。一个计算文件频率（DF）的方法是测定有多少份文件出现过"母牛"一词，然后除以文件集里包含的文件总数。所以，如果"母牛"一词在 1000 份文件出现过，而文件总数是 10,000,000 份的话，其文件频率就是 0.0001 (1000/10,000,000)。最后，TF-IDF 分数就可以由计算词频除以文件频率而得到。以上面的例子来说，"母牛"一词在该文件集的 TF-IDF 分数会是 300 (0.03/0.0001)。

9.9.2 词向量化工具

词向量化工具（Word2Vec）用于将词转换为分布式词向量格式，分布式格式的主要优点在于在向量空间中相似词相近，使得更易生成小说模式以及模型评估更加健壮。分布式向量格式在很多自然语言分析应用中很有用，如命名实体识别、消歧、解析、打标签和机器翻译。

MLlib 中使用 skip-gram 模型来实现 Word2Vec，在 skip-gram 模型中，训练目标是学习到同一句子中最能预测其环境的词向量表示。从数学上来说，给定一系列词 w1,w2,...,wT，skip-gram

模型最大化对数似然均值：

$$\frac{1}{T}\sum_{t=1}^{T}\sum_{j=-k}^{j=k}\log p(w_{t+j}|w_t)$$，其中，K 是训练窗口大小

在 skip-gram 模型中，每个词 W 都与两个向量 u_w 和 v_w 相关，u_w 和 v_w 表示词 W 自身及其上下文。在 softmax 模型中，给定词 w_j，正确预测词 w_i 的概率由下式决定：

$$p(w_i|w_j) = \frac{\exp(u_{w_i}^T v_{w_j})}{\sum_{l=1}^{V}\exp(u_l^T v_{w_j})}$$，其中，V 是词量大小

在 skip-gram 模型中使用 softmax 的代价很高，因为 logp(w_i|w_j)的计算量随着 V 线性增长，而 V 很容易就达到百万级。为了提高训练 Word2Vec 的速度，我们使用分层 softmax 技术，该方法可以将计算 logp(w_i|w_j)的复杂度降低到 O(log(V))。

9.9.3 标准化

特征标准化（StandardScaler）是在训练集上对每列使用统计分析技术，将数据调整为标准差的倍数以及（或）去均值。这是一种非常通用的预处理步骤。比如，支持向量机中的 RBF 内核或 L1/L2 正则化线性模型就在特征具有单位变化以及（或）零均值时效果更好。特征标准化能够在优化过程中加快收敛速度，也能够让特征免于被训练时的一个超大值发生剧烈影响。

StandardScaler 的构建函数具有下列参数。

- withMean：默认值为 False，在调整前将数据中心化处理。它的输出是密集的，不能工作于稀疏输入（抛出异常）。
- withStd：默认值为 True，将数据调整为标准差。

在特征标准化中，我们提供一个 fit 方法，它接收 RDD[Vector]格式的输入，进行统计分析，然后输出一个标准差的倍数以及（或）去均值化的模型，模型结果依赖于我们如何配置特征标准化。模型输出一个特征标准化模型，它既能够将一个 Vector 标准化，也能够将一个 RDD[Vector]标准化。

9.9.4 范数化

范数化（Normalizer）将独立的样本调整为具有 L^p 范数。这是在文本分类或聚类中广泛应用的一个操作。比如，两个 L^2 化的 TF-IDF 向量，其点积就是向量间的余弦相似度。

Normalizer 的构建函数具有参数 p：默认值 2，在 L^p 空间范数化。

在范数化中，提供转换方法，它既能够将一个 Vector 范数化，也能够将一个 RDD[Vector]范数化。注意：若输入的模数是 0，它将返回原值。

9.10 频繁模式挖掘

9.10.1 频繁模式增长

频繁项集挖掘算法用于挖掘经常一起出现的 item 集合（称为频繁项集），通过挖掘出这些频繁项集，当在一个事务中出现频繁项集的其中一个 item 时，则可以把该频繁项集的其他 item 作为推荐。比如经典的购物篮分析中啤酒、尿布故事，啤酒和尿布经常在用户的购物篮中一起出现，从而挖掘出啤酒、尿布这个项集，即当一个用户购买了啤酒的时候可以为他推荐尿布，这样用户购买尿布的可能性会比较大，从而达到组合营销的目的。

常见的频繁项集挖掘算法有两类，一类是 Apriori 算法，另一类是频繁模式增长算法（FP-growth）。Apriori 通过不断地构造候选集、筛选候选集来挖掘出频繁项集，需要多次扫描原始数据，当原始数据较大时，磁盘 I/O 次数太多，效率比较低下。频繁模式增长算法则只需扫描原始数据两遍，通过 FP-tree 数据结构对原始数据进行压缩，效率较高。

频繁模式增长算法主要分为两个步骤：FP-tree 构建和递归挖掘 FP-tree。FP-tree 构建通过两次数据扫描，将原始数据中的事务压缩到一个 FP-tree 树，该 FP-tree 类似于前缀树，相同前缀的路径可以共用，从而达到压缩数据的目的。接着通过 FP-tree 找出每个 item 的条件模式基、条件 FP-tree，递归的挖掘条件 FP-tree 得到所有的频繁项集。算法的主要计算瓶颈在 FP-tree 的递归挖掘上。

9.10.2 关联规则挖掘

关联规则挖掘（Association Rules）属于描述型模式，发现关联规则的算法属于无监督学习的方法。

9.10.3 PrefixSpan

PrefixSpan 算法韩家炜老师在 2004 年提出的序列模式算法，该算法和他在 2000 提出的频繁模式增长算法有很大的相似之处，都避免产生候选序列。PrefixSpan 算法的核心是产生前缀和对应的后缀，每次递归都将合适的后缀变为前缀。难点是类似：<a x>，<(a x)>和<(_ x)>，后两种可以做一类处理。

9.11 实例演示

9.11.1 K-means 聚类算法实例

在该实例中将先使用 K-Means 算法计算给出族簇中心点，然后通过 KMeansModel 类里提供了 computeCost 方法评估聚类的效果，最后对给出点位置和测试数据确定所属族簇。K-Means 属于基于平方误差的迭代重分配聚类算法，其核心思想十分简单：

- 随机选择 K 个中心点；
- 计算所有点到这 K 个中心点的距离，选择距离最近的中心点为其所在的簇；
- 简单地采用算术平均数（mean）来重新计算 K 个簇的中心；
- 重复步骤 2 和 3，直至簇类不再发生变化或者达到最大迭代值；
- 输出结果。

Spark MLlib 中 K-means 算法的实现类 (KMeans.scala) 有以下参数，具体如下：

```
class KMeans private (
    private var k: Int,
    private var maxIterations: Int,
    private var runs: Int,
    private var initializationMode: String,
    private var initializationSteps: Int,
    private var epsilon: Double,
    private var seed: Long) extends Serializable with Logging
```

通过下面默认构造函数，我们可以看到这些可调参数具有以下初始值。

```
def this() = this(2, 20, 1, KMeans.K_MEANS_PARALLEL, 5, 1e-4, Utils.random.
  nextLong())
```

参数的含义解释如下：

- k 表示期望的聚类的个数。
- maxInterations 表示方法单次运行最大的迭代次数。
- runs 表示算法被运行的次数。K-means 算法不保证能返回全局最优的聚类结果，所以在目标数据集上多次跑 K-means 算法，有助于返回最佳聚类结果。
- initializationMode 表示初始聚类中心点的选择方式，目前支持随机选择或者 K-means|| 方式。默认是 K-means||。
- initializationSteps 表示 K-means||方法中的部数。
- epsilon 表示 K-means 算法迭代收敛的阀值。
- seed 表示集群初始化时的随机种子。

通常应用时，会先调用 KMeans.train 方法对数据集进行聚类训练，返回 KMeansModel 类实例，然后使用 KMeansModel.predict 方法对新的数据点进行所属聚类的预测

1．测试数据

该实例使用的数据为 kmeanscluster.data，可以在附带资源/data/chapter9 目录中找到，在该文件中提供了 12 个点的空间位置坐标，把该文件上传到 HDFS 的/chapter9 目录中。

```
2.1 2.1 2.1
2.2 2.2 2.2
1.9 1.9 1.9
2.0 2.0 2.0
18.9 18.9 18.9
19.0 19.0 19.0
19.1 19.1 19.1
19.2 19.2 19.2
66.0 66.0 66.0
66.1 66.1 66.1
66.2 66.2 66.2
66.3 66.3 66.3
```

2．程序代码

```scala
import org.apache.log4j.{Level, Logger}
import org.apache.spark.{SparkConf, SparkContext}
import org.apache.spark.mllib.clustering.KMeans
import org.apache.spark.mllib.linalg.Vectors

object KMeansClustering {
  def main(args: Array[String]) {
    if (args.length < 3) {
      println("Usage:KMeansClustering trainingDataFilePath numClusters
        numIterations")
      sys.exit(1)
    }

    //屏蔽不必要的日志显示在终端上
    Logger.getLogger("org.apache.hadoop").setLevel(Level.ERROR)
    Logger.getLogger("org.apache.spark").setLevel(Level.ERROR)
    Logger.getLogger("org.eclipse.jetty.server").setLevel(Level.OFF)

    //设置运行环境
    val conf = new SparkConf().setAppName("KMeansClusting")
    val sc = new SparkContext(conf)
```

```scala
//装载数据集
val data = sc.textFile(args(0))
val parsedTrainingData = data.map(s => Vectors.dense(s.split('
   ').map(_.toDouble)))

//将数据集聚类,进行模型训练形成数据模型
val numClusters = args(1).toInt
val numIterations = args(2).toInt
val model = KMeans.train(parsedTrainingData, numClusters, numIterations)

//打印数据模型的中心点
println("Cluster centers:")
for (c <- model.clusterCenters) {
  println(" " + c.toString)
}

//使用误差平方之和来评估数据模型
val ks:Array[Int] = Array(1,2,3,4,5,6,7,8,9,10)
ks.foreach(cluster => {
  val model:KMeansModel = KMeans.train(parsedTrainingData, cluster,30)
  val ssd = model.computeCost(parsedTrainingData)
  println("Within Set Sum of Squared Errors When K=" + cluster + " -> "+ ssd)
})

//使用模型测试单点数据
println("Vectors 2.0 2.0 2.0 is belongs to clusters:" + model.
   predict(Vectors.dense("2.0 2.0 2.0".split(' ').map(_.toDouble))))
println("Vectors 20 20 20 is belongs to clusters:" + model.
   predict(Vectors.dense("20 20 20".split(' ').map(_.toDouble))))
println("Vectors 66.2 66.2 66.2 is belongs to clusters:" + model.
   predict(Vectors.dense("66.2 66.2 66.2".split(' ').map(_.toDouble))))

//交叉评估,返回数据集和结果
val result = data.map {
  line =>
    val linevectore = Vectors.dense(line.split(' ').map(_.toDouble))
    val prediction = model.predict(linevectore)
    line + " " + prediction
}
result.collect().foreach(line => println(line))

sc.stop()
}
}
```

3. 执行情况

对运行代码进行打包，生成打包文件 sparklearning.jar 并移动到 Spark 根目录下。在运行前需要启动 HDFS 和 Spark 集群，然后使用 spark-submit 脚本提交程序到集群进行运行，其运行命令和截图如图 9-10 所示。

```
$cd /app/spark/spark-2.0.0
$./bin/spark-submit --class chapter9.KMeansClustering \
  --master spark://master:7077 \
  --executor-memory 1024m \
  sparklearning.jar \
  hdfs://master:9000/chapter9/kmeanscluster.data 3 20
```

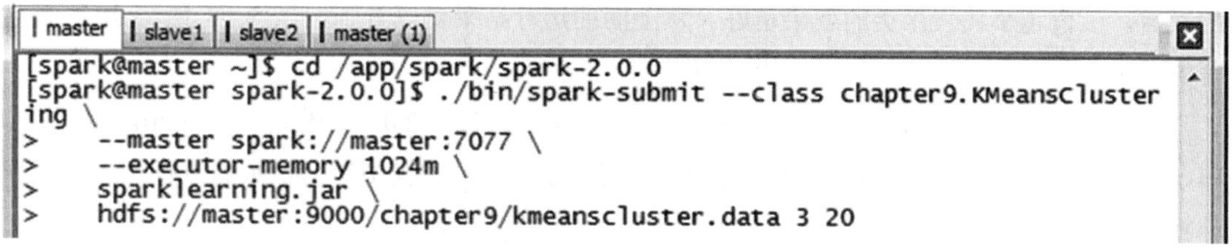

图 9-10 运行 K-means 聚类界面

该算法程序运行结果日志如下：

```
Cluster centers:
  [19.05,19.05,19.05]
  [66.15,66.15,66.15]
  [2.0500000000000003,2.0500000000000003,2.0500000000000003]
Within Set Sum of Squared Errors When K=1 -> 26465.329999999994
Within Set Sum of Squared Errors When K=2 -> 1734.4499999999937
Within Set Sum of Squared Errors When K=3 -> 0.4499999999951889
Within Set Sum of Squared Errors When K=4 -> 0.32999999999256957
Within Set Sum of Squared Errors When K=5 -> 0.31499999999315165
Within Set Sum of Squared Errors When K=6 -> 0.19499999999255735
Within Set Sum of Squared Errors When K=7 -> 0.0749999999909825
Within Set Sum of Squared Errors When K=8 -> 0.11999999999767041
Within Set Sum of Squared Errors When K=9 -> 0.07499999999214069
Within Set Sum of Squared Errors When K=10 -> 0.029999999992146087
Vectors 2.0 2.0 2.0 is belongs to clusters:2
Vectors 20 20 20 is belongs to clusters:0
Vectors 66.2 66.2 66.2 is belongs to clusters:1
2.1 2.1 2.1 2
2.2 2.2 2.2 2
1.9 1.9 1.9 2
2.0 2.0 2.0 2
```

```
18.9 18.9 18.9 0
19.0 19.0 19.0 0
19.1 19.1 19.1 0
19.2 19.2 19.2 0
66.0 66.0 66.0 1
66.1 66.1 66.1 1
66.2 66.2 66.2 1
66.3 66.3 66.3 1
```

在运行结果中可以看到,通过算法找出三个族簇中心点:(19.05, 19.05, 19.05)、(66.15, 66.15, 66.15) 和 (2.0500000000000003, 2.0500000000000003, 2.0500000000000003)。

K 值作为初始聚类中心,它是 K-means 算法的关键,一般来说,同样的迭代次数和算法跑的次数,该值越小代表聚类的效果越好。从上面的运行结果可以看到,当 K=3 的 cost 值有显著的减少,所以我们选择 3 这个临界点作为 K 的个数。

最后根据模型去判断给定点所属的族簇,其中(2.0, 2.0, 2.0)属于族簇 2,(20, 20, 20)属于族簇 0,(66.2, 66.2, 66.2)属于族簇 1。同时根据计算的模型对测试数据进行判定,计算并打印所属的族簇。

9.11.2 手机短信分类实例

在该实例中将对手机的短信的样本数据分类,通过训练的模型去检测发送的短信是否为垃圾短信。该过程是机器学习中经典的文本分类问题,该问题的主要目标是通过对已有语料库文本数据训练得到分类模型,进而对新文本进行类别标签的预测,该分类在垃圾邮件检测、新闻网站的新闻自动分类和非法信息过滤等领域经常使用。

在本实例中,首先将手机短信转化成单词数组,然后使用 ML 中 Word2Vec 工具将单词数组转化成一个 K 维向量,最后通过训练 K 维向量样本数据得到一个前馈神经网络模型,通过该模型对其他短信的标签预测。具体使用有 Spark MLlib 算法有词向量化工具 Word2Vec 和多层感知器分类器(Multiple Layer Perceptron Classifier),整体对数据的处理和模型训练采用的是 ML Pipeline 的方式。

Word2Vec 工具内容参见 9.10.2 节,Spark 在 Word2Vec.scala 类实现该方法,Word2Vec 有以下参数:

- inputCol:源数据 DataFrame 中存储文本词数组列的名称;
- outputCol:经过处理的数值型特征向量存储列名称;
- vectorSize:目标数值向量的维度大小,默认是 100;
- windowSize:上下文窗口大小,默认是 5;

- numPartitions：训练数据的分区数，默认是 1；
- maxIter：算法求最大迭代次数，小于或等于分区数，默认是 1；
- minCount：只有当某个词出现的次数大于或者等于 minCount 时，才会被包含到词汇表里，否则会被忽略掉；
- stepSize：优化算法的每一次迭代的学习速率。默认值是 0.025。

多层感知器（MLP, Multilayer Perceptron）是一种多层的前馈神经网络模型，所谓前馈型神经网络，指其从输入层开始只接收前一层的输入，并把计算结果输出到后一层，并不会给前一层有所反馈，整个过程可以使用有向无环图来表示。该类型的神经网络由三层组成，分别是输入层（Input Layer），一个或多个隐层（Hidden Layer），输出层（Output Layer），如图 9-11 所示。

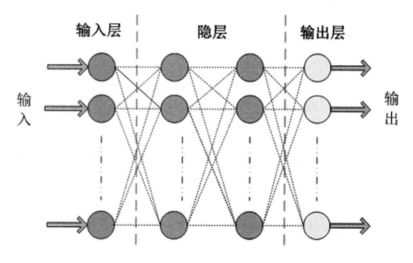

图 9-11　多层感知器模型

Spark 在 MultilayerPerceptronClassifier.scala 类中实现多层感知器分类器，支持以下参数：
- featuresCol：输入数据 DataFrame 中指标特征列的名称。
- labelCol：输入数据 DataFrame 中标签列的名称。
- layers：这个参数是一个整型数组类型，第一个元素需要和特征向量的维度相等，最后一个元素需要训练数据的标签取值个数相等，如 2 分类问题就写 2。中间的元素有多少个就代表神经网络有多少个隐层，元素的取值代表了该层的神经元的个数。例如 val layers = Array[Int](100,6,5,2)。
- maxIter：优化算法求解的最大迭代次数。默认值是 100。
- predictionCol：预测结果的列名称。

- tol：优化算法迭代求解过程的收敛阀值。默认值是 1e-4。不能为负数。
- blockSize：该参数被前馈网络训练器用来将训练样本数据的每个分区都按照 blockSize 大小分成不同组，并且每个组内的每个样本都会被叠加成一个向量，以便于在各种优化算法间传递。该参数的推荐值是 10-1000，默认值是 128。

1．测试数据说明

本示例测试的数据来自 UCI 的 SMS Spam Collection 数据集，该数据集包含两列数据：第一列是短信的标签，第二列是短信内容。它们之间使用制表符（Tab）进行分隔，以下为测试数据部分样例：

```
ham    Go until jurong point, crazy.. Available only in bugis n great world la e
  buffet... Cine there got amore wat...
ham    Ok lar... Joking wif u oni...
spam   Free entry in 2 a wkly comp to win FA Cup final tkts 21st May 2005. Text
  FA to 87121 to receive entry question(std txt rate)T&C's apply
  08452810075over18's
ham    U dun say so early hor... U c already then say...
ham    Nah I don't think he goes to usf, he lives around here though
spam   FreeMsg Hey there darling it's been 3 week's now and no word back! I'd like
  some fun you up for it still? Tb ok! XxX std chgs to send, £1.50 to rcv
ham    Even my brother is not like to speak with me. They treat me like aids patent.
......
```

2．程序代码

```scala
import org.apache.log4j.{Level, Logger}
import org.apache.spark.ml.Pipeline
import org.apache.spark.ml.classification.MultilayerPerceptronClassifier
import org.apache.spark.ml.evaluation.MulticlassClassificationEvaluator
import org.apache.spark.ml.feature.{IndexToString, Word2Vec, StringIndexer}
import org.apache.spark.sql.SQLContext
import org.apache.spark.{SparkContext, SparkConf}

object SpamMessageClassifier {
  def main(args: Array[String]) {
    //屏蔽不必要的日志显示在终端上
    Logger.getLogger("org.apache.spark").setLevel(Level.ERROR)
    Logger.getLogger("org.eclipse.jetty.server").setLevel(Level.OFF)

    //对输入参数个数进行判断，如果输入参数不为 1 则退出
    if (args.length != 1) {
      println("Usage: /path/to/spark/bin/spark-submit --master spark://master:
        9000 " +
```

```scala
    "--driver-memory 1g --class chapter9.SpamMessageClassifier " +
    "sparklearning.jar SMSDataFilePath")
  sys.exit(1)
}

//设置应用程序运行环境
val conf = new SparkConf().setAppName("SpamMessageClassifier").setMaster
  ("local[4]")
val sc = new SparkContext(conf)
val sqlCtx = new SQLContext(sc)

//从 HDFS 读取手机短信作为处理数据源,在此基础上创建 DataFrame,该 DataFrame 包含
  labelCol、contextCol 两个列
val messageRDD = sc.textFile(args(0)).map(_.split("\t")).map(line => {
  (line(0),line(1).split(" "))
})
val smsDF = sqlCtx.createDataFrame(messageRDD).toDF("labelCol",
  "contextCol")

//将原始的文本标签("Ham"或者"Spam")转化成数值型的表型
val labelIndexer = new StringIndexer()
  .setInputCol("labelCol")
  .setOutputCol("indexedLabelCol")
  .fit(smsDF)

//使用 Word2Vec 将短信文本转化成数值型词向量
val word2Vec = new Word2Vec()
  .setInputCol("contextCol")
  .setOutputCol("featuresCol")
  .setVectorSize(100)
  .setMinCount(1)

val layers = Array[Int](100,6,5,2)

//使用 MultilayerPerceptronClassifier 训练一个多层感知器模型
val mpc = new MultilayerPerceptronClassifier()
  .setLayers(layers)
  .setBlockSize(512)
  .setSeed(1234L)
  .setMaxIter(128)
  .setFeaturesCol("featuresCol")
  .setLabelCol("indexedLabelCol")
  .setPredictionCol("predictionCol")
```

```
//使用 LabelConverter 将预测结果的数值标签转化成原始的文本标签
val labelConverter = new IndexToString()
  .setInputCol("predictionCol")
  .setOutputCol("predictedLabelCol")
  .setLabels(labelIndexer.labels)

//将原始文本数据按照 8:2 的比例分成训练和测试数据集
val Array(trainingData, testData) = smsDF.randomSplit(Array(0.8, 0.2))

//使用 pipeline 对数据进行处理和模型的训练
val pipeline = new Pipeline().setStages(Array(labelIndexer, word2Vec, mpc,
  labelConverter))
val model = pipeline.fit(trainingData)
val preResultDF = model.transform(testData)

//使用模型对测试数据进行分类处理并在屏幕打印 20 笔数据
preResultDF.select("contextCol", "labelCol", "predictedLabelCol").show(20)

//测试数据集上测试模型的预测精确度
val evaluator = new MulticlassClassificationEvaluator()
  .setLabelCol("indexedLabelCol")
  .setPredictionCol("predictionCol")
val predictionAccuracy = evaluator.evaluate(preResultDF)
println("Testing Accuracy is %2.4f".format(predictionAccuracy * 100) + "%")
  sc.stop
  }
}
```

3. 执行情况

第一步　打包运行

对运行代码进行打包，生成打包文件 sparklearning.jar 并移动到 Spark 根目录下。在运行前需要启动 HDFS 和 Spark 集群，然后使用 spark-submit 脚本提交程序到集群中，其运行命令如下所示：

```
$cd /app/spark/spark-2.0.0
$./bin/spark-submit --class chapter9.SpamMessageClassifier \
  --master spark://master:7077 \
  --executor-memory 1024m \
  sparklearning.jar \
  hdfs://master:9000/chapter9/SMSSpamCollection
```

第二步　执行并观察输出，如图 9-12 所示。

```
| master | slave1 | slave2                                              |□|
+--------------------+------+----------------+
|         contextCol|labelCol|predictedLabelCol|
+--------------------+------+----------------+
|["Aww, you, must,...|   ham|              ham|
|["EY!, CALM, DOWN...|   ham|              ham|
|["Gimme, a, few",...|   ham|              ham|
|["Happy, valentin...|   ham|              ham|
|["Hey, sorry, I, ...|   ham|              ham|
|["alright, babe,,...|   ham|              ham|
|[&lt;#&gt;, , am,...|   ham|              ham|
|[&lt;#&gt;, , gre...|   ham|              ham|
|[&lt;#&gt;, %of,...|   ham|              ham|
|[*, was, a, nice,...|   ham|              ham|
|[1), Go, to, writ...|   ham|              ham|
|    [26th, OF, JULY]|   ham|              ham|
|[4, oclock, at, m...|   ham|             spam|
|[7, wonders, in, ...|   ham|             spam|
|[;-(, oh, well,,...|   ham|              ham|
|[A, boy, was, lat...|   ham|              ham|
|[A, gram, usually...|   ham|              ham|
|[Aaooooright, are...|   ham|              ham|
|[Abeg,, make, pro...|   ham|              ham|
|[After, the, drug...|   ham|              ham|
+--------------------+------+----------------+
only showing top 20 rows
Testing Accuracy is 93.4992%
```

图 9-12　手机短信分类运行结果

9.12　小结

在本章中先介绍了机器学习，一般认为机器学习是研究计算机怎样模拟或实现人类的学习行为，以获取新的知识或技能，重新组织已有的知识结构使之不断改善自身的性能。机器学习可以分为监督学习、半监督学习、无监督学习和强化学习五种方式。

然后介绍了 Spark MLlib，MLlib 是 MLBase 一部分，其中 MLBase 分为四部分：MLlib、MLI、ML Optimizer 和 MLRuntime。它提供了常用机器学习算法的实现，包括分类、回归、聚集、协同过滤和降维等。Spark 机器学习库从 1.2 版本以后被分为两个包 spark.mllib 和 spark.ml，从官方文档来看，Spark ML Pipeline 虽然是被推荐的机器学习方式，但是并不会在短期内替代原始的 MLlib 库，因为 MLlib 已经包含了丰富稳定的算法实现。

接着几个小节中，按照不同的机器学习类型介绍详细内容。最后通过使用 K-means 聚类算法和手机短信分类两个实例演示如何利用 Spark MLlib 进行数据分析，在 K-means 聚类算法中使用了 K-Means 算法对坐标数据进行分类，而在手机短信分类中使用词向量化工具 Word2Vec 和多层感知器分类器对手机的短信进行分类。

第 10 章

Spark GraphX

10.1　GraphX 介绍

10.1.1　图计算

"图计算"是以"图论"为基础对现实"图"结构的抽象，以及在该结构上进行的计算。在图计算中，图一般表示为 G=（V，E，D），其中，V 为顶点或节点（Vertex），E 为边（Edge），D 为权重（Data）。比如在电子购物中可以分为两类节点，用户和所购买的商品，边指的是用户购买商品的行为，而权重则是购买的次数、价格和购买的时间等。在现实社会中，很多行为可以用图结构来进行抽象处理，比如微信、微博、社交网络、地图导航和网页链接关系，甚至消费者网上的购物、评论内容和评论标签等。

图结构很好地表示了数据间的关联性，其是大数据计算的核心，通过关联性可以从海量的数据中分析挖掘有用的信息。比如，通过对社交网络的建模，可以进行社交圈子识别；或者通过构建用户兴趣图谱为用户推荐如新闻、图片、群组、广告等信息。但是由于图结构中数据内部存在较高的关联性，在图计算时会引入大量的连接和聚集等操作，严重消耗计算资源，迫切需要对这些算法进行优化。

针对图计算这些问题，大量的图并行计算框架被开发出来，这些计算框架根据图结构的特点，对数据进行更好的组织和存储，优化计算框架以达到更好的分布式计算性能。这些计算框架一般使用分布式数据存储，在各个计算节点上对图数据进行计算、传输和更新等操作。通过图顶点的数据及与顶点相连的边上的数据，产生发送给与其相连的顶点的消息，消息通过网络或其他方式传输给指定的顶点，顶点在收到消息或状态后进行更新，生成新的图数据。

10.1.2 GraphX 介绍

GraphX 是构建于 Spark 核心基础上的并行图计算框架，基于 Spark 提供的 RDD 抽象，一方面依赖于 RDD 的容错性实现了高效健壮，另一方面 GraphX 可以与 Spark SQL、MLLib 等无缝地结合使用，例如使用 Spark SQL 收集的数据可以交由 GraphX 进行处理，而 GraphX 在计算时可以和 MLLib 结合完成深度数据挖掘等操作，这些是其他图计算框架所没有的。类似于 GraphLab，GraphX 基于 GAS 模型，以边为中心，对点进行切割存储的。在该模型中顶点数据将分配个集群中个顶点进行存储，从而增大并行度解决常遇到的高出度顶点的情况。

GraphX 将图计算和数据计算集成到一起，数据不仅可以作为图进行操作，同样可以作为表进行操作。这样可以解决传统的图计算往往需要不同的系统支持不同的 View，例如在 Table View 这种视图下可能需要 Spark 的支持或者 Hadoop 的支持，而在 Graph View 这种视图下可能需要 Pregel 或者 GraphLab，如图 10-1 所示。

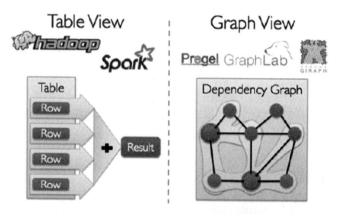

图 10-1　GraphX 将图计算和数据计算集成一起

上面所描述的图计算处理方式是传统的计算方式，当然现在除了 Spark GraphX 之外的图计算框架也在考虑这个问题；不同的系统带来的问题是之一是需要学习、部署和管理不同的系统，例如要同时学习、部署和管理 Hadoop、Hive、Spark、Giraph 和 GraphLab 等。

其实最关键的还是效率问题，因为在不同的转换中间每步都要落地的话，数据转换和复制带来的开销也非常大，包括序列化带来的开销。同时中间结果和相应的结构无法重用，特别是一些结构性的东西，比如说顶点或者边的结构一直没有变，这种情况下结构内部的 Structure 是不需要改变的，而如果每次都重新构建的话，就算不变也无法重用，这会使性能非常差，如图 10-2 所示。

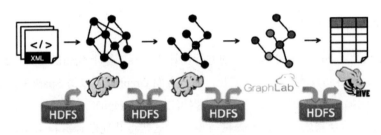

图 10-2　转换的效率问题

解决方案就是 Spark GraphX。GarphX 实现了 Unified Representation，GraphX 统一了 Table View 和 Graph View，基于 Spark 可以非常轻松地进行 pipeline 的操作，如图 10-3 所示。

图 10-3　基于 GarphX 的解决方案

GraphX 不仅被广泛应用于用户网络的社区发现、用户影响力、能量传播、标签传播等，而且也越来越多地应用到推荐领域的标签推理、人群划分、年龄段预测、商品交易时序跳转等。另外，GraphX 非常适合于微信、微博、社交网络、电子商务、地图导航等类型的产品。

10.1.3　发展历程

- 早在 Spark 0.5 版本，Spark 就带了一个基本的 Bagel 模块，提供了类似 Pregel 的功能，所有图处理任务由一系列超步组成，每个超步中，用户可以通过 Bagel 接口指定顶点更新和处理的函数，并指定该顶点与其他顶点通信的消息。当然，这个版本还非常原始，性能和功能都比较弱，属于实验型产品。

- 到 Spark 0.8 版本时，鉴于业界对分布式图计算的需求日益见涨，Spark 开始独立一个分支 Graphx-Branch，作为独立的图计算模块，借鉴 GraphLab，开始设计开发 GraphX。GraphX 在 Spark 定位为新的面向图数据和并行图计算的接口，提供丰富的图操作接口，扩展和改善 Bagel 的处理性能和应用范围。

- 在 Spark 0.9 版本中，这个模块被正式集成到主干，虽然是 Alpha 版本，但已可以试用。Spark 1.0 版本时，对 GraphX 进行了性能优化，EdgeRDD 可以通过临界点属性构建边点三元组 Triplet 对象。

- 在 Spark 1.2 版本中，推出了 GraphX 正式版。在该版本 EdgeRDD 和 VertexRDD 加入了 StorageLevel 参数能够更加灵活控制图数据存储方式。

值得注意的是，GraphX 依然处于快速发展中，每个版本代码都有不少的改进和重构。虽然和 GraphLab 的性能还有一定差距，但凭借 Spark 整体上的一体化流水线处理，社区热烈的活跃度及快速改进速度，GraphX 具有强大的竞争力。

10.2 GraphX 实现分析

如同 Spark 核心实现，GraphX 代码非常简洁，其实现架构大致分为 3 层，如图 10-4 所示。

- 实现层：该层定义了 GraphX 由最基本的数据结构，包括顶点、边和边三元组；介绍了 GraphX 不同的顶点切分策略以及数据存储方式；介绍了图计算过程使用的数据存储结构，如路由表、重复顶点视图等。
- 操作层：主要包括抽象类 GraphX 及其实现类 GraphImpl，在这两个类中定义了构建图操作、转换操作、结构操作、聚合操作和缓存操作等。另外，在 GraphOps 类中也实现了图基本属性操作和连接操作等。
- 算法层：在 GraphX 根据实现层和操作层实现了常用的算法，如 PageRank、三角关系统计、最短路径等。

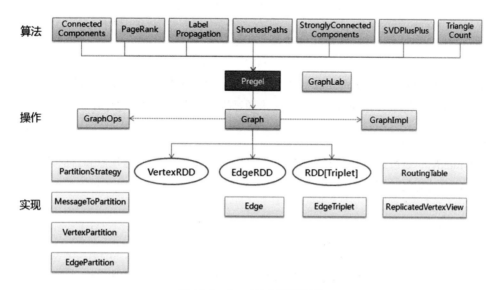

图 10-4　GraphX 实现架构图

10.2.1 GraphX 图数据模型

GraphX 中图定义为 G(P)=(V, E, P) 来表示。其中 V 表示顶点（Vertex）；E 表示边（Edge），由边的两个顶点 ID 构成的元组表示，在有向图中（i, j）代表由顶点 i 指向顶点 j 的边，在无向图中（i, j）代表由顶点 i 和顶点 j 之间的边；P（Property）表示边属性，为顶点的数据和边的数据。

GraphX 最基本的数据结构为顶点（Vertex）、边（Edge）和边三元组（EdgeTriplet）。

- 顶点（Vertex）：包含顶点 ID 和顶点数据（VD），可以表示为（顶点 ID，顶点数据 VD）。
- 边（Edge）：包含源顶点和目标顶点 ID 以及边数据（ED），可以表示为（源顶点 ID，目标顶点 ID，边数据 ED）。
- 边三元组（EdgeTriplet）：它是边的子类，在边的基础上存储了边的源顶点和目标顶点的数据，可以表示为（源顶点，目标顶点，边数据）。

1. Vertex

在 GraphX 中顶点数据抽象为 VertexRDD，它是对 RDD[（VertexID, VD）]的继承和扩展，RDD 的类型是 VertexId 和 VD，其中的 VD 是属性的类型，即 VertexRDD 有 ID 和点属性。VertexRDD 由若干个顶点分区组成，其顶点分区的定义如下所示：

```
private[graphx] class VertexPartition[VD: ClassTag](
    val index: VertexIdToIndexMap,   //顶点 ID 与顶点数据在顶点数组中的索引
    val values: Array[VD],           //存放顶点数据的数组
    val mask: BitSet)                //表示过滤 index 中顶点的掩码
  extends VertexPartitionBase[VD]
```

顶点使用顶点 ID 进行哈希分区，每个分区中顶点的数据存储在一个数据组中，顶点的 ID 存在在一个哈希列表中，该哈希列表记录了顶点 ID 与顶点数据在顶点数组中的索引的映射。每个顶点分区中包含掩码，该掩码用于计算时，通过掩码的设置过滤分区中不需要的顶点。

2. Edge

在 GraphX 中边数据抽象为 EdgeRDD，在边定义中包含了边的源顶点和目的顶点的 ID 以及边数据，其边定义如下：

```
case class Edge[ED] (
    var srcId: VertexId = 0,
    var dstId: VertexId = 0,
    var attr: ED = null.asInstanceOf[ED])
```

类似于顶点，EdgeRDD 由若干个边分区组成，其中边分区的定义如下：

```
class EdgePartition[@specialized(Char, Int, Boolean, Byte, Long, Float, Double)
```

```
    ED: ClassTag, VD: ClassTag](
    localSrcIds: Array[Int],              //源顶点数组
    localDstIds: Array[Int],              //目标顶点数组
    data: Array[ED],                      //所有边数据
    index:GraphXPrimitiveKeyOpenHashMap[VertexId, Int],//顶点ID在源ID数组中的索引
    global2local: GraphXPrimitiveKeyOpenHashMap[VertexId, Int],
    local2global: Array[VertexId],        //顶点数组
    vertexAttrs: Array[VD],               //顶点属性数组
    activeSet: Option[VertexSet])         //活跃顶点的ID
```

为了快速查找，在一个分区内的所有边根据源顶点聚集在一起，并使用哈希映射存储源顶点 ID 和其所在源顶点 ID 数组中的起始索引的映射。在某个边分区中查找某条边时，先通过边的源顶点 ID 查找到该顶点在源顶点数组中的起始索引，然后使用该索引为起点，在源顶点数组和目的顶点数组中进行搜索，直到找到对应的源顶点 ID 和目的顶点 ID。

3. Triplet

边点三元体 Triplet 是 Graphx 提出的特有的数据结构，它的属性有源顶点 ID、源顶点属性、边属性、目标顶点 ID 和目标顶点属性，Triplets 其实是对 Vertices 和 Edges 做了连接操作，如图 10-5 所示。

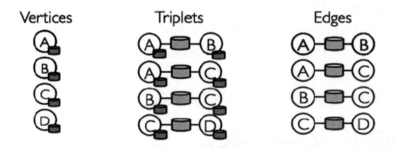

图 10-5　Triplets 对 Vertices 和 Edges 进行的连接操作

从源码中我们看到，GraphX 使用 EdgeTriplet 表示 Triplet，EdgeTriplet 继承自 Edge[ED]，加入了 srcAttr 表示源顶点的属性和 dstAttr 表示目的顶点的属性。Vertices 具有顶点 ID 和属性，Edges 具有源顶点 ID、目的顶点 ID 和自身的属性，Triplets 对 Vertices 和 Edges 的连接操作使得 Triplets 具备源顶点的 ID 和属性、目的顶点 ID 和属性以及自身的属性，EdgeTriplet 为图的数据遍历提供了方便。其源码实现如下：

```
class EdgeTriplet[VD, ED] extends Edge[ED] {
  var srcAttr: VD = _
  var dstAttr: VD = _
```

```
protected[spark] def set(other: Edge[ED]): EdgeTriplet[VD, ED] = {
  srcId = other.srcId
  dstId = other.dstId
  attr = other.attr
  this
}
......
}
```

10.2.2　GraphX 图数据存储

图存储一般有边分割和点分割两种存储方式。2013 年，GraphLab 2.0 将其存储方式由边分割变为点分割，在性能上取得重大提升，目前基本上被业界广泛接受并使用。

- **边分割（Edge-Cut）**：每个顶点都存储一次，但有的边会被打断分到两台机器上。这样做的好处是节省存储空间；坏处是对图进行基于边的计算时，对于一条两个顶点被分到不同机器上的边来说，要跨机器通信传输数据，内网通信流量大。如图 10-6 中左边图分为 3 个分区，其分区 1 中包含顶点 A 和顶点 C，分区 2 中包含顶点 B，而分区 3 中包含顶点 D，这些顶点在集群中只存储一次。

- **点分割（Vertex-Cut）**：每条边只存储一次，都只会出现在一台机器上。邻居多的点会被复制到多台机器上，增加了存储开销，同时会引发数据同步问题。点分割的好处是可以大幅减少内网通信量。如图 10-6 中右边图分为 3 个分区，其中分区 1 包含顶点 A、B 和边 AB，分区 2 包含顶点 B、C 和边 BC，分区 3 包含顶点 C、D 和边 CD，这些边在集群中只存储一次，而顶点可能重复存储。

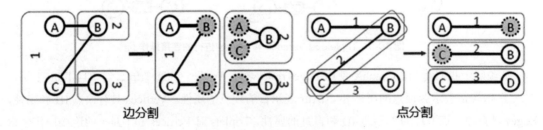

图 10-6　边分割和点分割方式

虽然两种方法互有利弊，但现在是点分割占上风，各种分布式图计算框架都将自己底层的存储形式变成了点分割。其主要原因有以下两个。

（1）磁盘价格下降，存储空间不再是问题，而内网的通信资源没有突破性进展，集群计算时内网带宽是宝贵的,时间比磁盘更珍贵。这点就类似于常见的空间换时间的策略。

(2)在当前的应用场景中,绝大多数网络都是"无尺度网络",遵循幂律分布,不同点的邻居数量相差非常悬殊。而边分割会使那些多邻居的点所相连的边大多数被分到不同的机器上,这样的数据分布会使得内网带宽更加捉襟见肘,于是边分割存储方式被渐渐抛弃了。

Graphx 使用的是点分割方式存储图,用 VerterRDD、EdgeRDD 和 RoutingTable 三个 RDD 存储图数据信息。存储实现如图 10-7 所示。

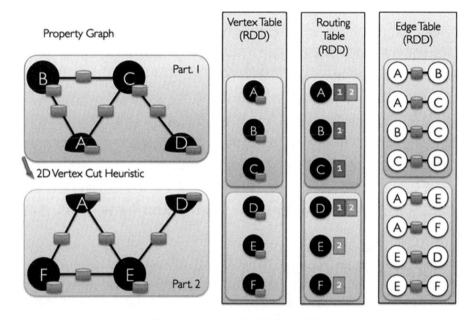

图 10-7　Graphx 点分割方式存储图

其中,路由表中记录了图在进行运算时,将顶点数据从顶点 RDD 发给边 RDD 的路由数据。收集的数据包括由每条边分区中源顶点和目的顶点 ID 以及边分区的 ID 组成的键值对,并对这些数据根据顶点分区数据进行重新分区,生成针对每个顶点分区的路由数据,生成一个新的 RDD,其定义如下:

```
private val routingTable: Array[(Array[VertexId], BitSet, BitSet)])
```

将路由数据进行这样的分类时,为了根据图计算的实际需要选择合适的路由数据,例如,若计算仅需要源顶点的数据,那么可以选择仅包含源顶点数据的路由数据,在运算时只会将源顶点的数据发送给边分区,从而减少了数据传输量。

另外在图存储中重复顶点视图用于记录顶点数据根据路由表的路由数据被发送到边时的状态,定义如下:

```
class ReplicatedVertexView[VD: ClassTag, ED: ClassTag](
    var edges: EdgeRDDImpl[ED, VD],
    var hasSrcId: Boolean = false,
    var hasDstId: Boolean = false)
```

重复顶点视图与边 RDD 是同步分区（co-partition）的，即分区个数相同且分区方法相同，在图计算时可以对重复顶点视图和边 RDD 的分区进行拉链（zipPartition）操作，即将重复顶点视图和边 RDD 的分区一一对应地组合起来。在整个运算过程中，只有在创建重复顶点视图时需要移动顶点数据，边数据不需要移动。由于顶点数据一般比边数据要少得多，而且随着迭代次数的增加需要更新的顶点数据也越来越少，这样可以大大减少数据的移动量。重复顶点视图在创建之后会缓存在内存中，用于多次使用，如果程序不需要使用该重复顶点视图，需要手动调用 GraphImpl 的 unpersistVertices 将其清除出内存。

生成重复顶点视图时，在边 RDD 的每个分区中创建集合，存储该分区包含的源顶点和目的奠定的 ID 集合，该集合被称作本地顶点 ID 映射。在生成重复顶点视图时，若顶点视图时第一次被创建，则把本地顶点 ID 映射和发送给边 RDD 各分区的顶点数据组合起来，在每个分区中以分区中的本地顶点 ID 映射为索引存储顶点数据，生成新的顶点分区，最后得到一个新的顶点 RDD。若重复顶点视图不是第一次创建，则使用之前重复顶点视图创建的顶点 RDD 与发送给边 RDD 各分区的顶点更新数据进行连接操作，更新顶点 RDD 中的顶点数据，生成新的顶点 RDD。

10.2.3　GraphX 图切分策略

图的切分策略是指如何将图切分成不同的分区进行存储，GraphX 中的图数据是以 RDD 的方式进行存储的，对图的切分也就是对图的 RDD 进行切分。切分策略由 PartitionStrategy 特质 (trait)进行定义，在该特质中定义了 getPartition 方法。

在 GraphX 中内置实现了 4 种图切分策略，不过用户也可以通过实现 PartitionStrategy 接口来自定义分区方法，其中内置切分策略如图 10-8 所示。

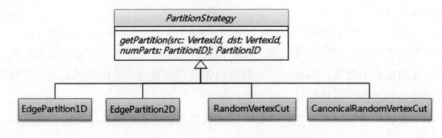

图 10-8　GraphX 的 4 种图切分策略类图

1. 1D 分区方法

在 1D 分区方法中，仅使用边的源顶点 ID 计算分区值，这样可以将所有源顶点相同的边放到同一个分区中。

```
case object EdgePartition1D extends PartitionStrategy {
  override def getPartition(src: VertexId, dst: VertexId, numParts: PartitionID):
    PartitionID = {
    val mixingPrime: VertexId = 1125899906842597L
    (math.abs(src * mixingPrime) % numParts).toInt
  }
}
```

2. 2D 分区方法

2D 分区方法是二维划分方法，它能够将边划分至 n 个节点，保证顶点在集群中的复制份数小于 $2^{\sqrt{numParts}}$，同时尽可能保持边分布的均匀性，从而保证图计算负载的均衡。该分区方法定义如下：

```
case object EdgePartition2D extends PartitionStrategy {
  override def getPartition(src: VertexId, dst: VertexId, numParts:
    PartitionID): PartitionID = {
    //获取分配矩阵的平方根因子，它能够保证对源顶点和目的顶点随机化的过程中，
    //复制份数小于 2^sqrt(numParts)
    val ceilSqrtNumParts: PartitionID = math.ceil(math.sqrt(numParts)).toInt
    val mixingPrime: VertexId = 1125899906842597L
    if (numParts == ceilSqrtNumParts * ceilSqrtNumParts) {
      //引入 mixingPrime 常量增加 col 和 row 分配的随机性和均匀性
      val col:PartitionID=(math.abs(src*mixingPrime)%ceilSqrtNumParts).toInt
      val row:PartitionID=(math.abs(dst*mixingPrime)%ceilSqrtNumParts).toInt
      //均衡分配到 numParts 个分区
      (col * ceilSqrtNumParts + row) % numParts
    } else {
      val cols = ceilSqrtNumParts
      val rows = (numParts + cols - 1) / cols
      val lastColRows = numParts - rows * (cols - 1)
      val col = (math.abs(src * mixingPrime) % numParts / rows).toInt
      val row = (math.abs(dst * mixingPrime) % (if (col < cols - 1) rows else
        lastColRows)).toInt
      col * rows + row
    }
  }
}
```

3. 随机顶点分区方法

随机顶点切分使用边的源顶点和目的顶点 ID 计算哈希值得到分区值，这样分区可以将两个顶点之间相同的边放到同一个分区中。

```
case object RandomVertexCut extends PartitionStrategy {
  override def getPartition(src: VertexId, dst: VertexId, numParts:
    PartitionID): PartitionID = {
   math.abs((src, dst).hashCode()) % numParts
  }
}
```

4. 正则随机顶点分区方法

与随机顶点分区方法类似，只是在计算哈希值前进行一次顶点排序。在源顶点和目的顶点中，ID 小的排前面，ID 大的排后面，然后对边进行计算哈希值并除余计算得到相应的分区数，这样分区可以将两个顶点之间的边都放到同一个分区中，无论边的方向如何。

```
case object CanonicalRandomVertexCut extends PartitionStrategy {
  override def getPartition(src: VertexId, dst: VertexId, numParts:
    PartitionID): PartitionID = {
   if (src < dst) {
    math.abs((src, dst).hashCode()) % numParts
   } else {
    math.abs((dst, src).hashCode()) % numParts
   }
  }
}
```

10.2.4　GraphX 图操作

GraphX 可以使用 Spark 所支持的数据并行操作，同时 GraphX 提供了丰富的图并行操作。

1. 构建图操作（见表 10-1）

表 10-1　构建图操作

接　　口	描　　述
fromEdgeTuples[VD: ClassTag](rawEdges: RDD[(VertexId, VertexId)], defaultValue: VD, uniqueEdges: Option[PartitionStrategy] = None, edgeStorageLevel: StorageLevel =	通过一组边构建图，其中边由源顶点和目的顶点组成，在参数中提供了顶点和边的存储级别，默认情况下均存储在内存中 ● defaultValue 为顶点的默认数据，用于当顶点在边 RDD 存在但在顶点 RDD 中不存在为顶点提供默认值

接口	描述
StorageLevel.MEMORY_ONLY, vertexStorageLevel: StorageLevel = StorageLevel.MEMORY_ONLY): Graph[VD, Int]	• uniqueEdges 参数用于提供一个分区策略，对生成的图结构进行分区操作，若提供了该参数，图中重复的边会被合并，重复边的属性进行相加得到合并后的属性。若不提供该参数，则图中的重复边不会做任务处理 • edgeStorageLevel 参数为边的存储级别，默认情况为存储在内存中 • vertexStorageLevel 参数为顶点的存储级别，默认情况为存储在内存中
fromEdges[VD: ClassTag, ED: ClassTag](edges: RDD[Edge[ED]], defaultValue: VD, edgeStorageLevel: StorageLevel = StorageLevel.MEMORY_ONLY, vertexStorageLevel: StorageLevel = StorageLevel.MEMORY_ONLY): Graph[VD, ED]	通过一组边构建图，其中边的数据包含了顶点数据和边的值，该接口提供了顶点的默认值，在参数中提供了顶点和边的存储级别，默认情况下均存储在内存中 • defaultValue 为顶点的默认数据，用于当顶点在边 RDD 存在但在顶点 RDD 中不存在为顶点提供默认值 • edgeStorageLevel 参数为边的存储级别，默认情况为存储在内存中 • vertexStorageLevel 参数为顶点的存储级别，默认情况为存储在内存中
apply[VD: ClassTag, ED: ClassTag](vertices: RDD[(VertexId, VD)], edges: RDD[Edge[ED]], defaultVertexAttr: VD = null.asInstanceOf[VD], edgeStorageLevel: StorageLevel = StorageLevel.MEMORY_ONLY, vertexStorageLevel: StorageLevel = StorageLevel.MEMORY_ONLY): Graph[VD, ED]	该方法是 Graph 在创建图时使用的默认方法，在参数中提供了顶点和边的存储级别，默认情况下均存储在内存中 • RDD[VertexId, VD] 顶点 RDD 包含顶点的 ID 和数据 • RDD[Edge[ED]]边 RDD 包含边的源顶点和目标顶点 ID 以编数据 • edgeStorageLevel 参数为边的存储级别，默认情况为存储在内存中 • vertexStorageLevel 参数为顶点的存储级别，默认情况为存储在内存中

2. 转换操作（见表 10-2）

表 10-2　转换操作

接口	描述
partitionBy(partitionStrategy: PartitionStrategy): Graph[VD, ED]	根据分区策略对图的边进行分区操作，关于分区策略详细信息参见 9.2.3 中的图切分策略

续表

接口	描述
partitionBy(partitionStrategy: PartitionStrategy, numPartitions: Int): Graph[VD, ED]	根据分区策略对图的边进行分区操作，在参数中指定了分区数目，关于分区策略详细信息参见 9.2.3 中的图切分策略
mapVertices[VD2: ClassTag](map: (VertexId, VD) => VD2): Graph[VD2, ED]	用于对图中每个顶点的数据 VD 进行转换，生成新的顶点数据 VD2，从而生成一个新的图，新的图与原图具有相同的结构。例如：可以将顶点中的字符串类型的数据转换成对应的整型数据
mapEdges[ED2: ClassTag](map: Edge[ED] => ED2): Graph[VD, ED2]	和 mapVertices 类似，mapEdges 方法是对图中的每条边的数据 ED 进行转换，生成新的边数据 ED2，从而生成一个新的图，新的图与原图具有相同的结构。不过在转换的过程中并没有把顶点的数据传递给邻边的顶点
mapEdges[ED2: ClassTag](map: (PartitionID, Iterator[Edge[ED]]) => Iterator[ED2]): Graph[VD, ED2]	与上面的 mapEdges 功能一致，不同之处在于上面接口方法每次只能处理一个边的数据，而该方法则批量处理一个分区所有边的数据。在该方法中传入的是该分区编号 partitionID 和针对该分区边的 Iterator，返回的是包含新的边值（ED）的分区的 Iterator，并且新分区中的表与原始分区中的边是一一对应的
mapTriplets[ED2: ClassTag](map: EdgeTriplet[VD, ED] => ED2) : Graph[VD, ED2]	使用该接口方法能够进行每条边的数据（ED）及边所连接的两个顶点的数据（VD）进行计算，同时生成新的边数据（ED2）。该变换不会改变图结构或改变图的值
mapTriplets[ED2: ClassTag](map: EdgeTriplet[VD, ED] => ED2, tripletFields: TripletFields) : Graph[VD, ED2]	与上面的 mapTriplets 功能上一致，不同之处在于增加了参数 TripletFields，该参数用于过滤不需要进行转换的 Triplet 对象，这样有助于提交处理效率
mapTriplets[ED2: ClassTag](map: (PartitionID, Iterator[EdgeTriplet[VD, ED]]) => Iterator[ED2],tripletFields: TripletFields): Graph[VD, ED2]	与上面的 mapTriplets 功能上一致，不同之处在于上面接口方法每次只能处理一个 EdgeTriplet 对象，而该方法则批量处理一个分区所有 EdgeTriplet 对象。在该方法中传入的是该分区编号 partitionID 和针对该分区 EdgeTriplet 对象的 Iterator，返回的是包含新的 Iterator[ED2]和 TripletFields，并且新分区中的数据与原始分区的数据是一一对应的

3. 结构操作（见表 10-3）

表 10-3 结构操作

接口	描述
reverse: Graph[VD, ED]	将图中所有边进行反转，即若某条边连接两个顶点 a 和 b，初始时边的方向为由 a 指向 b，经过反转操作后边的方向由 b 指向 a
subgraph(epred: EdgeTriplet[VD, ED] => Boolean = (x => true), vpred: (VertexId, VD) => Boolean = ((v, d) => true)): Graph[VD, ED]	在图中对顶点和边数据按照要求进行过滤生成一个新图。首先生成图的边三元组，然后根据对图中源顶点、目的顶点和边的判定条件过滤边三元组，从而生成新的边数据，顶点数据可以通过对原图的顶点过滤得到，最后使用新的边数据和顶点数据生成新图
mask[VD2: ClassTag, ED2: ClassTag](other: Graph[VD2, ED2]): Graph[VD, ED]	在当前图中获取其他图中同样存在的顶点和边，获取的新图并保持顶点和边的数据与原图一致
groupEdges(merge: (ED, ED) => ED): Graph[VD, ED]	将两个顶点之间多条边合并成一条，在合并过程中先使用 partitionBy 方法进行分区操作，以保证获取正确的结果

4. 连接操作（见表 10-4）

表 10-4 连接操作

接口	描述
outerJoinVertices[U: ClassTag, VD2: ClassTag](other: RDD[(VertexId, U)])(mapFunc: (VertexId, VD, Option[U]) =>VD2)(implicit eq: VD =:= VD2 = null): Graph[VD2, ED]	通过该接口可以实现当前图和其他图进行连接操作，并在连接结果上使用 mapFunc 函数进行计算，形成一个新图。该方法为 GraphX 核心操作接口

5. 聚合操作（见表 10-5）

表 10-5 聚合操作

接口	描述
mapReduceTriplets[A: ClassTag](mapFunc: EdgeTriplet[VD, ED] => Iterator[(VertexId, A)],reduceFunc: (A, A) => A,activeSetOpt: Option[(VertexRDD[_],	该接口为 GraphX 模型的核心操作接口，首先从某个顶点相邻的边和顶点获取数据组合成 EdgeTriplet[VD,ED]，然后使用用户定义的 mapFunc 对这些 EdgeTriplet 执行计算，生成该顶点发送给其他顶点的消息，最后使用用户定义的 reduceFunc 将不同顶点发送给同一顶点的消息 A 进行归并，生成发送给该顶点的消息。由于 mapFunc 只生成发送给源或目的节点的消息，从而确保了数据移动是沿着边进行的。计算是按照边进行的，

续表

接　　口	描　　述
EdgeDirection)] = None): VertexRDD[A]	生成发送给顶点的消息后再进行聚集操作（aggregation），从而减少高出入度的顶点在计算和消息传递过程中造成的负载不均衡的影响
aggregateMessages[A: ClassTag](sendMsg : EdgeContext[VD, ED, A] => Unit, mergeMsg: (A, A) => A, tripletFields: TripletFields = TripletFields.All): VertexRDD[A]	在 Spark1.2 版本中加入该方法，用于替换 mapReduce- Triplets 方法。该方法成为 GraphX 最重要的图操作方法，主要用来高效解决相邻边或相邻顶点之间的通信问题，例如：将与顶点相邻的边或顶点的数据聚集在顶点上、将顶点数据散发在相邻边上，它能够简单、高效地解决 PageRank 等图迭代应用。 该方法计算分为三步：①由边三元组生成消息；②向边三元组的顶点发送消息；④顶聚合收到的消息。它实现分为 Map 阶段和 Reduce 阶段两个阶段 ● Map 阶段中 GraphX 使用顶点 RDD 更新重复顶点视图。重复顶点视图与边 RDD 进行分区拉链（zipPartitions）操作，将顶点数据传往边 RDD 分区，实现边三元组视图 ● 对边 RDD 进行 map 操作，依据用户提供的函数为每个边三元组产生一个消息（Msg），生成以顶点 ID、消息为元素的 RDD，其类型为 RDD[(VertexId, Msg)] ● Reduce 阶段中，首先对第一步中的消息 RDD 按顶点分区方式进行分区（使用顶点 RDD 的分区函数），分区后的消息 RDD 与该图的顶点 RDD 元素分布状况将完全相同。在分区时，GraphX 会使用用户提供的聚合函数合并相同顶点的消息，最终形成类似顶点 RDD 的消息 RDD
aggregateMessagesWithActiveSet[A: ClassTag](sendMsg: EdgeContext[VD, ED, A] => Unit, mergeMsg: (A, A) => A, tripletFields: TripletFields, activeSetOpt: Option[(VertexRDD[_], EdgeDirection)]): VertexRDD[A]	相对前一个方法 aggregateMessages 方法增加了 activeSetOpt 过滤参数，功能和计算过程类似

6. 缓存操作（见表 10-6）

表 10-6　缓存操作

接　　口	描　　述
persist(newLevel: StorageLevel = StorageLevel.MEMORY_ONLY): Graph[VD, ED]	使用指定的存储级别存储顶点和边的数据，忽略在此之前数据所指定的存储级别
cache(): Graph[VD, ED]	将图缓存到内存中。由于在图的计算过程中，RDD 并不是一直都保存在内存中，然而在计算过程中，可能会多次用到图数据，为了避免开销，将图缓存到内存中
unpersist(blocking: Boolean = true): Graph[VD, ED]	释放存储中的顶点和边的数据，该方法多用于迭代生成新图之前对旧数据的清理
unpersistVertices(blocking: Boolean = true): Graph[VD, ED]	释放内存中缓存的顶点数据，适用于只修改点的属性值，但会重复使用边进行计算的迭代操作。此方法可以释放先前迭代的顶点属性（当其不再需要的时候），提高 GC 性能
checkpoint(): Unit	对图计算过程中的结果进行检查点操作，这些结果会暂时保存在可靠的存储中

7. 基本属性操作（见表 10-7）

表 10-7　基本属性操作

接　　口	描　　述
numVertices: Long	整个图顶点总数
numEdges: Long	整个图边总数
inDegrees: VertexRDD[Int]	整个图所有顶点的入度，若顶点无入度，则不出现在结果中
outDegrees: VertexRDD[Int]	整个图所有顶点的出度，若顶点无出度，则不出现在结果中
degrees: VertexRDD[Int]	计算图中各顶点的度
degreesRDD(edgeDirection: EdgeDirection): VertexRDD[Int]	计算相邻顶点的度，edgeDirection 参数控制收集方向
collectNeighborIds(edgeDirection: EdgeDirection): VertexRDD[Array[VertexId]]	收集每个顶点的相邻顶点的 ID 数据，edge-Direction 用于控制收集的方向
collectNeighbors(edgeDirection: EdgeDirection): VertexRDD[Array[(VertexId, VD)]]	收集每个顶点的相邻顶点的数据，当图中顶点的出入度较大时，可能会占用很大的存储空间，参数 edgeDirection 用于控制收集的方向

续表

接 口	描 述
collectEdges(edgeDirection: EdgeDirection): VertexRDD[Array[Edge[ED]]]	收集每个顶点边的数据,参数 edgeDirection 用于控制收集的方向
joinVertices[U: ClassTag](table: RDD[(VertexId, U)])(mapFunc: (VertexId, VD, U) => VD): Graph[VD, ED]	使用输入的顶点数据更新生成新的顶点数据。将当前图的数据和输入的顶点数据做内连接操作,过滤输入数据中不存在的顶点,并对过滤的结果数据使用自定义函数进行计算,如果输入数据中没有包含图中某些顶点数据,则在新图中使用原图的顶点数据
filter[VD2: ClassTag, ED2: ClassTag](preprocess: Graph[VD, ED] => Graph[VD2, ED2], epred: (EdgeTriplet[VD2, ED2]) => Boolean = (x: EdgeTriplet[VD2, ED2]) => true,vpred: (VertexId, VD2) => Boolean = (v: VertexId, d: VD2) => true): Graph[VD, ED]	根据条件对图进行过滤操作,首先使用预处理函数(preprocess)对图进行转换操作生成新的顶点和边的数据,然后在新的图数据上使用 epred 和 vpred 函数分别对边和顶点进行过滤操作,最后返回过滤后的过结果数据
pickRandomVertex(): VertexId	在图中随机获取一个顶点并返回该顶点的 ID

10.3 实例演示

10.3.1 图例演示

1. 实例介绍

图 10-9 中有 6 个人,每个人有名字和年龄,这些人根据社会关系形成 8 条边,每条边有其属性。在以下例子演示中将构建顶点、边和图,打印图的属性、转换操作、结构操作、连接操作、聚合操作,并结合实际要求进行演示。

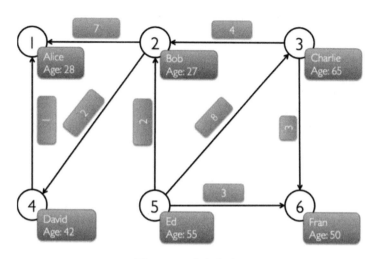

图 10-9　6 个人关系图

2．程序代码

其代码如下：

```
import org.apache.log4j.{Level, Logger}
import org.apache.spark.{SparkContext, SparkConf}
import org.apache.spark.graphx._
import org.apache.spark.rdd.RDD

object GraphXExample {
  def main(args: Array[String]) {
    //屏蔽日志
    Logger.getLogger("org.apache.spark").setLevel(Level.WARN)
    Logger.getLogger("org.eclipse.jetty.server").setLevel(Level.OFF)

    //设置运行环境
    val conf = new SparkConf().setAppName("SimpleGraphX").setMaster("local")
    val sc = new SparkContext(conf)

    //设置顶点和边,注意顶点和边都是用元组定义的Array
    //顶点的数据类型是VD:(String,Int)
    val vertexArray = Array(
      (1L, ("Alice", 28)),
      (2L, ("Bob", 27)),
      (3L, ("Charlie", 65)),
      (4L, ("David", 42)),
      (5L, ("Ed", 55)),
      (6L, ("Fran", 50))
```

```scala
)
//边的数据类型 ED:Int
val edgeArray = Array(
  Edge(2L, 1L, 7),
  Edge(2L, 4L, 2),
  Edge(3L, 2L, 4),
  Edge(3L, 6L, 3),
  Edge(4L, 1L, 1),
  Edge(5L, 2L, 2),
  Edge(5L, 3L, 8),
  Edge(5L, 6L, 3)
)

//构造 vertexRDD 和 edgeRDD
val vertexRDD: RDD[(Long, (String, Int))] = sc.parallelize(vertexArray)
val edgeRDD: RDD[Edge[Int]] = sc.parallelize(edgeArray)

//构造图 Graph[VD,ED]
val graph: Graph[(String, Int), Int] = Graph(vertexRDD, edgeRDD)

//*******************************************************************
//*****************************图的属性*********************************
//*******************************************************************
println("*************************************************")
println("属性演示")
println("*************************************************")
println("找出图中年龄大于 30 的顶点:")
graph.vertices.filter { case (id, (name, age)) => age > 30 }.collect.foreach {
  case (id, (name, age)) => println(s"$name is $age")
}

//边操作:找出图中属性大于 5 的边
println("找出图中属性大于 5 的边:")
graph.edges.filter(e => e.attr > 5).collect.foreach(e => println(s"${e.srcId}
  to ${e.dstId} att ${e.attr}"))
println

//triplets 操作, ((srcId, srcAttr), (dstId, dstAttr), attr)
println("列出边属性>5 的tripltes: ")
for (triplet <- graph.triplets.filter(t => t.attr > 5).collect) {
  println(s"${triplet.srcAttr._1} likes ${triplet.dstAttr._1}")
}
println
```

```scala
//Degrees 操作
println("找出图中最大的出度、入度、度数：")
def max(a: (VertexId, Int), b: (VertexId, Int)): (VertexId, Int) = {
  if (a._2 > b._2) a else b
}
println("max of outDegrees:" + graph.outDegrees.reduce(max) + " max of
  inDegrees:" + graph.inDegrees.reduce(max) + " max of Degrees:" +
  graph.degrees.reduce(max))
println

//*******************************************************************
//******************************转换操作******************************
//*******************************************************************
println("*******************************************************")
println("转换操作")
println("*******************************************************")
println("顶点的转换操作，顶点age + 10: ")
graph.mapVertices{ case (id, (name, age)) => (id, (name,
  age+10))}.vertices.collect.foreach(v => println(s"${v._2._1} is
  ${v._2._2}"))
println
println("边的转换操作，边的属性*2: ")
graph.mapEdges(e=>e.attr*2).edges.collect.foreach(e => println(s"${e.srcId}
  to ${e.dstId} att ${e.attr}"))
println

//*******************************************************************
//******************************结构操作******************************
//*******************************************************************
println("*******************************************************")
println("结构操作")
println("*******************************************************")
println("顶点年纪>30 的子图：")
val subGraph = graph.subgraph(vpred = (id, vd) => vd._2 >= 30)
println("子图所有顶点：")
subGraph.vertices.collect.foreach(v => println(s"${v._2._1} is
  ${v._2._2}"))
println
println("子图所有边：")
subGraph.edges.collect.foreach(e => println(s"${e.srcId} to ${e.dstId} att
  ${e.attr}"))
println

//*******************************************************************
```

```
//***********************************结构操作*********************************
//****************************************************************************
println("****************************************************************")
println("连接操作")
println("****************************************************************")
val inDegrees: VertexRDD[Int] = graph.inDegrees
case class User(name: String, age: Int, inDeg: Int, outDeg: Int)

//创建一个新图，顶点VD的数据类型为User，并从graph做类型转换
val initialUserGraph: Graph[User, Int] = graph.mapVertices { case (id, (name,
    age)) => User(name, age, 0, 0)}

//initialUserGraph 与 inDegrees、outDegrees（RDD）进行连接,
//并修改 initialUserGraph 中 inDeg 值、outDeg 值
val userGraph = initialUserGraph.outerJoinVertices(initialUserGraph.
    inDegrees) {
  case (id,u,inDegOpt)=>User(u.name,u.age,inDegOpt.getOrElse(0),u.outDeg)
}.outerJoinVertices(initialUserGraph.outDegrees) {
    case (id, u, outDegOpt) => User(u.name, u.age, u.inDeg,
        outDegOpt.getOrElse(0))
}

println("连接图的属性：")
userGraph.vertices.collect.foreach(v => println(s"${v._2.name}  inDeg:
    ${v._2.inDeg}  outDeg: ${v._2.outDeg}"))
println

println("出度和入读相同的人员：")
userGraph.vertices.filter {
    case (id, u) => u.inDeg == u.outDeg
}.collect.foreach {
    case (id, property) => println(property.name)
}
println

//****************************************************************************
//**********************************实用操作***********************************
//****************************************************************************
println("****************************************************************")
println("实用操作")
println("****************************************************************")
println("找出5到各顶点的最短：")
val sourceId: VertexId = 5L  //定义源点
val initialGraph = graph.mapVertices((id, _) => if (id == sourceId) 0.0 else
```

```
        Double.PositiveInfinity)
    val sssp = initialGraph.pregel(Double.PositiveInfinity)(
      (id, dist, newDist) => math.min(dist, newDist),
      triplet => {    //计算权重
        if (triplet.srcAttr + triplet.attr < triplet.dstAttr) {
          Iterator((triplet.dstId, triplet.srcAttr + triplet.attr))
        } else {
          Iterator.empty
        }
      },
      (a,b) => math.min(a,b)    //最短距离
    )
    println(sssp.vertices.collect.mkString("\n"))

    sc.stop()
  }
}
```

3. 运行结果

在 IDEA 中首先对 GraphXExample.scala 代码进行编译，编译通过后进行执行，执行结果如下：

```
************************************************************
属性演示
************************************************************
```

找出图中年龄大于 30 的顶点：

```
David is 42
Fran is 50
Charlie is 65
Ed is 55
```

找出图中属性大于 5 的边：

```
2 to 1 att 7
5 to 3 att 8
```

列出边属性>5 的 tripltes：

```
Bob likes Alice
Ed likes Charlie
```

找出图中最大的出度、入度、度数：

```
max of outDegrees:(5,3) max of inDegrees:(2,2) max of Degrees:(2,4)
```

```
************************************************************
转换操作
```

```
*************************************************************
顶点的转换操作，顶点 age + 10：
4 is (David,52)
1 is (Alice,38)
6 is (Fran,60)
3 is (Charlie,75)
5 is (Ed,65)
2 is (Bob,37)
边的转换操作，边的属性*2：
2 to 1 att 14
2 to 4 att 4
3 to 2 att 8
3 to 6 att 6
4 to 1 att 2
5 to 2 att 4
5 to 3 att 16
5 to 6 att 6

*************************************************************
结构操作
*************************************************************
顶点年纪>30 的子图：
子图所有顶点：
David is 42
Fran is 50
Charlie is 65
Ed is 55
子图所有边：
3 to 6 att 3
5 to 3 att 8
5 to 6 att 3

*************************************************************
连接操作
*************************************************************
连接图的属性：
David inDeg: 1  outDeg: 1
Alice inDeg: 2  outDeg: 0
Fran inDeg: 2  outDeg: 0
Charlie inDeg: 1  outDeg: 2
Ed inDeg: 0  outDeg: 3
```

```
Bob inDeg: 2  outDeg: 2
```
出度和入读相同的人员：
```
David
Bob
```

```
**************************************************
实用操作
**************************************************
```
找出 5 到各顶点的最短：
```
(4,4.0)
(1,5.0)
(6,3.0)
(3,8.0)
(5,0.0)
(2,2.0)
```

10.3.2 社区发现演示

1. 例子介绍

现实中存在着各种网络，比如社交网络人际关系网、交易网等，对于这些网络进行社区的发现有重要的意义。在人际关系网中，可以发现不同兴趣、背景的社会团体，方便进行不同的宣传策略；在交易网中，不同的社区代表不同购买力的客户群体，方便运营为他们推荐合适的商品；在店铺网络中，社区发现可以检测出商帮、价格联盟等，对商家进行指导等。总之，社区发现在各种网络中都能有应用场景，图 10-10 展示了基于图的拓扑结构进行社区发现的例子。

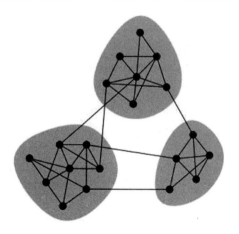

图 10-10　基于图的拓扑结构进行社区发现

2. 测试数据

该测试数据为某社交网站关注情况，其中 users.txt 为社交网站用户和 followers.txt 为这些用户的关注情况，需要注意的是这些关注形成的图是有向的。，这两个文件可以在本系列附带资源 /data/chapter10 目录中找到，在测试前把这两个文件上传到 HDFS 上，其中内容为：

（1）用户数据（users.txt）

该用户数据总共三个字段，分别为序号、昵称和名字全程，字段之间使用逗号分隔，该文件内容如下：

```
1,BarackObama,Barack Obama
2,ladygaga,Goddess of Love
3,jeresig,John Resig
4,justinbieber,Justin Bieber
6,matei_zaharia,Matei Zaharia
7,odersky,Martin Odersky
8,anonsys
```

（2）用户关系数据（followers.txt）

每笔用户关系数据包含两个用户的联系，它们之间使用空格进行分隔，具体内容如下：

```
2 1
4 1
1 2
6 3
7 3
7 6
6 7
3 7
```

3. 程序代码

```scala
import org.apache.log4j.{Level, Logger}
import org.apache.spark.graphx.{GraphLoader, PartitionStrategy}
import org.apache.spark.sql.SparkSession

object TriangleCounting {
  def main(args: Array[String]): Unit = {
    //屏蔽不必要的日志显示在终端上
    Logger.getLogger("org.apache.hadoop").setLevel(Level.ERROR)
    Logger.getLogger("org.apache.spark").setLevel(Level.ERROR)
    Logger.getLogger("org.eclipse.jetty.server").setLevel(Level.OFF)

    //创建运行环境
    val spark = SparkSession.builder
```

```scala
    .appName(s"${this.getClass.getSimpleName}")
    .getOrCreate()
  val sc = spark.sparkContext

  //加载用户关系数据，对这些数据采用随机顶点分区方法
  val graph = GraphLoader.edgeListFile(sc, "/home/spark/work/chapter10/
     followers.txt", true)
    .partitionBy(PartitionStrategy.RandomVertexCut)

  //通过顶点进行社区发现
  val triCounts = graph.triangleCount().vertices

  //读取用户数据，读取用户的编号和昵称
  val users = sc.textFile("/home/spark/work/chapter10/users.txt").map { line
     =>
    val fields = line.split(",")
    (fields(0).toLong, fields(1))
  }

  //对发现的结果与用户数据进行关联
  val triCountByUsername = users.join(triCounts).map { case (id, (username, tc))
     =>
    (username, tc)
  }

  //打印发现社区中包含的成员数
  triCountByUsername.map(x=>(x._2, 1)).reduceByKey(_+_).foreach {
    case (communication, count) => println("The Communication:" + communication
       +" have " + count + " members.")
  }
  println

  //按照社区的顺序，打印详细信息
  println("All communication's members:")
  println(triCountByUsername.map(x=>(x._2, x._1)).sortByKey(true).collect().
     mkString("\n"))

  spark.stop()
  }
}
```

4. 运行结果

（1）打包运行

对运行代码进行打包,生成打包文件 sparklearning.jar 并移动到 Spark 根目录下。在运行前需要启动 HDFS 和 Spark 集群,然后使用 spark-submit 脚本提交程序到集群中,其运行命令如下所示:

```
$cd /app/spark/spark-2.0.0
$./bin/spark-submit --class chapter10.TriangleCounting \
  --master spark://master:7077 \
  --executor-memory 1024m \
  sparklearning.jar \
  hdfs://master:9000/chapter10/followers.txt \
  hdfs://master:9000/chapter10/users.txt
```

提交 jar 包运行需要两个参数,分别为用户关系数据和用户数据文件路径:

(2)执行并观察输出

通过社区发现算法运行结果如下:

```
The Communication:1 have 3 members.
The Communication:0 have 3 members.
All communication's members:
(0,justinbieber)
(0,ladygaga)
(0,BarackObama)
(1,matei_zaharia)
(1,jeresig)
(1,odersky)
```

该结果可以得出这些人之间有两个社区,分别为娱乐和技术两个社区,关系图如图 10-11 所示。

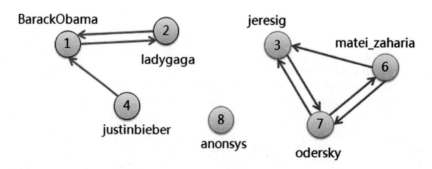

图 10-11　社区发现结果图

10.4　小结

本章中先介绍了图计算、图计算框架和 Spark GraphX。其中"图计算"是以"图论"为基础对现实"图"结构的抽象，以及在该结构上进行的计算；图的并行计算框架已经有较多，重点介绍比如来自 Google 的 Pregel 和最为著名的 GraphLab；而 GraphX 是构建于 Spark 核心基础上的并行图计算框架，基于 Spark 提供的 RDD 抽象，依赖于 RDD 的容错性实现了高效健壮，并可以与 Spark SQL、MLlib 等无缝地结合使用。

然后对 Spark GraphX 实现进行了深入分析，分别从 GraphX 图数据模型、图数据存储、图切分策略和图操作进行描述。

最后介绍了图例演示和社区发现演示两个实例。在图例演示中演示中将构建顶点、边和图并打印图的属性、转换、结构、连接和聚合等操作，而在社区发现实例中演示基于图的拓扑结构进行社区发现的例子。

第 11 章

SparkR

11.1 概述

11.1.1 R 语言介绍

R 是 S 语言的一种实现。S 语言是由 AT&T 贝尔实验室开发的一种用来进行数据探索、统计分析、作图的解释型语言。最初 S 语言的实现版本主要是 S-Plus。S-Plus 是一个商业软件，它基于 S 语言，并由 MathSoft 公司的统计科学部进一步完善。后来 Auckland 大学的 Robert Gentleman 和 Ross Ihaka 及其他志愿人员开发了一个 R 系统。R 的使用与 S-Plus 有很多相似之处，两个软件有一定的兼容性。

R 是用于统计分析、绘图的语言和操作环境。R 是属于 GNU 系统的一个自由、免费、源代码开放的软件，它是一个用于统计计算和统计制图的优秀工具。R 是一套完整的数据处理、计算和制图软件系统，其功能包括数据存储和处理系统；数组运算工具（其在向量、矩阵运算方面功能尤其强大），完整连贯的统计分析工具，优秀的统计制图功能，简便而强大的编程语言，可操纵数据的输入和输出，可实现分支、循环，用户可自定义功能。

R 是一个免费的自由软件，它有 UNIX、Linux、MacOS 和 Windows 版本，都是可以免费下载和使用的，还可以下载到 R 的安装程序、各种外挂程序和文档。在 R 的安装程序中只包含了 8 个基础模块，其他的外在模块可以通过 CRAN 获得。

R 语言是一种面向对象的语言，所有的对象都有两个内在属性：元素类型和长度。

元素类型是对象内元素的基本类型，包括数值（numeric）、字符型（character）、复数型（complex）、逻辑型（logical）、函数（function）等，通过 mode()函数可以查看一个对象的类型。

长度是对象中元素的数目，通过函数 length()可以查看对象的长度。

除了元素类型外，对象本身也有不同的"类型"，表示不同的数据结构（struct）。R 中的对象类型主要包括以下几种。

- 向量（vector）：由一系列有序元素构成。
- 因子（factor）：对同长的其他向量元素进行分类（分组）的向量对象。R 同时提供有序（ordered）和无序（unordered）因子。
- 数组（array）：带有多个下标的类型相同的元素的集合。
- 矩阵（matrix）：矩阵仅仅是一个带有双下标的数组。R 提供了一个函数专门处理二维数组（矩阵）。
- 数据帧（data frame）：和矩阵类似的一种结构。在数据帧中，列可以是不同的对象。
- 时间序列（time series）：包含一些额外的属性，如频率和时间。
- 列表（list）：是一种泛化（general form）的向量。它没有要求所有元素是同一类型，许多时候就是向量和列表类型。列表为统计计算的结果返回提供了一种便利的方法。

11.1.2 SparkR 介绍

SparkR 应该被看做是 R 版 Spark 的轻量级前端，这意味着它不会拥有像 Scala 或 Java 那样广泛的 API，但它还是能够在 R 里运行 Spark 任务和操作数据。它其中的一项关键特性就是有能力序列化闭包，从而能依次透明地将变量副本传入需要参与运算的 Spark 集群。SparkR 还通过内置功能的形式集成了其他的 R 模块，这一功能会在需要某些模块参与运算的时候通知 Spark 集群加载特定的模块，但是不同于闭包，这个需要手动设置。

SparkR 的出现解决了 R 语言中无法级联扩展的难题，同时也极大地丰富了 Spark 在机器学习方面能够使用的 Lib 库，SparkR 和 Spark MLlib 将共同构建出 Spark 在机器学习方面的优势地位。使用 SparkR 能让用户同时使用 Spark RDD 提供的丰富的 API，也可以调用 R 语言中丰富的 Lib 库。

SparkR 的运行原理如图 11-1 所示，在运行前需要加载 R 语言包和 rJava 包,然后通过 SparkR 初始化 Spark Context。如果是本地模式，直接在本地启动 Spark Executor 与 RScript 进行交互执行；如果是集群模式，则任务提交给资源管理器进行调度，由资源管理器将任务分配到 Worker 节点上进行执行。

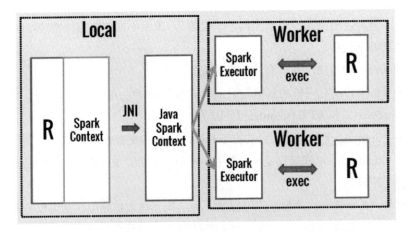

图 11-1　SparkR 的运行原理

11.2　SparkR 与 DataFrame

11.2.1　DataFrames 介绍

DataFrame 是带有列名称的分布式数据集,在概念上和关系型数据库中的表类似,或者和 R 语言中的 Data Frame 相似,相对两者 DataFrame 进行了深入的优化。DataFrame 可以通过广义的数组类型数据源进行构造,例如结构化文件、Hive 表、外部数据库或 R Data Frame 等。

SparkContext 是 SparkR 的入口,由它进行 R 程序和 Spark 集群互通。用户可以通过 sparkR.init 来创建 SparkContext,在创建的过程中根据需要传入 Spark 访问地址、应用程序名等参数。如果要使用 DataFrames,需要创建 SQLContext,它可以通过 SparkContext 来构造。如果使用 SparkR 命令行,SQLContext 和 SparkContext 会自动地构建好,不需要显示创建 sparkR.init。

```
R>sc <- sparkR.init()
R>sqlContext <- sparkRSQL.init(sc)
```

下面介绍通过本地的 R Data Frame、Hive 表或者是其他数据源来创建 DataFrames。

1. 通过本地 data frame 构造

最简单地创建 DataFrames 是将 R 的 Data Frame 转换成 SparkR DataFrames,我们可以通过 createDataFrame 来创建,并传入本地 R 的 Data Frame 以此来创建 SparkR DataFrames,下面例子就是这种方法。

```
R>df <- createDataFrame(sqlContext, faithful)
```

```
#显示DataFrame数据内容
R>head(df)
##   eruptions waiting
##1  3.600     79
##2  1.800     54
##3  3.333     74
```

2．通过数据源构造

通过 DataFrame 接口，SparkR 支持操作多种数据源。通过数据源创建 DataFrames 的一般是使用 read.df 方法，在该方法中需要传入 SQLContext、需要加载的文件路径以及数据源的类型 3 个参数。SparkR 内置支持读取 JSON 和 Parquet 文件，通过 Spark 数据读取依赖包可以读取很多类型的数据，比如 CSV 和 Avro 等文件，代码示例如下：

```
R>sc <- sparkR.init(sparkPackages="com.databricks:spark-csv_2.11:1.0.3")
R>sqlContext <- sparkRSQL.init(sc)
```

下面是介绍如何读取 JSON 文件，不过需要注意的是，这里使用的文件不是典型的 JSON 文件，而是每行必须包含一个分隔符、自包含有效的 JSON 对象。

```
R>people <- read.df(sqlContext, "./examples/src/main/resources/people.json",
   "json")
R>head(people)
##   age  name
##1  NA   Michael
##2  30   Andy
##3  19   Justin

# SparkR 可自动识别 JSON 文件的数据结构
R>printSchema(people)
# root
#  |-- age: integer (nullable = true)
#  |-- name: string (nullable = true)
```

数据源的 API 还可以将 DataFrames 保存成多种的文件格式，比如可以通过 write.df 方法将上面的 DataFrame 保存成 Parquet 文件。

```
R>write.df(people, path="people.parquet", source="parquet", mode="overwrite")
```

3．通过 Hive tables 构造

用户也可以通过 Hive 表来创建 SparkR DataFrames，为了达到这个目的，需要创建 HiveContext，从而访问 Hive MetaStore 中的表。注意：Spark 内置就对 Hive 提供了支持，SQLContext 和 HiveContext 的区别可以参见 Spark SQL 章节内容。

```r
#sc 默认存在，不需要创建
hiveContext <- sparkRHive.init(sc)

#创建表 src，并导入数据
sql(hiveContext, "CREATE TABLE IF NOT EXISTS src (key INT, value STRING)")
sql(hiveContext, "LOAD DATA LOCAL INPATH 'examples/src/main/resources/kv1.txt '
   INTO TABLE src")

#使用 HiveQL 进行查询
results <- sql(hiveContext, "FROM src SELECT key, value")

#显示查询结果数据
head(results)
##  key   value
## 1 238 val_238
## 2  86 val_86
## 3 311 val_311
```

11.2.2　与 DataFrame 的相关操作

SparkR DataFrames 中提供了大量操作结构化数据的函数，这里仅仅列出其中一小部分，详细的 API 可以参见 SparkR 编程的 API 文档。

1．选择行和列

```r
#创建 DataFrame
df <- createDataFrame(sqlContext, faithful)

#获取 DataFrame 的基本信息
df
## DataFrame[eruptions:double, waiting:double]

#显示"eruptions"列数据
head(select(df, df$eruptions))
##  eruptions
##1    3.600
##2    1.800
##3    3.333

#通过字符串的列名获取数据
head(select(df, "eruptions"))

#过滤出等待时间小于 50min 的数据
head(filter(df, df$waiting < 50))
```

```
##  eruptions waiting
##1    1.750      47
##2    1.750      47
##3    1.867      48
```

2. 归组和聚集

```
#使用'n'操作符用于计算等待时间出现的次数
head(summarize(groupBy(df, df$waiting), count = n(df$waiting)))
##  waiting count
##1      81    13
##2      60     6
##3      68     1

#可以显示排序好的聚集的数据
waiting_counts <- summarize(groupBy(df, df$waiting), count = n(df$waiting))
head(arrange(waiting_counts, desc(waiting_counts$count)))
##  waiting count
##1      78    15
##2      83    14
##3      81    13
```

3. 列操作

SparkR 提供了大量的函数用于直接对列进行数据处理的操作。

```
#可对列进行计算,把显示时间由时转化为s
df$waiting_secs <- df$waiting * 60
head(df)
##  eruptions waiting waiting_secs
##1    3.600      79         4740
##2    1.800      54         3240
##3    3.333      74         4440
```

11.3 编译安装 SparkR

11.3.1 编译安装 R 语言

根据 SparkR 运行架构图可以知道，在 Spark 的每个节点运行时会调用本地的 R，所以需要在每个节点上安装 R 语言环境。

1. 下载 R 语言源代码

用户可以在 R 语言的官方网站 http://www.r-project.org/ 选择对应的版本进行下载，不过建议在该页面选择最佳的 CRAN 镜像站点下载，下载后把解压包解压，移动到 /app/soft 目录下。

```
$cd /home/spark/work
$wget http://mirror.bjtu.edu.cn/cran/src/base/R-3/R-3.2.3.tar.gz
$tar -zxf R-3.2.3.tar.gz
$mv R-3.2.3 /app/soft
```

2. 安装依赖包

R 语言运行依赖相关程序包，可以通过 root 身份进行安装，命令如下：

```
#yum install gcc -y
#yum install gcc-c++ -y
#yum install gcc-gfortran -y
#yum install readline-devel -y
#yum install libXt-devel -y
#yum install libpng-devel -y
```

3. 编译安装

使用 make 方法对 R 语言进行编译安装。

```
#cd /app/soft/R-3.2.3
#./configure --prefix=$prefix --enable-R-shlib --with-x --with-libpng
   --with-jpeglib
#make && make install
```

4. 配置工作路径

打开 /etc/profile 配置文件，命令如下：

```
#sudo vi /etc/profile
```

加入 R 语言的工作目录，最后使用 $source /etc/profile 生效该配置：

```
export R_HOME=/app/soft/R-3.2.3
export PATH=$PATH:$R_HOME/bin
```

5. 启动 R Shell

编译完成后，在终端输入 R 命令进入 R 语言命令界面，如图 11-2 所示，表示 R 语言安装成功。

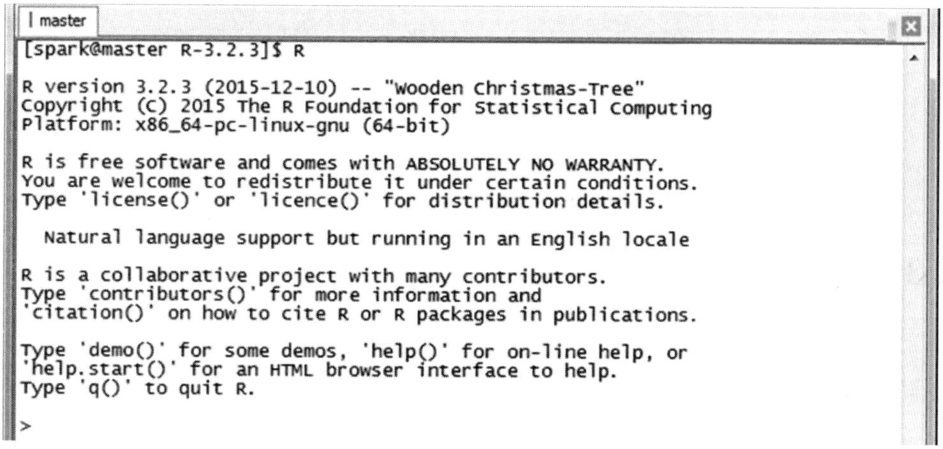

图 11-2　验证 R 语言安装结果

11.3.2　安装 SparkR 运行环境

1. 安装依赖包

以 root 身份安装 SparkR 依赖包，否则在安装依赖包会提示错误。

```
#yum install libcurl-devel -y
#yum install openssl-devel -y
#yum install libxml2-devel -y
```

2. 在 R Shell 中安装相关依赖包

在安装依赖包的时候，如 rJava，会提示选择最佳的镜像站点，可以根据界面提示选项选择。

```
#R
R>install.packages("rJava")
R>install.packages("devtools")
R>install.packages("git2r")
R>install.packages("xml2")
R>install.packages("rversions")
```

由于后面实例中需要画图，需要加入 png、jpeg 等支持，如图 11-3 所示。

```
R>install.packages("png")
R>install.packages("jpeg")
R>capabilities()
```

```
> capabilities()
     jpeg      png     tiff    tcltk      X11     aqua
    FALSE     TRUE    FALSE    FALSE     TRUE    FALSE
 http/ftp  sockets   libxml     fifo   cledit    iconv
     TRUE     TRUE     TRUE     TRUE     TRUE     TRUE
      NLS  profmem    cairo      ICU long.double libcurl
     TRUE    FALSE    FALSE    FALSE     TRUE    FALSE
```

图 11-3　查看依赖包支持功能

11.3.3　安装 SparkR

1．安装方法一：在线安装

如果安装机器联网并且网速较快可以采用在线方式安装 SparkR，如图 11-4 所示。

```
R>library(devtools)
R>install_github("amplab-extras/SparkR-pkg", subdir="pkg")
```

```
| master
[info] SHA-1: 692ca80aa1db3f5d0432c5b9c2ef1aa29d9fabbd
[info] Packaging /tmp/RtmpNqnnPM/devtools65c13270c660/amplab-extras-SparkR-pkg
[info] Done packaging.
[success] Total time: 798 s, completed Apr 28, 2016 9:11:40 PM
cp -f target/scala-2.10/sparkr-assembly-0.1.jar ../inst/
R CMD SHLIB -o SparkR.so string_hash_code.c
make[1]: Entering directory `/tmp/RtmpNqnnPM/devtools65c13270c660/amplab-extra
gcc -std=gnu99 -I/app/soft/R-3.2.3/include -DNDEBUG -I/usr/local/include -
gcc -std=gnu99 -shared -L/app/soft/R-3.2.3/lib -L/usr/local/lib64 -o SparkR.so
make[1]: Leaving directory `/tmp/RtmpNqnnPM/devtools65c13270c660/amplab-extras
installing to /app/soft/R-3.2.3/library/SparkR/libs
** R
** inst
** tests
** preparing package for lazy loading
Creating a generic function for 铌apply铌?from package 铌ase铌?in packag
Creating a generic function for 铌ilter铌?from package 铌ase铌?in packag
** help
*** installing help indices
** building package indices
** testing if installed package can be loaded
* DONE (SparkR)
>
```

图 11-4　使用在线方式安装 SparkR

2．安装方法二：源代码编译安装

（1）下载 SparkR 源代码：用户既可以从 SparkR 官方网站下载源代码，也可以从源代码托管服务器进行下载，下载地址如下：

http://amplab-extras.github.io/SparkR-pkg/

https://github.com/amplab-extras/SparkR-pkg/tarball/master

下载后把解压包解压，移动到 /app/soft 目录下：

```
$cd /home/spark/work
$tar -zxf amplab-extras-SparkR-pkg-385fbe5.tar.gz
$mv amplab-extras-SparkR-pkg-385fbe5 /app/spark/sparkr
```

（2）下载 sbt：从 sbt 官网下载其安装包 http://www.scala-sbt.org/download.html，下载完毕后

把安装包解压，移动/app/soft 目录下。

```
$cd /home/spark/work
$tar -zxf sbt-0.13.11.tgz
$mv sbt-0.13.11 /app/soft/
```

把 sbt 加入到 PATH 路径中，首先使用#sudo vi /etc/profile 命令打开/etc/profile 配置文件，在该文件中加入 sbt 的 bin 目录：

```
export SBT_HOME=/app/soft/sbt-0.13.11
export PATH=$PATH:$R_HOME/bin
```

最后使用$source /etc/profile 生效该配置。

（3）修改 sbt 编译配置文件内容。打开 SparkR 源代码包的 sbt 编译配置文件：

```
$cd /app/spark/sparkr/pkg/src/sbt
$sudo vi sbt
```

在该文件中，把 sbt 的路径修改为绝对路径：

```
JAR=/app/soft/sbt-0.13.11/bin/sbt-launch.jar
```

（4）安装 SparkR。在 SparkR 根目录下，指定 Spark 和 Hadoop 版本进行编译，编译过程需要下载依赖包，所以耗费时间和网速关系较大：

```
$cd /app/spark/sparkr/
$SPARK_VERSION=1.4.0 HADOOP_VERSION=2.6.0 ./install-dev.sh
```

11.3.4 启动并验证安装

首先启动 R Shell，然后加载 SparkR。

```
$R
#加载 SparkR 包
R>library(SparkR)
在 R Shell 中，初始化 SparkContext 及执行 WordCount，代码如下：
#初始化 Spark Context，并读取 Spark 根目录下文件
sc <- sparkR.init(master="local", "RWordCount")
lines <- textFile(sc, "file:///app/spark/spark-2.0.0/README.md")

#按分隔符拆分每一行为多个元素，返回结果序列
words <- flatMap(lines,
  function(line) {
    strsplit(line, " ")[[1]]
  })
#使用 lapply 来定义对应每一个 RDD 元素的运算，这里返回一个（K, V）对
wordCount <- lapply(words, function(word) { list(word, 1L) })

#对（K, V）对进行聚合计算
```

```
counts <- reduceByKey(wordCount, "+", 2L)
#以数组的形式，返回数据集的所有元素
output <- collect(counts)
#按格式输出结果
for (wordcount in output) {
  cat(wordcount[[1]], ": ", wordcount[[2]], "\n")
}
```
验证 SparkR 安装结果显示页面如图 11-5 所示。

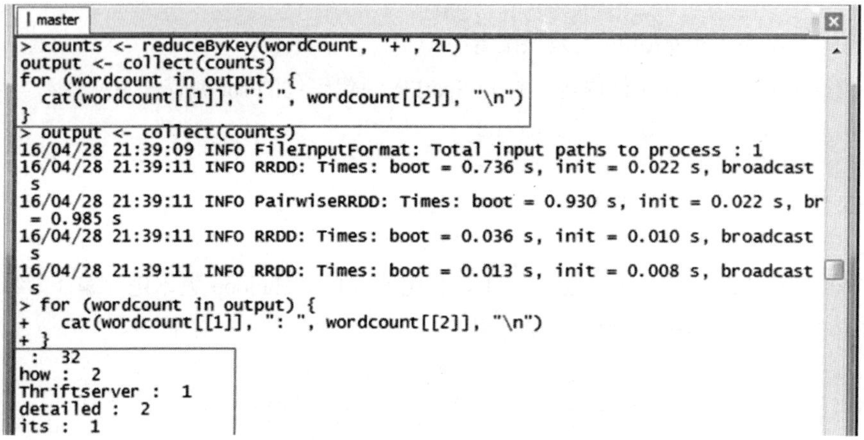

图 11-5　验证 SparkR 安装结果

如果想将 SparkR 运行于集群环境中，则只需要将 master=local 换成 Spark 集群的监听地址即可。

11.4　实例演示

1. 例子介绍

在该 SparkR 实例中，将对 2011 年所有 Houston 的航班数据进行操作，这些数据包含 14 个列，227496 行数据。实例将先读入该航班 CSV 格式的数据，打印头部数据并计算数据总的行数等，然后选择目的地和取消两个列，把这两个列数据注册 flightsTable 临时表，使用 SQL 语句对该表进行查询并显示前 6 行，最后对这些航班数据进行统计每日平均航班延误时间。

2. 程序代码

```
#加载 SparkR 包
library(SparkR)
```

```r
#对输入参数进行判断，参数个数不为 1 则退出
args <- commandArgs(trailing = TRUE)
if (length(args) != 1) {
  print("Usage: DataManipulation.R <path-to-flights.csv>")
  q("no")
}

#初始化 Spark 运行环境
sparkR.session(appName = "DataManipulation ")

#读取航班数据并创建 R 的 DataFrame
flightsCsvPath <- args[[1]]
flightsDF <- read.df(flightsCsvPath, source = "csv", header = "true")

#打印该 DataFrame 的结构，并缓存该 DataFrame
printSchema(flightsDF)
cache(flightsDF)

#显示 DataFrame 前 6 行数据
showDF(flightsDF, numRows = 6) ## Or
head(flightsDF)

#显示 DataFrame 总行数
count(flightsDF)

#选择目的地和取消两个列，并把这两个列数据注册 flightsTable 表
destDF <- select(flightsDF, "dest", "cancelled")
createOrReplaceTempView(flightsDF, "flightsTable")

#使用 SQL 语句查询这些数据生成 DataFrame 并显示
destDF <- sql("SELECT dest, cancelled FROM flightsTable")
local_df <- collect(destDF)
head(local_df)

#下面两种方式可以过滤出目的地为 JFK 的数据
jfkDF <- filter(flightsDF, "dest = \"JFK\"") ##OR
jfkDF <- filter(flightsDF, flightsDF$dest == "JFK")
showDF(jfkDF, numRows = 6)

#按照日期进行统计，计算每次航班延误平均时间，并打印这些数据
if("magrittr" %in% rownames(installed.packages())) {
  library(magrittr)
  groupBy(flightsDF, flightsDF$date) %>%
```

```
    summarize(avg(flightsDF$dep_delay), avg(flightsDF$arr_delay)) ->
      dailyDelayDF
  head(dailyDelayDF)
}

#停止Spark运行环境
sparkR.session.stop()
```

3. 运行结果

把程序代码保存为 **DataManipulation.R** 文件，并放在在 Spark 根目录下。由于该程序中使用了 **magrittr** 包，在运行前需要安装，命令如下：

```
#R
R> install.packages('magrittr')
R> library(magrittr)
```

在运行程序文件之前启动 **HDFS**、**Hive** 和 **Spark** 集群，然后使用如下命令进行执行：

```
$cd /app/spark/spark-2.0.0
$./bin/spark-submit --master spark://master:7077  DataManipulation.R
    file:///home/spark/work/chapter11/flights.csv
```

执行的结果如下：

```
root
 |-- date: string (nullable = true)
 |-- hour: string (nullable = true)
 |-- minute: string (nullable = true)
 |-- dep: string (nullable = true)
 |-- arr: string (nullable = true)
 |-- dep_delay: string (nullable = true)
 |-- arr_delay: string (nullable = true)
 |-- carrier: string (nullable = true)
 |-- flight: string (nullable = true)
 |-- dest: string (nullable = true)
 |-- plane: string (nullable = true)
 |-- cancelled: string (nullable = true)
 |-- time: string (nullable = true)
 |-- dist: string (nullable = true)

   SparkDataFrame[date:string, hour:string, minute:string, dep:string, arr:string,
dep_delay:string, arr_delay:string, carrier:string, flight:string, dest:string,
plane:string, cancelled:string, time:string, dist:string]


+------------------+---+---+----+----+--+---+---+----+----+------+--+---+----+
|              date| h| m| dep| arr|de|a_d|car|flig|est|plane|ca|tim|dist|
+------------------+---+---+----+----+--+---+---+----+----+------+--+---+----+
```

```
|2011-01-01 12:00:00|  14|  0|1400|1500|  0|-10|     AA|   428|  DFW|N576AA|  0|  40| 224|
|2011-01-02 12:00:00|  14|  1|1401|1501|  1| -9|     AA|   428|  DFW|N557AA|  0|  45| 224|
|2011-01-03 12:00:00|  13| 52|1352|1502| -8| -8|     AA|   428|  DFW|N541AA|  0|  48| 224|
|2011-01-04 12:00:00|  14|  3|1403|1513|  3|  3|     AA|   428|  DFW|N403AA|  0|  39| 224|
|2011-01-05 12:00:00|  14|  5|1405|1507|  5| -3|     AA|   428|  DFW|N492AA|  0|  44| 224|
|2011-01-06 12:00:00|  13| 59|1359|1503| -1| -7|     AA|   428|  DFW|N262AA|  0|  45| 224|
+-------------------+----+---+----+----+---+---+-------+------+-----+------+---+----+----+
only showing top 6 rows


         date  h  m   dep  arr  d_de  a_d  carrier  flight dest   plane  can  time dist
1 2011-01-01 12:00:00 14  02  1400 1500   -0  -10       AA     428  DFW  N576AA    0   40  224
2 2011-01-02 12:00:00 14  12  1401 1501   -1   -9        A     428  DFW  N557AA    0   45  224
3 2011-01-03 12:00:00 13  52  1352 1502   -8   -8       AA     428  DFW  N541AA    0   48  224
4 2011-01-04 12:00:00 14  32  1403 1513   -3   -3       AA     428  DFW  N403AA    0   39  224
5 2011-01-05 12:00:00 14  52  1405 1507   -5   -3       AA     428  DFW  N492AA    0   44  224
6 2011-01-06 12:00:00 13  59  1359 1503   -1   -7       AA     428  DFW  N262AA    0   45  224

[1] 227496

    dest   cancelled
1   DFW    0
2   DFW    0
3   DFW    0
4   DFW    0
5   DFW    0
6   DFW    0

+-------------------+---+---+----+----+--+---+---+----+----+------+--+---+----+
|               date|  h|  m| dep| arr|de|a_d|car|flig| est|plane |ca|tim|dist|
+-------------------+---+---+----+----+--+---+---+----+----+------+--+---+----+
|2011-01-01 12:00:00|  6| 54|1654|1124|-6|165| B6| 620| JFK|N324JB| 0|181|1428|
|2011-01-01 12:00:00| 16| 39|1639|2110|54|161| B6| 622| JFK|N324JB| 0|188|1428|
|2011-01-02 12:00:00|  7|  3|1703|1113| 3|186| B6| 620| JFK|N324JB| 0|172|1428|
|2011-01-02 12:00:00| 16|  4|1604|2040|19|131| B6| 622| JFK|N324JB| 0|176|1428|
|2011-01-03 12:00:00|  6| 59|1659|1100|-1|-19| B6| 620| JFK|N229JB| 0|166|1428|
|2011-01-03 12:00:00| 18|  1|1801|2200|136|111| B6| 622| JFK|N206JB| 0|165|1428|
+-------------------+---+---+----+----+--+---+---+----+----+------+--+---+----+
only showing top 6 rows
```

11.5 小结

在本章中先介绍 R 语言和 SparkR，R 是用于统计分析、绘图的语言和操作环境，而 SparkR 应该被看作是 R 版 Spark 的轻量级前端，SparkR 的出现解决了 R 语言中无法级联扩展的难题，同时也极大地丰富了 Spark 在机器学习方面能够使用的 Lib 库。

然后介绍了 SparkR 相关操作，在新版的 SparkR 使用了 DataFrame，它是带有列名称的分布式数据集，可以通过广义的数组类型数据源进行构造：例如结构化文件、Hive 表、外部数据库或 R Data Frame 等。接着介绍了 SparkR 的编译和安装，安装 SparkR 需要通过编译安装 R 语言，再通过在线或者编译的方式安装 SparkR。

最后通过数据操作和结合机器学习实例进行演示，在数据操作实例中，将对 2011 年所有 Houston 的航班数据进行操作。

第 12 章

Alluxio

12.1 Alluxio 简介

12.1.1 Alluxio 介绍

随着实时计算的需求日益增多，分布式内存计算也持续升温，怎样将海量数据近乎实时地处理，或者说怎样把离线批处理的速度再提升到一个新的高度是当前研究的重点。近年来，内存的吞吐量成指数倍增长，而磁盘的吞吐量增长缓慢，那么将原有计算框架中文件落地磁盘替换为文件落地内存，也是提高效率的优化点。

目前已经使用基于内存计算的分布式计算框架有 Spark、Impala 及 SAP 的 HANA 等。但是其中不乏一些还是有文件落地磁盘的操作，如果能让这些落地磁盘的操作全部落地到一个共享的内存中，那么这些基于内存的计算框架的效率会更高。

Alluxio 是原伯克利分校 AMP 实验室的李浩源所开发的一个分布式内存文件系统，原来名字叫 Tachyon，在他毕业后成立了 Alluxio 公司，并在 2016 年 2 月份发布了 Alluxio1.0 版本。作为世界上首款以内存为中心的虚拟分布式存储系统，它能够统一数据访问并成为连接计算框架和底层存储系统的桥梁，应用程序只需要连接 Alluxio 便能够访问底层任意存储系统中的数据。除此之外，Alluxio 以内存为中心的架构使得数据访问比现有的解决方案能快若干个数量级。

Alluxio 是架构在最底层的分布式文件存储和上层的各种计算框架之间的一种中间件，其主要职责是将那些不需要落地的文件落地到分布式内存文件系统中来达到共享内存，从而提高效率。同时可以减少内存冗余、GC 时间等，Alluxio 的在大数据中层次关系如图 12-1 所示。

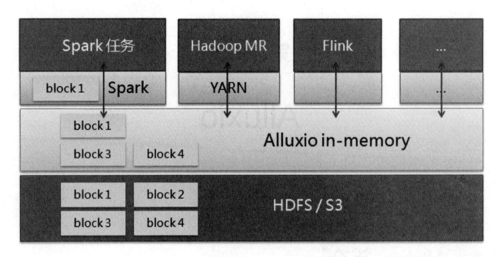

图 12-1　Alluxio 的在大数据中层次关系

Alluxio 允许文件以内存的速度在集群框架中进行可靠的共享，就像 Spark、MapReduce 和 Flink 等那样，通过利用信息继承、内存侵入，Alluxio 获得了高性能。Alluxio 工作集文件缓存在内存中，并且让不同的作业/查询以及框架都能以内存的速度来访问缓存文件。因此，Alluxio 可以减少那些需要经常使用数据集通过访问磁盘来获得的次数。

12.1.2　Alluxio 系统架构

1. 系统架构

Alluxio 在 Spark 平台的部署：总的来说，Alluxio 有 3 个主要的部件，即 Master、Client 和 Worker。在集群中，每个 Worker 节点上部署了一个 Alluxio Worker，所有的 Alluxio Worker 都由 Alluxio Master 所管理，Alluxio Master 通过 Alluxio Worker 定时发出的心跳来判断 Worker 是否已经崩溃以及每个 Worker 剩余的内存空间量。图 12-2 为 Alluxio 高可用架构。

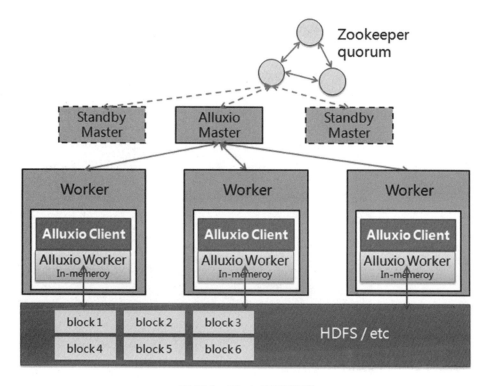

图 12-2　Alluxio 高可用架构

2．Alluxio Master 的结构

Alluxio Master 结构的主要功能如下：

（1）Alluxio Master 是一个主管理器，处理从各个 Client 发出的请求，这一系列的工作由 Service Handler 来完成。这些请求包括获取 Worker 的信息，读取 File 的 Block 信息，创建 File 等。

（2）Alluxio Master 是一个 Name Node，存放着所有文件的信息，每个文件的信息都被封装成一个 Inode，每个 Inode 都记录着属于这个文件的所有 Block 信息。在 Alluxio 中，Block 是文件系统存储的最小单位。假设每个 Block 是 256MB，如果有一个文件的大小是 1GB，那么这个文件会被切为 4 个 Block。每个 Block 可能存在多个副本，存储在多个 Alluxio Worker 中，因此 Master 里面也必须记录每个 Block 所存储的 Worker 地址。

（3）Alluxio Master 同时管理着所有的 Worker，Worker 会定时向 Master 发送心跳通知本次活跃状态以及剩余存储空间。Master 是通过 Master Worker Info 去记录每个 Worker 的上次心跳时间、已使用的内存空间以及总存储空间等信息。

图 12-3 为 Alluxio Master 的结构示意图。

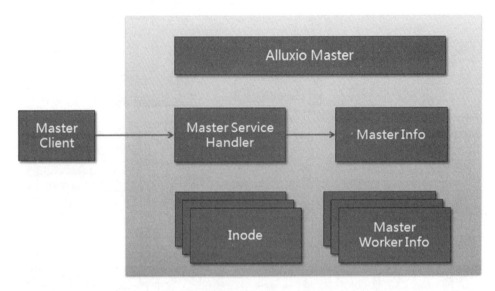

图 12-3　Alluxio Master 结构示意图

3．Alluxio Worker 的结构

Alluxio Worker 主要负责存储管理，主要功能如下：

（1）Alluxio Worker 的 Service Handler 处理来自 Client 发来的请求，这些请求包括读取某个 Block 的信息、缓存某个 Block、锁住某个 Block、向本地内存存储要求空间等。

（2）Alluxio Worker 的主要部件是 Worker Storage，其作用是管理 Local Data（本地的内存文件系统）以及 Under File System（Alluxio 以下的磁盘文件系统，比如 HDFS）。

（3）Alluxio Worker 有一个 Data Server，以便处理其他的 Client 对其发起的数据读/写请求。当请求达到时，Alluxio 会先在本地的内存存储找数据，如果没有找到，则会尝试去其他的 Alluxio Worker 的内存存储中进行查找。如果数据完全不在 Alluxio 里，则需要通过 Under File System 的接口去磁盘文件系统（比如 HDFS）中读取。

图 12-4 为 Alluxio Worker 结构示意图。

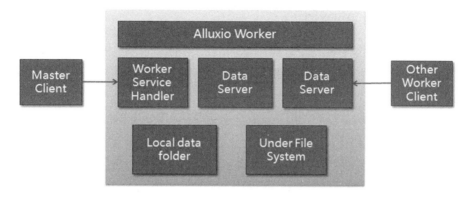

图 12-4　Alluxio Worker 结构示意图

4．Alluxio Client 的结构

Alluxio Client 的主要功能是向用户抽象一个文件系统接口，以屏蔽掉底层实现细节。Alluxio Client 会通过 Master Client 部件跟 Alluxio Master 交互，比如可以向 Alluxio Master 查询某个文件的某个 Block 在哪里。Alluxio Client 也会通过 Worker Client 部件与 Alluxio Worker 交互，比如向某个 Alluxio Worker 请求存储空间。在 Alluxio Client 实现中最主要的是 Alluxio File 部件。在 Alluxio File 下，实现了 Block Out Stream，主要用于写本地内存文件，实现了 Block In Stream 主要负责读内存文件。在 Block In Stream 内包含了两个不同的实现：Local Block In Stream 主要是用来读本地的内存文件，而 Remote Block In Stream 主要是读非本地的内存文件。注意，非本地既可以是在其他的 Alluxio Worker 的内存文件里，也可以是在 Under File System 的文件里。

图 12-5 为 Alluxio Client 结构示意图。

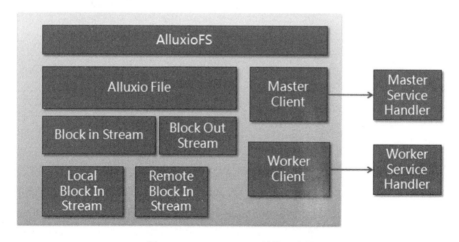

图 12-5　Alluxio Client 结构示意图

12.1.3　HDFS 与 Alluxio

　　HDFS 是一个分布式文件系统，它具有高容错性（fault-tolerant）的特点，并且设计用来部署在低廉的硬件上，而且它提供高吞吐量来访问应用程序的数据，适合那些有着超大数据集的应用程序。

　　HDFS 采用 Master/Slave 架构，HDFS 集群是由一个名称节点（NameNode）和一定数目的数据节点（DataNode）组成的。名称节点是一台中心服务器，负责管理文件系统的名字空间以及客户端对文件的访问。集群中的数据节点一般是一个节点一个，负责管理它所在节点上的存储。HDFS 暴露了文件系统的名字空间，用户能够以文件的形式在上面存储数据。从内部看，一个文件其实被分成一个或多个数据块，这些块存储在一组数据节点上。名称节点执行文件系统的名字空间操作，比如打开、关闭、重命名文件或目录，它也负责确定数据块到具体数据节点的映射。数据节点负责处理文件系统客户端的读/写请求，在名称节点的统一调度下对数据块进行创建、删除和复制。HDFS 的读/写架构如图 12-6 所示。

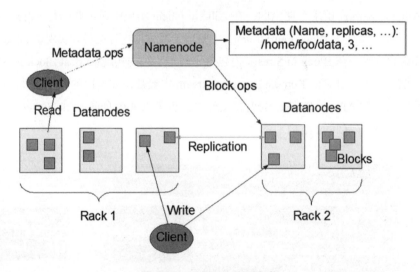

图 12-6　HDFS 的读/写架构

　　名称节点和数据节点被设计成可以在普通的商用机器上运行，这些机器一般运行着 GNU/Linux 操作系统。HDFS 采用 Java 语言开发，因此任何支持 Java 的机器都可以部署名称节点或数据节点。由于采用了可移植性极强的 Java 语言，使得 HDFS 可以部署到多种类型的机器上。一个典型的部署场景是一台机器上只运行一个名称节点实例，而集群中的其他机器则分别运行一个数据节点实例。这种架构并不排斥在一台机器上运行多个数据节点，只不过这样的情况比较少见。集群中单一名称节点的结构大大简化了系统的架构。名称节点是所有 HDFS 元数据的仲裁者和管理者。

对比 HDFS 和 Alluxio，首先从两者的存储结构来看，HDFS 设计为用来存储海量文件的分布式系统，Alluxio 设计为用来缓存常用数据的分布式内存文件系统。从这点来看，Alluxio 可以认为是操作系统层面上的缓存，HDFS 可以认为是磁盘。

在可靠性方面，HDFS 采用副本技术来保证出现系统宕机等意外情况时，文件访问的一致性以及可靠性；而 Alluxio 是依赖于底层文件系统的可靠性来实现自身文件的可靠性的。由于相对于磁盘资源来说，内存是非常宝贵的，所以 Alluxio 通过在其 underfs（一般使用 HDFS）上写入 CheckPoint 日志信息来实现对文件系统的可恢复性。

从文件的读取以及写入方式来看，Alluxio 可以更好地利用本地模式来读取文件信息，当文件读取客户端和文件所在的 Worker 位于一台机器上时，客户端会直接绕过 Worker 直接读取对应的物理文件，减少了本机的数据交互。而 HDFS 在遇到这样的情况时，会通过本地 Socket 进行数据交换，这也会有一定的系统资源开销。在写入文件时，HDFS 只能写入磁盘，而 Alluxio 却提供了多种数据写入模式用以满足不同需求。

对于 Alluxio 更为高级的特性可以参见官网提供的文档，如 http://www.alluxio.org/docs/master/cn/File-System-API.html 链接中讲述了 Alluxio 文件系统客户端 API，在 http://www.alluxio.org/docs/master/cn/Unified-and-Transparent-Namespace.html 中介绍了 Alluxio 统一透明命名空间。

12.2 Alluxio 编译部署

Alluxio 文件系统有 3 种部署方式：单机模式、集群模式和高可用集群模式，集群模式相比于高可用集群模式区别在于多个 Master 节点。下面将介绍单机和集群环境下安装、配置和使用 Alluxio，如果读者想成为 Alluxio 开源社区并成为其中一员，可以参考 http://www.alluxio.org/docs/master/cn/ Contributing-Getting-Started.html，在该页面中有成为 Alluxio 开发者上手的详细指南。

12.2.1 编译 Alluxio

为了更好地契合用户的本地环境，如 Java 版本、Hadoop 版本或其他一些软件包的版本，可以下载 Alluxio 源码自行编译。Alluxio 开源在 GitHub 上，可以很方便地获得其不同版本的源码。

1. 下载源代码

对于已经发布的版本可以直接从 Github 下载 Alluxio 编译好的安装包并解压，下载地址为 https://github.com/Alluxio/alluxio/releases。下面演示下载的是 alluxio-1.3.0.tar.gz 源代码包，下载的源代码包在 master 节点上解压缩：

```
$cd /home/spark/work/
$tar -xzf alluxio-1.3.0.tar.gz
```

把 alluxio-1.3.0 改名并移动到 /app/complied 目录下：

```
$mv alluxio-1.3.0 /app/compile/alluxio-src-1.3.0
$ll /app/complied
```

2. 编译代码

Alluxio 项目采用 Maven 进行管理，因此可以采用 mvn package 命令进行编译打包。编译 Alluxio 源代码的时候，需要从网上下载依赖包，所以整个编译过程机器必须保证在联网状态。编译执行如下脚本：

```
$cd /app/compile/alluxio-src-1.3.0
$export MAVEN_OPTS="-Xmx2g -XX:MaxPermSize=512M -XX:ReservedCodeCacheSize=512m"
$mvn clean package -Pspark -Djava.version=1.7 -Dhadoop.version=2.7.0 -DskipTests
```

整个编译过程编译了约 24 个任务，整个过程耗时大约 17min，如图 12-7 所示。

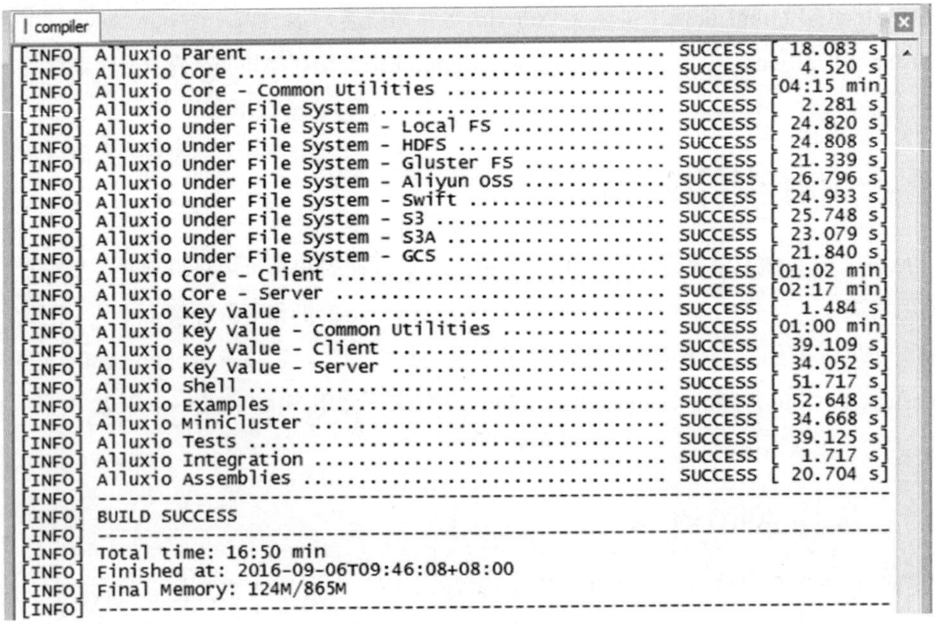

```
[INFO] Alluxio Parent ............................... SUCCESS [ 18.083 s]
[INFO] Alluxio Core ................................. SUCCESS [  4.520 s]
[INFO] Alluxio Core - Common Utilities .............. SUCCESS [04:15 min]
[INFO] Alluxio Under File System .................... SUCCESS [  2.281 s]
[INFO] Alluxio Under File System - Local FS ......... SUCCESS [ 24.820 s]
[INFO] Alluxio Under File System - HDFS ............. SUCCESS [ 24.808 s]
[INFO] Alluxio Under File System - Gluster FS ....... SUCCESS [ 21.339 s]
[INFO] Alluxio Under File System - Aliyun OSS ....... SUCCESS [ 26.796 s]
[INFO] Alluxio Under File System - Swift ............ SUCCESS [ 24.933 s]
[INFO] Alluxio Under File System - S3 ............... SUCCESS [ 25.748 s]
[INFO] Alluxio Under File System - S3A .............. SUCCESS [ 23.079 s]
[INFO] Alluxio Under File System - GCS .............. SUCCESS [ 21.840 s]
[INFO] Alluxio Core - Client ........................ SUCCESS [01:02 min]
[INFO] Alluxio Core - Server ........................ SUCCESS [02:17 min]
[INFO] Alluxio Key Value ............................ SUCCESS [  1.484 s]
[INFO] Alluxio Key Value - Common Utilities ......... SUCCESS [01:00 min]
[INFO] Alluxio Key Value - Client ................... SUCCESS [ 39.109 s]
[INFO] Alluxio Key Value - Server ................... SUCCESS [ 34.052 s]
[INFO] Alluxio Shell ................................ SUCCESS [ 51.717 s]
[INFO] Alluxio Examples ............................. SUCCESS [ 52.648 s]
[INFO] Alluxio MiniCluster .......................... SUCCESS [ 34.668 s]
[INFO] Alluxio Tests ................................ SUCCESS [ 39.125 s]
[INFO] Alluxio Integration .......................... SUCCESS [  1.717 s]
[INFO] Alluxio Assemblies ........................... SUCCESS [ 20.704 s]
[INFO] ------------------------------------------------------------
[INFO] BUILD SUCCESS
[INFO] ------------------------------------------------------------
[INFO] Total time: 16:50 min
[INFO] Finished at: 2016-09-06T09:46:08+08:00
[INFO] Final Memory: 124M/865M
[INFO] ------------------------------------------------------------
```

图 12-7　Alluxio 编译成功结果

使用如下命令查看编译后该 Alluxio 项目大小：
```
$cd /app/compile/alluxio-src-1.3.0
$du -s /app/compile/alluxio-src-1.3.0
```
完成这一步后，我们就得到了能够运行在用户本地环境的 Alluxio，下面分别介绍如何在单机和分布式环境下配置和启动 Alluxio，在进行部署之前先把编译好的文件复制到/app/spark 下，并把文件夹命名为 alluxio-1.3.0：
```
$cd /app/compile
$cp -r alluxio-src-1.3.0 /app/spark/alluxio-1.3.0
$ll /app/spark
```

12.2.2　单机部署 Alluxio

部署文件既可以使用前一节编译好的安装包，也可以在 Alluxio 官网下载对应的安装包，官网下载链接为 http://alluxio.org/downloads/files/。这里要注意一点，Alluxio 在单机模式下启动时会自动挂载 RamFS，所以请保证使用的账户具有 sudo 权限。

1. 配置 Alluxio

Alluxio 相关配置文件在$ALLUXIO_HOME/conf 目录下，在 workers 文件中配置需要启动 Alluxio Worker 的节点，默认是 localhost，所以在单机模式下不用更改。在这里需要修改 alluxio-env.sh 配置文件，具体操作是将 alluxio-env.sh.template 复制为 alluxio-env.sh：
```
$cd /app/spark/alluxio-1.3.0/conf
$cp alluxio-env.sh.template alluxio-env.sh
$sudo vi alluxio-env.sh
```
在 alluxio-env.sh 中替换如下配置。
```
export JAVA_HOME=/app/soft/jdk1.7.0_55
export JAVA="$JAVA_HOME/bin/java"
export ALLUXIO_MASTER_HOSTNAME=localhost
export ALLUXIO_UNDERFS_ADDRESS=$ALLUXIO_HOME/underFSStorage
export ALLUXIO_WORKER_MEMORY_SIZE=1GB
```
下面列举了一些重要的配置项。
- JAVA_HOME：系统中 Java 的安装路径。
- ALLUXIO_MASTER_ADDRESS：启动 Alluxio Master 的地址，默认为 localhost，所以在单机模式下不用更改。
- ALLUXIO_UNDERFS_ADDRESS：Alluxio 使用的底层文件系统的路径，在单机模式下可以直接使用本地文件系统，如"/tmp/alluxio"，也可以使用 HDFS，如

"hdfs://host:port"。

- ALLUXIO_WORKER_MEMORY_SIZE：每个 Alluxio Worker 使用的 RamFS 大小。

2．格式化 Alluxio

完成配置后即可单机模式启动 Alluxio，启动前需要格式化存储文件。格式化和启动 Alluxio 的命令分别如下：

```
$cd /app/spark/alluxio-1.3.0/bin
$alluxio format
```

存储文件在$ALLUXIO_HOME/underfs/tmp/alluxio 目录下。

3．启动 Alluxio

使用如下命令启动 Alluxio，可以看到在/nmt/ramdisk 目录下格式化 RamFS，启动界面如图 12-8 所示。

```
$cd /app/spark/alluxio-1.3.0/bin
$./alluxio-start.sh local
```

```
[spark@master bin]$ ./alluxio-start.sh local
/app/spark/alluxio-1.3.0/bin
Killed 1 processes on master
Killed 1 processes on master
Waiting for WORKERS tasks to finish...
All WORKERS tasks finished, please analyze the log at /app/spark/alluxio-1.3.0/logs/task.log.
Starting master @ localhost. Logging to /app/spark/alluxio-1.3.0/logs
ALLUXIO_RAM_FOLDER was not set. Using the default one: /mnt/ramdisk
Formatting RamFS: /mnt/ramdisk (1gb)
Starting worker @ master. Logging to /app/spark/alluxio-1.3.0/logs
[spark@master bin]$
```

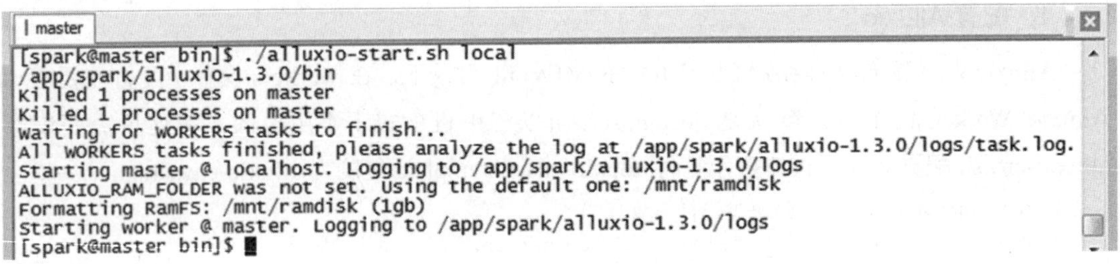

图 12-8　启动 Alluxio 界面

4．验证启动

使用 JPS 命令查看 Alluxio 进程，分别为 Alluxio Worker 和 Alluxio Master。查看 Alluxio 监控页面，访问地址为 http://master:19999，如图 12-9 所示。

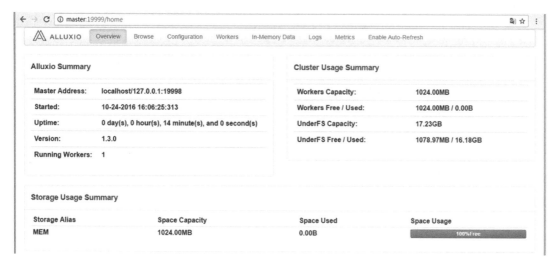

图 12-9　单机部署 Alluxio 监控页面

5. 停止 Alluxio

停止 Alluxio 的命令如下：

```
$cd /app/spark/alluxio-1.3.0/bin
$alluxio-stop.sh all
```

12.2.3　集群模式部署 Alluxio

1. 配置 conf/workers

Alluxio 相关配置文件在 $ALLUXIO_HOME/conf 目录下，对 workers 文件中配置需要启动 Alluxio Worker 的节点，在这里需要设置 master、slave1 和 slave2 三个节点。

```
$cd /app/spark/alluxio-1.3.0/conf
$sudo vi workers
```

2. 配置 conf/alluxio-env.sh

在 $ALLUXIO_HOME/conf 目录下，将 alluxio-env.sh.template 复制为 alluxio-env.sh，并在 alluxio-env.sh 中修改具体配置。不同于单机模式，这里需要修改 Alluxio Master 地址以及底层文件系统路径。

```
$cd /app/spark/alluxio-1.3.0/conf
$cp alluxio-env.sh.template alluxio-env.sh
$sudo vi alluxio-env.sh
```

在该文件中修改一下两个参数，这里使用底层文件系统为 HDFS。

```
export ALLUXIO_MASTER_HOSTNAME=master
export ALLUXIO_UNDERFS_ADDRESS=hdfs://master:9000
```

3. 向各个节点分发 Alluxio

使用如下命令把 alluxio 文件夹复制到 slave1 和 slave2 节点上。

```
$cd /app/spark/
$scp -r alluxio-1.3.0 spark@slave1:/app/spark/
$scp -r alluxio-1.3.0 spark@slave2:/app/spark/
```

4. 格式化 Alluxio

格式化前启动 HDFS。

```
$cd /app/spark/hadoop-2.7.2/sbin
$./start-dfs.sh
```

启动前需要格式化存储文件，格式化命令如下：

```
$cd /app/spark/alluxio-1.3.0/bin
$./alluxio format
```

可以看到在 HDFS 的/tmp 创建了 Alluxio 文件夹。

5. 启动 Alluxio

在这里使用 SudoMout 参数，需要在启动过程中输入 spark 用户的密码，具体过程如下：

```
$cd /app/spark/alluxio-1.3.0/bin
$./alluxio-start.sh all SudoMount
```

启动 Alluxio 有了更多的选项，具体如下。

- **./alluxio-start.sh all Mount**：在启动前自动挂载 Alluxio Worker 所使用的 RamFS，然后启动 Alluxio Master 和所有 Alluxio Worker。由于直接使用 mount 命令，所以需要用户为 root。
- **./alluxio-start.sh all SudoMount**：在启动前自动挂载 Alluxio Worker 所使用的 RamFS，然后启动 Alluxio Master 和所有 Alluxio Worker。由于使用 sudo mount 命令，所以需要用户有 sudo 权限。
- **./alluxio-start.sh all NoMount**：认为 RamFS 已经挂载好，不执行挂载操作，只启动 AlluxioMaster 和所有 Alluxio Worker。

因此，如果不想每次启动 Alluxio 都挂载一次 RamFS，可以先使用命令"./alluxio-mount.sh Mount workers"或"./alluxio-mount.sh SudoMount workers"挂载好所有 RamFS，然后使用"./alluxio-start.sh all NoMount"命令启动 Alluxio。

单机和集群式模式的区别就在于节点配置和启动步骤，事实上，也可以在集群模式下只设置一个 Alluxio Worker，此时就成为伪分布模式。

6．验证启动

使用 JPS 命令查看 Alluxio 进程，分别为 Alluxio Worker 和 Alluxio Master。用户可以在浏览器内打开 Alluxio 的 WebUI，如 http://master:19999，查看整个 Alluxio 的状态，各个 Alluxio Worker 的运行情况、各项配置信息和浏览文件系统等，如图 12-10 所示。

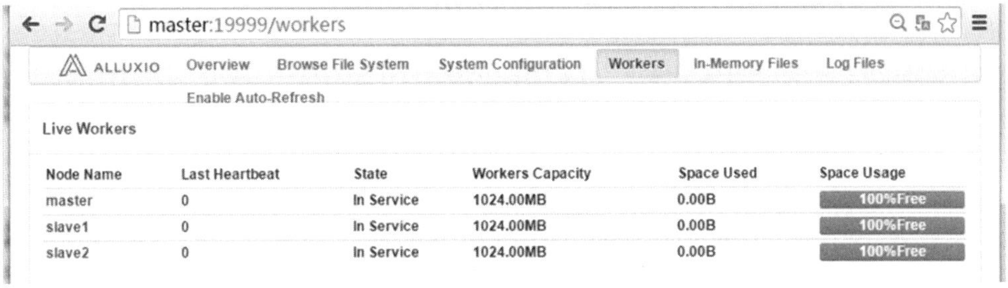

图 12-10　集群部署 Alluxio 监控页面

使用如下命令测试样例程序，如果所有测试样例显示"Passed the test!"表示启动正常。

```
$cd /app/spark/alluxio-1.3.0/bin
$./alluxio runTests
```

12.3　Alluxio 命令行使用

Alluxio 的命令行接口和配置等在其官网有中英文帮助文档，具体链接地址为 http://alluxio.org/documentation。Alluxio 命令行界面让用户可以对文件系统进行基本的操作，调用命令行工具使用以下脚本：

```
$alluxio fs
```

文件系统访问的路径格式如下：

```
alluxio://<master node address>:<master node port>/<path>
```

在 Alluxio 命令行使用中 alluxio://<master node address>:<master node port>前缀可以省略，该信息从配置文件中读取。

12.3.1　接口说明

用户可以通过如下命令查看 Alluxio 所有接口命令。

```
$cd /app/spark/alluxio-1.3.0/bin
```

```
$alluxio fs -help
```
其中大部分的命令含义可以参考 Linux 下的同名命令，命令含义见表 12-1。

表 12-1　Alluxio 命令说明

命　令	语　法	含　义
cat	cat "path"	将 Alluxio 中的一个文件内容打印在控制台中
chgrp	chgrp "group" "path"	修改 Alluxio 中的文件或文件夹的所属组
chmod	chmod "permission" "path"	修改 Alluxio 中的文件或文件夹的访问权限
chown	chown "owner" "path"	修改 Alluxio 中的文件或文件夹的所有者
copyFromLocal	copyFromLocal "source path" "remote path"	将"source path"指定的本地文件系统中的文件复制到 Alluxio 中"remote path"指定的路径，如果"remote path"已经存在，则该命令失败
copyToLocal	copyToLocal "remote path" "local path"	将"remote path"指定的 Alluxio 中的文件复制到本地文件系统中
count	count "path"	输出"path"中的所有名称匹配一个给定前缀的文件及文件夹的总数
du	du "path"	输出一个指定的文件或文件夹的大小
fileInfo	fileInfo "path"	输出指定的文件的数据块信息
free	free "path"	将 Alluxio 中的文件或文件夹移除，如果该文件或文件夹存在于底层存储中，那么仍然可以在那里访问
getCapacityBytes	getCapacityBytes	获取 Alluxio 文件系统的容量
getUsedBytes	getUsedBytes	获取 Alluxio 文件系统已使用的字节数
load	load "path"	将底层文件系统的文件或者目录加载到 Alluxio 中
loadMetadata	loadMetadata "path"	将底层文件系统的文件或者目录的元数据加载到 Alluxio 中
location	location "path"	输出包含某个文件数据的主机
ls	ls "path"	列出给定路径下的所有直接文件和目录的信息，例如大小
mkdir	mkdir "path1"… "pathn"	在给定路径下创建文件夹，以及需要的父文件夹，多个路径用空格或者 Tab 键分隔。如果其中的任何一个路径已经存在，该命令失败
mount	mount "path" "uri"	将底层文件系统的"uri"路径挂载到 Alluxio 命名空间中的"path"路径下，"path"路径事先不能存在并由该命令生成。没有任何数据或者元数据从底层文件系统加载。当挂载完成后，对该挂载路径下的操作会同时作用于底层文件系统的挂载点

续表

命令	语法	含义
mv	mv "source" "destination"	将"source"指定的文件或文件夹移动到"destination"指定的新路径。如果"destination"已经存在，则该命令失败。
persist	persist "path"	将仅存在于 Alluxio 中的文件或文件夹持久化到底层文件系统中
pin	pin "path"	将给定文件锁定到内容中以防止剔除。如果是目录，递归作用于其子文件以及里面新创建的文件
report	report "path"	向 master 报告一个文件已经丢失
rm	rm "path"	删除一个文件。如果输入路径是一个目录，则该命令失败
setTtl	setTtl "time"	设置一个文件的 TTL 时间，单位为 ms
tail	tail "path"	将指定文件的最后 1KB 内容输出到控制台
touch	touch "path"	在指定路径创建一个空文件
unmount	unmount "path"	卸载挂载在 Alluxio 中"path"指定路径上的底层文件路径，Alluxio 中该挂载点的所有对象都会被删除，但底层文件系统会将其保留
unpin	unpin "path"	将一个文件解除锁定，从而可以对其剔除，如果是目录，则递归作用
unsetTtl	unsetTtl	删除文件的 ttl 值

12.3.2 接口操作示例

为了在命令行中使用 alluxio 命令，需要把$ALLUXIO_HOME/bin 配置到/etc/profile 配置文件的 PATH 参数中，并通过#source /etc/profile 生效。

1. copyFromLocal

将本地$ALLUXIO_HOME/conf 目录复制到 Alluxio 文件系统的根目录下的 conf 子目录。

```
$cd /app/spark/alluxio-1.3.0/bin
$alluxio fs copyFromLocal ../conf alluxio://master:19998/conf
Copied ../conf to alluxio://master:19998/conf
$alluxio fs ls /conf
5.26KB 05-08-2016 22:47:26:403  In Memory  /conf/alluxio-env.sh.template
95.00B 05-08-2016 22:47:26:612  In Memory  /conf/workers
5.43KB 05-08-2016 22:47:26:627  In Memory  /conf/alluxio-swift-env.sh.template
3386.00B 05-08-2016 22:47:26:641  In Memory  /conf/log4j.properties
5.46KB   05-08-2016 22:47:26:687  In Memory  /conf/alluxio-env.sh
3218.00B 05-08-2016 22:47:26:744  In Memory  /conf/core-site.xml.template
```

```
4077.00B  05-08-2016 22:47:26:779  In Memory  /conf/metrics.properties.template
2579.00B  05-08-2016 22:47:26:793  In Memory  /conf/alluxio-glu…-env.sh.template
```

2. cat

查看指定文件的内容，代码如下：

```
$alluxio fs cat alluxio://master:19998/conf/workers
master
slave1
slave2
$alluxio fs cat /conf/workers
master
slave1
slave2
```

3. copyToLocal

把 Alluxio 文件系统文件复制到本地，需要注意的是，命令中的 src 必须是 Alluxio 文件系统中的文件不支持目录复制，否则报错无法复制。图 12-11 为 copyToLocal 命令操作界面。

```
$mkdir -p /home/spark/upload/chapter12/conflocal
$alluxio fs copyToLocal /conf /home/spark/upload/chapter12/conflocal
$ll /home/spark/upload/chapter12/conflocal
```

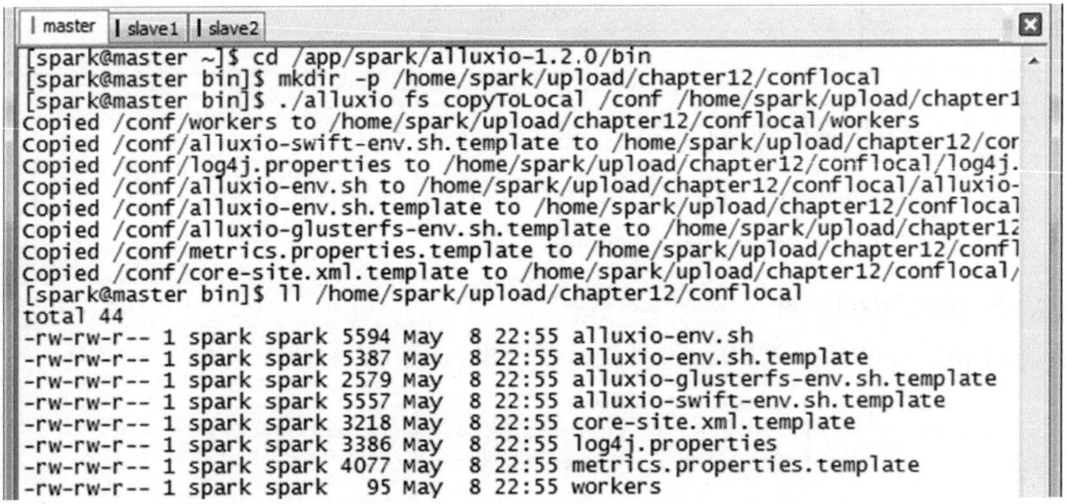

图 12-11　copyToLocal 命令操作界面

4. count

统计当前路径下的目录、文件信息，包括文件数、目录树以及总的大小。

```
$alluxio fs count /
File Count              Folder Count            Total Bytes
32                      3                       31861
```

5. ls

使用 ls 和 lsr 命令查看 Alluxio 文件系统下的文件信息，其中 lsr 命令可以递归地查看子目录。

```
$alluxio fs ls /conf
$alluxio fs ls alluxio://master:19998/conf
$alluxio fs lsr /
```

6. mkdir、rm 和 touch

- **mkdir**：创建目录，支持自动创建不存在的父目录。
- **rm**：删除文件，不能删除目录，注意，递归删除根目录是无效的。
- **touch**：创建文件，不能创建已经存在的文件。

7. pin 和 unpin

pin 命令将指定的路径常驻在内存中，如果指定的是一个文件夹会递归地包含所有文件以及任何在这个文件夹中新创建的文件。unpin 命令撤销指定路径的常驻内存状态。

pin 执行前或 unpin 执行后的 Web Interface 界面，分别如图 12-12 和图 12-13 所示。

```
$alluxio fs pin /conf/log4j.properties
$alluxio fs unpin /conf/log4j.properties
```

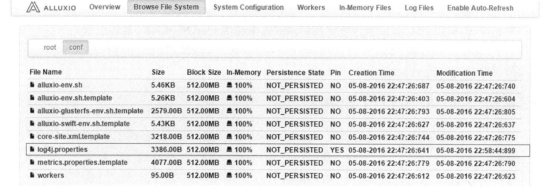

图 12-12　pin 执行前的界面

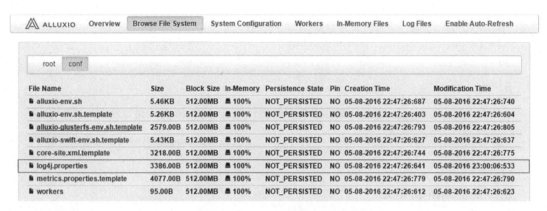

图 12-13　unpin 执行后的界面

12.4　实例演示

12.4.1　启动环境

1. 启动 HDFS

```
$cd /app/spark/hadoop-2.7.2/sbin
$./start-dfs.sh
```

2. 启动 Alluxio

在这里使用 SudoMout 参数，需要在启动过程中输入 spark 的密码，具体过程如下：

```
$cd /app/spark/alluxio-1.3.0/bin
$alluxio-start.sh all SudoMount
```

12.4.2　Alluxio 上运行 Spark

Alluxio Client 需要在编译时指定 Spark 选项。在顶层 alluxio 目录中执行如下命令构建 Alluxio：

```
$mvn clean package -Pspark -DskipTests
```

1. 修改 spark-env.sh

修改 $SPARK_HOME/conf 目录下 spark-env.sh 文件：

```
$cd /app/spark/spark-2.0.0/conf
$sudo vi spark-env.sh
```

在该配置文件中添加如下内容：

```
export SPARK_CLASSPATH=/app/spark/alluxio-1.3.0/core/client/target/alluxio-
    core-client-1.3.0-jar-with-dependencies.jar:$SPARK_CLASSPATH
```

2．添加 core-site.xml

如果 Alluxio 运行 Hadoop 1.x 集群之上（如果使用 Hadoop 2.X 该步骤可以省略），需要在 $SPARK_HOME/conf 目录下创建 core-site.xml 包含以下内容，否则无法识别以 alluxio://开头的文件系统：

```
$cd /app/spark/spark-2.0.0/conf
$touch core-site.xml
$sudo vi core-site.xml
```

在该配置文件中添加如下内容：

```
<configuration>
  <property>
    <name>fs.alluxio.impl</name>
    <value>alluxio.hadoop.FileSystem</value>
  </property>
</configuration>
```

3．启动 Spark 集群

```
$cd /app/spark/spark-2.0.0/sbin
$./start-all.sh
```

4．读取文件并保存

（1）准备测试数据文件：使用 Alluxio 命令行准备测试数据文件。

```
$cd /app/spark/alluxio-1.3.0/bin
$alluxio fs copyFromLocal ../conf/alluxio-env.sh /alluxio-env.sh
Copied ../conf/alluxio-env.sh to /alluxio-env.sh
$alluxio fs ls /alluxio-env.sh
5.46KB    05-09-2016 10:58:16:461   In Memory     /alluxio-env.sh
```

启动 Spark-Shell：

```
$cd /app/spark/spark-2.0.0/bin
$./spark-shell--master spark://master:7077 --executor-memory 1024m
```

对测试数据文件进行计数并另存，对前面放入到 Alluxio 文件系统的文件进行计数：

```
scala> val s = sc.textFile("alluxio://master:19998/alluxio-env.sh")
scala> s.count()
```

把前面的测试文件另存为 alluxio-env-bak.sh 文件：

```
scala> s.saveAsTextFile("alluxio://master:19998/alluxio-env-bak.sh")
```

（2）在 Alluxio 的 UI 界面查看。

用户可以查看到该文件在 Alluxio 文件系统中保存成 alluxio-env-bak.sh 文件夹，如图 12-14 所示。

File Name	Size	Block Size	In-Memory	Persistence State	Pin	Creation Time	Modification Time
alluxio-env-bak.sh				NOT_PERSISTED	NO	05-09-2016 11:06:49:846	05-09-2016 11:06:51:800
alluxio-env.sh	5.46KB	512.00MB	100%	NOT_PERSISTED	NO	05-09-2016 10:58:16:461	05-09-2016 10:58:16:748
conf				NOT_PERSISTED	NO	05-08-2016 22:47:26:354	05-08-2016 22:47:26:793
default_tests_files				PERSISTED	NO	05-08-2016 22:29:47:184	05-08-2016 22:31:20:752

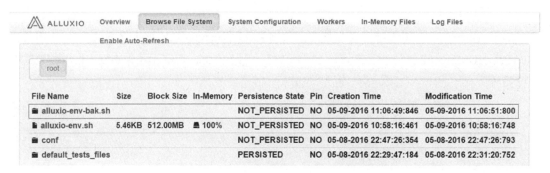

图 12-14　查看保存 alluxio-env-bak.sh 结果界面

该文件夹中包含两个文件，分别为 part-00000 和 part-00001，如图 12-15 所示。

File Name	Size	Block Size	In-Memory	Persistence State	Pin	Creation Time	Modification Time
_SUCCESS	0.00B	512.00MB	100%	NOT_PERSISTED	NO	05-09-2016 11:06:51:800	05-09-2016 11:06:51:814
part-00000	2826.00B	512.00MB	100%	NOT_PERSISTED	NO	05-09-2016 11:06:50:670	05-09-2016 11:06:51:121
part-00001	2768.00B	512.00MB	100%	NOT_PERSISTED	NO	05-09-2016 11:06:51:426	05-09-2016 11:06:51:443

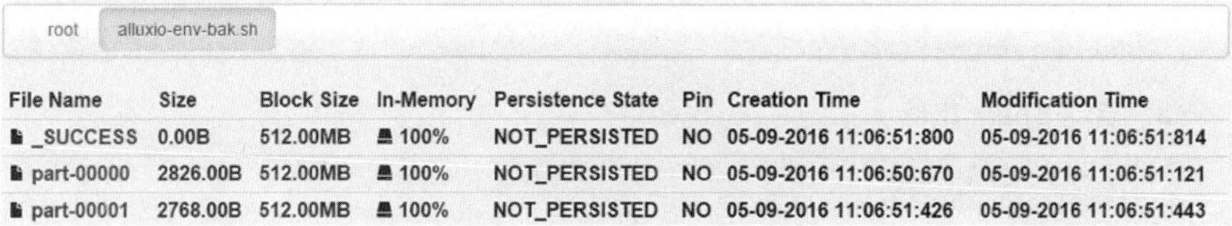

图 12-15　查看 alluxio-env-bak.sh 目录内容

另外通过内存存在文件的监控页面可以观测到，这几个操作文件在内存中，如图 12-16 所示。

File Path	Size	Block Size	Pin	Creation Time	Modification Time
/alluxio-env-bak.sh/_SUCCESS	0.00B	512.00MB	NO	05-09-2016 11:06:51:800	05-09-2016 11:06:51:814
/alluxio-env-bak.sh/part-00000	2826.00B	512.00MB	NO	05-09-2016 11:06:50:670	05-09-2016 11:06:51:121
/alluxio-env-bak.sh/part-00001	2768.00B	512.00MB	NO	05-09-2016 11:06:51:426	05-09-2016 11:06:51:443
/alluxio-env.sh	5.46KB	512.00MB	NO	05-09-2016 10:58:16:461	05-09-2016 10:58:16:748

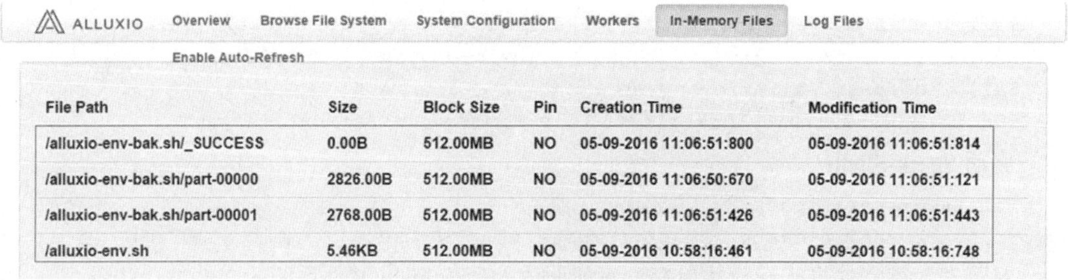

图 12-16　通过监控页面查看内存中的文件

12.4.3　Alluxio 上运行 MapReduce

1. 修改 core-site.xml

该配置文件为 $Hadoop_HOME/conf 目录下的 core-site.xml 文件：

```
$cd /app/spark/hadoop-2.7.2/etc/hadoop
$sudo vi core-site.xml
```

修改 core-site.xml 文件配置，添加如下配置项：

```
<property>
  <name>fs.alluxio.impl</name>
  <value>alluxio.hadoop.FileSystem</value>
  <description>The Alluxio FileSystem (Hadoop 1.x and 2.x)</description>
</property>
<property>
  <name>fs.alluxio-ft.impl</name>
  <value>alluxio.hadoop.FaultTolerantFileSystem</value>
  <description>The Alluxio FileSystem (Hadoop 1.x and 2.x) with fault tolerant
   support</description>
</property>
<property>
  <name>fs.AbstractFileSystem.alluxio.impl</name>
  <value>alluxio.hadoop.AlluxioFileSystem</value>
  <description>The Alluxio AbstractFileSystem (Hadoop 2.x)</description>
</property>
```

2. 启动 YARN

```
$cd /app/spark/hadoop-2.7.2/sbin
$./start-yarn.sh
```

3. 运行 MapReduce 例子

（1）创建结果保存目录：

```
$hadoop fs -mkdir /chapter12
$hadoop fs -ls /chapter12
```

第一步运行 MapReduce 例子

```
$cd /app/spark/hadoop-2.7.2/bin
$hadoop jar ../share/hadoop/mapreduce/hadoop-mapreduce-examples-2.7.2.jar
  wordcount -libjars $ALLUXIO_HOME/core/client/target/alluxio-core-client-
  1.2.0-jar-with-dependencies.jar  alluxio://master:19998/alluxio-env.sh
  hdfs://master:9000/chapter12/output
```

（2）查看结果：查看 HDFS，可以看到在/chapter12 中创建了 output 目录，查看 part-r-0000 文件内容，为 alluxio-env.sh 单词计数，如图 12-17 所示。

```
$hadoop fs -ls /chapter12
Found 1 items
drwxr-xr-x   - spark supergroup          0 2016-05-09 11:21 /chapter12/output
$hadoop fs -cat /chapter12/output/part-r-00000
```

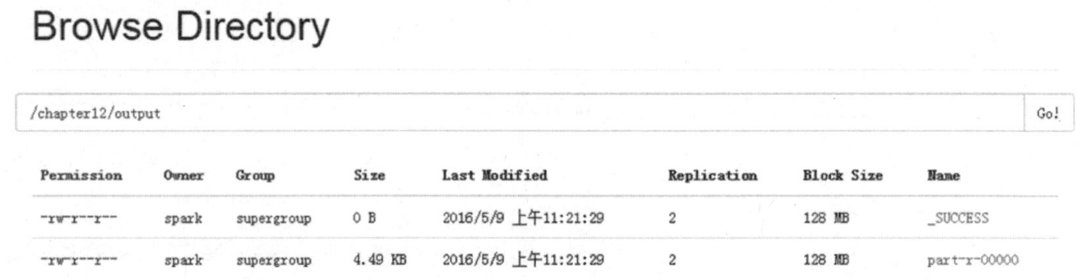

Browse Directory

Permission	Owner	Group	Size	Last Modified	Replication	Block Size	Name
/chapter12/output							Go!
-rw-r--r--	spark	supergroup	0 B	2016/5/9 上午11:21:29	2	128 MB	_SUCCESS
-rw-r--r--	spark	supergroup	4.49 KB	2016/5/9 上午11:21:29	2	128 MB	part-r-00000

图 12-17　查看 HDFS 单词计数结果

12.5　小结

在本章中先介绍了 Alluxio，前身是 Tachyon，它是世界上首款以内存为中心的虚拟分布式存储系统。Alluxio 有三个主要的部件：Master、Client 和 Worker。Alluxio Worker 部署在 Worker 节点上，Worker 通过 Alluxio 的 Client 访问 Alluxio 进行数据读写。所有的 Alluxio Worker 都被 Alluxio Master 所管理。然后介绍了 Alluxio 编译和部署，部署分为单机部署和集群部署两种方式，接着介绍了 Alluxio 命令行的使用。最后在 Spark 集群中使用 Alluxio 并进行文件的读取，另外也演示了 MapReduce 中的操作。